Smart Phone and Next-Generation Mobile Computing

Pei Zheng

Lionel Ni

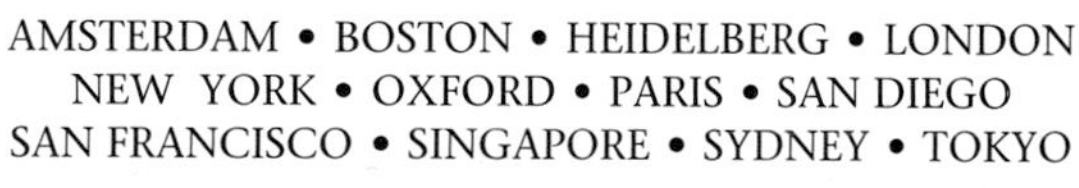

AMSTERDAM • BOSTON • HEIDELBERG • LONDON
NEW YORK • OXFORD • PARIS • SAN DIEGO
SAN FRANCISCO • SINGAPORE • SYDNEY • TOKYO

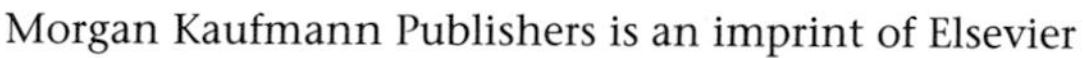

Morgan Kaufmann Publishers is an imprint of Elsevier

Senior Editor	Rick Adams
Acquisitions Editor	Rick Adams
Assistant Editor	Rachel Roumeliotis
Publishing Services Manager	Simon Crump
Senior Production Editor	Paul Gottehrer
Cover Design	Side-by-Side Design
Composition	Cepha Imaging Pvt. Ltd.
Interior printer	Maple-Vail Book Manufacturing Group
Cover printer	Phoenix Color Corp.

Morgan Kaufmann Publishers is an imprint of Elsevier.
500 Sansome Street, Suite 400, San Francisco, CA 94111

This book is printed on acid-free paper.

Library of Congress Cataloging-in-Publication Data

ISBN 13: 978-0-12-088560-2
ISBN 10: 0-12-088560-3

For information on all Morgan Kaufmann publications,
visit our Web site at www.mkp.com or www.books.elsevier.com

Printed in the United States of America
05 06 06 07 08 5 4 3 2 1

Smart Phone and Next-Generation Mobile Computing

The Morgan Kaufmann Series in Networking
Series Editor, David Clark, M.I.T.

Smart Phone and Next-Generation Mobile Computing
Pei Zheng and Lionel Ni

GMPLS: Architecture and Applications
Adrian Farrel and Igor Bryskin

Network Security: A Practical Approach
Jan L. Harrington

Content Networking: Architecture, Protocols, and Practice
Markus Hofmann and Leland R. Beaumont

Network Algorithmics: An Interdisciplinary Approach to Designing Fast Networked Devices
George Varghese

Network Recovery: Protection and Restoration of Optical, SONET-SDH, IP, and MPLS
Jean Philippe Vasseur, Mario Pickavet, and Piet Demeester

Routing, Flow, and Capacity Design in Communication and Computer Networks
Michal Pióro and Deepankar Medhi

Wireless Sensor Networks: An Information Processing Approach
Feng Zhao and Leonidas Guibas

Communication Networking: An Analytical Approach
Anurag Kumar, D. Manjunath, and Joy Kuri

The Internet and Its Protocols: A Comparative Approach
Adrian Farrel

Modern Cable Television Technology: Video, Voice, and Data Communications, 2e
Walter Ciciora, James Farmer, David Large, and Michael Adams

Bluetooth Application Programming with the Java APIs
C. Bala Kumar, Paul J. Kline, and Timothy J. Thompson

Policy-Based Network Management: Solutions for the Next Generation
John Strassner

Computer Networks: A Systems Approach, 3e
Larry L. Peterson and Bruce S. Davie

Network Architecture, Analysis, and Design, 2e
James D. McCabe

MPLS Network Management: MIBs, Tools, and Techniques
Thomas D. Nadeau

Developing IP-Based Services: Solutions for Service Providers and Vendors
Monique Morrow and Kateel Vijayananda

Telecommunications Law in the Internet Age
Sharon K. Black

Optical Networks: A Practical Perspective, 2e
Rajiv Ramaswami and Kumar N. Sivarajan

Internet QoS: Architectures and Mechanisms
Zheng Wang

TCP/IP Sockets in Java: Practical Guide for Programmers
Michael J. Donahoo and Kenneth L. Calvert

TCP/IP Sockets in C: Practical Guide for Programmers
Kenneth L. Calvert and Michael J. Donahoo

Multicast Communication: Protocols, Programming, and Applications
Ralph Wittmann and Martina Zitterbart

MPLS: Technology and Applications
Bruce Davie and Yakov Rekhter

High-Performance Communication Networks, 2e
Jean Walrand and Pravin Varaiya

Internetworking Multimedia
Jon Crowcroft, Mark Handley, and Ian Wakeman

Understanding Networked Applications: A First Course
David G. Messerschmitt

Integrated Management of Networked Systems: Concepts, Architectures, and their Operational Application
Heinz-Gerd Hegering, Sebastian Abeck, and Bernhard Neumair

Virtual Private Networks: Making the Right Connection
Dennis Fowler

Networked Applications: A Guide to the New Computing Infrastructure
David G. Messerschmitt

Wide Area Network Design: Concepts and Tools for Optimization
Robert S. Cahn

For further information on these books and for a list of forthcoming titles, please visit our Web site at http://www.mkp.com.

Contents

6 Mobile Security and Privacy 335

About the Authors

Pei Zheng was an assistant professor in the Computer Science Department at Arcadia University and a consultant working in the areas of mobile computing and distributed systems during the writing of this book. Dr. Zheng received his Ph.D. degree in Computer Science from Michigan State University in 2003. He was a Member of Technical Staff in Bell Laboratories, Lucent Technologies. He joined Microsoft in 2005. His research interests include distributed systems, network simulation and emulation, and mobile computing.

Lionel M. Ni is Professor and Head of the Computer Science Department at the Hong Kong University of Science and Technology. Dr. Ni earned his Ph.D. degree in electrical and computer engineering from Purdue University, West Lafayette, IN, in 1981. He was professor in Computer Science and Engineering Department at Michigan State University, where he started his academic career in 1981. He has been involved in many projects related to wireless technologies, 2.5G/3G cellular phones, and embedded systems. He is the co-author of the book *Interconnection Networks: An Engineering Approach* published by Morgan Kaufmann in 2002.

Preface

What will be the next wave of computing? What will be the driving force behind the constantly changing mobile wireless systems and applications? To answer these questions, let us first look into the history of computing. The first wave of computing was the wide use of Personal Computers (PC) that enabled the general public to take advantage of computing power for business operations or personal use. Before the first wave, computers were generally used by large companies and governments for mission critical purposes. The second wave of computing was the amazing Internet, which greatly empowered people with unprecedented levels of information sharing and data access via computer networks. Many believe the third wave of computing will be the advent of mobile computing that eventually liberates people from relying on a computing or communication device at a fixed location to access the network. Thanks to the remarkable advances in a broad set of wireless technologies, in the microprocessor and storage technologies, in the convergence of network infrastructure, and in the pervasive computing facilities, people will soon be able to access and consume information anytime, anywhere, from any device. As the most widely used computing device, a cell phone or a new-generation smart phone, will likely play a significant role in future mobile computing paradigm. This vision also encompasses a broad range of emerging mobile wireless technologies and novel services and applications. Needless to say, in order to realize this vision there are still many challenges—and opportunities—in almost every aspect of the mobile computing realm.

Scope and Outline of the Book

The book provides an in-depth coverage of next-generation mobile computing paradigm, including mobile wireless technologies, mobile services and applications, and research and development challenges surrounding backend systems, network infrastructure, and mobile terminals including smart phones and other mobile devices. It substantially emphasizes the following components.

Part I: A Detailed Survey of Emerging Mobile Wireless Technologies

As the ancient Greek philosopher Heraclitus of Ephesus pointed out, "Change is the only thing in the world that is unchanging." This is downright true especially in the mobile computing domain. New mobile wireless technologies continue to show up; even the old voice-centric cellular networks are undergoing fundamental changes. This book will provide state-of-the-art coverage of these technologies that show a strong potential to deeply change people's life, such as 3G cellular networks, wireless LAN, Bluetooth, Ultra Wide Band, WiMax, ZigBee, sensor networks, battery for mobile devices, mobile display technology, among others. Moreover, the coexistence and integration issues of these technologies are so important that a significant portion of the book is devoted to it.

Chapter 2 to Chapter 4 discusses the trend of convergence and integration in mobile wireless networks and emerging wireless technologies. Major software platforms such as Symbian, Microsoft Windows Mobile for Smartphone and Palm OS are also introduced briefly. Topics covered in Part I is the basis for readers to move forward to the remaining chapters.

Chapter 2, "The Next Wave of Computing," is concerned with the trend in mobile computing and related technical issues. The chapter first looks back into the evolution of the cell phone and PDA, as well as some handheld computers. Based on that, the chapter introduces the trend of the convergence between the cell phone and PDA, and between the cell phone and a mobile entertainment device. A smart phone, a converged mobile device, will likely be used as the universal mobile terminal alongside with the trend of convergence. Hence the chapter continues to identify a range of smart phone applications and services for consumers and businesses. The chapter moves forward beyond next-generation mobile devices to some fundamental challenges and opportunities that will turn the vision of mobile computing and pervasive computing into everyday reality.

Chapter 3, "Supporting Wireless Technologies," provides an extensive study of existing mobile wireless technologies. Much of the emphasis is on the highly anticipated 3G cellular networks and widely deployed wireless LANs, as the next-generation smart phone will likely offer at least these two types of wireless connectivity. Other wireless technologies, which either have already been commercialized, or still are in the intensive research and standardization stage, will be covered as well. Additionally, in order to explain how standards and politics play important roles in wireless industry, a summary of worldwide partnership groups, major competitors, and governmental policy makers is presented.

Chapter 4, "Mobile Terminal Platforms," focuses on the platforms for mobile terminal devices, including mobile embedded processors, memory and storage, extension interfaces, and display. On top of mobile hardware, a wide range of mobile software platforms are competing with each other. These mobile platforms are, to name a few, Symbian, Palm OS, Microsoft Windows Mobile, and embedded Linux, as well as middleware runtime such as J2ME and BREW. An overview of each competing operating system and corresponding software development platform is presented. The chapter also introduces various software emulators of those platforms and mobile markup languages.

Part II: An In-Depth Discussion of Mobile Computing Challenges and Approaches

To fulfill the need for mobile and pervasive computing, researchers and practitioners have to face many challenges encountered in both design and implementation of a viable application or service. To help the readers quickly gain an insightful understanding of the mobile computing paradigm, the book identifies a list of compelling and crucial research and technical issues, and then discusses some of the novel approaches to these problems, as well as highlighting future research directions in those fields. For each issue, the role of a smart phone will always be taken into consideration. As mentioned, the book will primarily focus on networking- and application design-related issues.

The history of the computer industry shows that services and applications utilizing whatever advanced technologies must satisfy business or consumer needs in a timely manner in order to survive and become widespread. Therefore, the utmost concern of service providers is, what do the users really want? The last chapter of Part II tries to answer this question by looking at a collection of mobile wireless application scenarios in different industries and trying to find the general guidelines to designing and implementing a viable application or service. Common software building blocks of mobile applications and services are briefly discussed as well.

Chapter 5 to Chapter 7 emphasizes the challenging issues for the future of mobile computing. These issues mainly fall into three categories: mobile network infrastructure design issues, mobile terminal hardware and software design issues, and application-specific issues. Each covered issue is concisely described and then thoroughly discussed with some real-world examples. The authors' view and ideas on these issues are also presented.

Chapter 5, "Mobile Networking Challenges," concerns a number of critical issues in the research and development of heterogeneous mobile wireless networks that a smart phone will interface, including next generation mobile networks, integration of wireless LAN and cellular networks, Mobile IPv6, wireless TCP, mobile ad hoc networks, mobile Quality of Services, and so on. The chapter provides a glimpse into the latest development of those challenges in both academia and the wireless and computing industry. Overall we intend to identify the motivation of building those wireless systems, and then discuss surrounding technical challenges and some prototype or commercial systems if any.

Chapter 6, "Mobile Security and Privacy," discusses the security problems and solutions in mobile computing, with emphasis on the issues of heterogeneous network authentication, authorization, accounting, and wireless LAN security. Privacy is probably another major concern that may impede the deployment of mobile technologies using smart phones, especially when pervasive mobile devices are becoming part of the environment. The chapter will describe keys issues of mobile privacy, and then highlight looming privacy-related issues in next-generation mobile computing paradigm.

Chapter 7, "Mobile Applications Challenges," introduces some of the interesting problems in designing next-generation smart phone-based mobile services, such as wireless Web, m-commerce, location-aware computing, and mobile wireless multimedia. These examples represent state of the art in this broad area. It talks about the planning and architectural design of smart phone mobile applications for enterprise and consumers. It does so by highlighting available technical building blocks and general guidelines to mobile application design, and by discussing some examples of such existing applications in the real world. It covers a full spectrum of software and hardware components of mobile applications, ranging from XML web services, to wireless application gateway, to mobile enterprise solutions, as well as a list of potential applications of smart phone in various scenarios. These may motivate the yet-to-come "killer app" in the wireless world and open up huge business opportunities in the next several years. We also introduce some commercial systems and prototypes in each category as examples to show how it can help business and the end user.

On the other hand, the book is NOT a mobile software programming guide, nor is a reference book of wireless communication concepts and theories. It does highlight existing software platforms and development tools, but only on a very high level. The book will introduce some wireless communication fundamentals, but clearly the focus is on the computing side, especially on networking and application issues. Readers interested with those areas can easily find many books with excellent coverage.

Audience

The book's audience may belong to diverse groups in industry and academia. Potential readers are industrial professionals, including technical managers, software architects and developers, and academic researchers who are willing to know more about mobile computing and smart phone technologies, potential services, applications, and future technical trends. The book should also appeal to students taking senior- or graduate-level mobile computing courses, assuming they have a general knowledge of computer networks.

Industry Technical Managers and Software Architects and Engineers

Technical managers in the mobile wireless industry may find the book useful in providing a thorough coverage of emerging mobile and wireless technologies that can be incorporated into the design of next-generation mobile applications and services targeting smart phones. Even for software developers working on the programming with smart phones and other handheld devices, a comprehensive and solid understanding of the concepts, technologies and architectural design of the whole system would definitely be invaluable. Moreover, the book's coverage and analysis on common software building blocks of mobile applications will help architects design and implement a new mobile application.

Technology Evangelists, Strategists, and Analysts

In conjunction with an in-depth discussion of technical details and research issues surrounding mobile computing and smart phones, the book also explores conceivable and potentially viable mobile applications and services based on technical and market trend in the mobile wireless industry. In this book, the coverage of mobile wireless technologies mostly concludes with a brief summary of its business perspective, status quo and trend of the market, and a standardization process if any. Technical analysts, strategists and evangelists who want to find out how the wireless technologies and applications will evolve and what the new market for growth is will particularly benefit from reading these sections.

Academic Researchers

For academic researchers such as faculties and students in the mobile wireless research community, this book serves as a full-coverage guide by examining almost all the important aspects of mobile computing and pervasive computing. Those academic researchers who have been actively involved in the research of specific areas in mobile computing may read corresponding sections of the book to grasp major ideas and approaches, as well as future research directions, in those areas. Students and instructors will find the book an excellent text or reference for mobile computing courses.

In all cases, to assist readers to probe further, the "References" and "Further Reading" sections at the end of each chapter offer pointers to a collection of carefully selected research projects, papers, standards, and market analyses. Please also see the URL www.books.elsevier.com/computerscience?isbn=0120885603.

Acknowledgments

Many of our colleagues in academics and industry have made significant contributions to the writing of this book. We would like to thank Anand Tripathi (University of Minnesota), Baijian Yang (Ball State University), Dapeng Wu (University of Florida), and Paul Huang (Sun Microsystems) for their insightful suggestions.

We would like to thank our employers, Arcadia University, USA, and Hong Kong University of Science and Technology for their support on the book project. We gratefully thank our manuscript reviewers, Lisa Phifer and Tim Thompson, for their insightful and detailed comment and suggestions on the structure of the book, topics to be covered, and presentations.

Our special thanks go to Rick Adams for his generous support throughout the writing and production of the book. Thanks are also due to Paul Gottehrer and Rachel Roumeliotis at Morgan Kaufman/Elsevier for their wonderful work.

Pei Zheng and Lionel M. Ni

Hong Kong
December 2005

1

Introduction to Smart Phone and Mobile Computing

You have probably heard the term *smart phones* when people talk about some fancy gadgets that look like a combination of a cell phone and a personal digital assistant (PDA), and you must have heard the term *wireless* many times, mostly from advertisements of cell phones and service plans by big wireless service providers. If you are working in the wireless industry or computer industry, then you should be familiar with the notion of *mobile computing* and should have some idea of the concept of *pervasive computing*. What exactly do these terms refer to? What are the similarities and differences between them? Furthermore, what are the cutting-edge technologies and emerging applications of the next generation of mobile computing? What are some of the challenging issues in this domain? Before we embark on an exciting journey into the fantastic world of mobile computing, we should answer these questions.

The first chapter serves as an introduction to the remainder of the book. It begins by defining a number of key terms and concepts used throughout the book, such as smart phone, wireless communication, nomadic computing, and mobile computing, and continues with an overall discussion of the next wave of computing and the role of smart phones in that context. The first chapter briefly introduces a range of mobile wireless technologies and applications, as well as key issues and challenges in this field.

1.1 Smart Phone and Mobile Computing

The next generation of mobile computing will foster the convergence of communication, computing, and consumer electronics, three traditionally distinct industries with quite low interoperability. On the front end, a smart phone is likely to become a universal mobile terminal carrying integrated functionalities augmented by mobility and ubiquitous network access.

1.1.1 Definitions

First off, what is a *smart phone*? What are the differences between a smart phone and a cell phone? What are the implications of mobile computing? How does mobile computing relate to wireless communication? People hear these terms, but very often they are being used by the media quite vaguely. A significant amount of literature from both academia and industry discusses research and development topics regarding smart phones and mobile computing, but some of these articles have confused the key terms in this domain, especially *mobile* and *wireless*. In many cases, these two terms are regarded as being the same and are used interchangeably; however, they are not identical.

1.1.1.1 Mobile Computing

Mobile computing refers to a broad set of computing operations that allow a user to access information from portable devices such as laptop computers, PDAs, cell phones, handheld computers, music players, portable game devices, and so on. The two operational modes in mobile computing are disconnected mode and connected mode. In the disconnected mode, information access on a mobile device is local, such as when someone uses a PDA to manage a schedule and an address book. In connected mode, the mobile device supplies one or more types of wireless or wired network connectivity to enable network access. In the disconnected mode, a user can synchronize data on a mobile device with a computer. The synchronization may involve both download from and upload to the host computer.

In the connected mode, applications on a mobile device are able to communicate directly with other mobile devices or back-end systems via a network connection. As mobile wireless technologies mature and provide a higher data rate at a lower cost, we will be able to take advantage of a new range of mobile devices that allow network mode operations, fulfilling the need for ubiquitous mobile access anywhere, anytime.

In wired mobile computing, network access from a mobile device is enabled via a wired network. For example, people bring their laptop computers to conferences, offices, and hotels and then connect to wired networks, such as a local area network (LAN), or they connect to the Internet via asymmetric digital subscriber line (ADSL), cable modem, or dialup. Wireless mobile computing (or mobile wireless computing) emerged and proliferated at a staggering pace as a result of the dramatic advancements of an assortment of wireless technologies, such as cellular wireless networks, wireless LAN, Bluetooth, infrared, and wireless sensor networks. These wireless technologies essentially make it possible for a user to enjoy such mobility that the location of a mobile device user no longer has to be fixed to maintain a network connection. The advantage is obvious: Tasks that had to be performed on a computer at a fixed location can now be done on the move.

The flexibility offered by mobile computing and its natural integration with heterogeneous network access have given rise to a large number of novel business and consumer applications and services utilizing voice, messaging, and data communication. Looking a few years down the road, as wireless data technologies continue to evolve, mobile network operators and service providers have begun to leverage powerful, wireless-enabled, always-on mobile devices and high-bandwidth, ubiquitous network access to provide a new spectrum of applications and services that effectively provide an unprecedented level of information access and assistance anywhere, anytime, from any device.

The notion of *nomadic computing* refers to a special case of mobile computing — using a mobile device to connect to a wired or wireless network intermittently from place to place with support for high-level mobility. A typical application of nomadic computing would be traveling salespersons and technical professionals who connect to a corporate network from office, homes, customer sites, hotels, airports, and so on, being able to consume the same set of services provided by the network transparently and conveniently regardless of location. An always-on, active network connection is not required in nomadic computing.

1.1.1.2 Wireless Communication

As the term *wireless* implies, no wires are attached to devices in the wireless domain; instead, various radio transmission techniques are utilized to connect either two devices or a device and a shared infrastructure. Numerous wireless technologies have been utilized by mobile computing to enhance a user's experience with regard to both user mobility and device mobility. Wireless does not necessarily mean mobile, however, because some wireless technologies, notably cable-replacement wireless technologies, can only be used when one or both of the parties are fixed or barely move during the wireless signal transmission, such as the short-range Bluetooth or infrared and long-range WiMax.

To summarize, some mobile applications do not require wireless technologies but some wireless applications are not mobile. Clearly, wireless communication places an emphasis on the radio layer and link layer of the network protocol stack (e.g., how the signal is transmitted and received), whereas mobile computing takes advantage of ubiquitous wireless communications to build applications and services. In reality, the notion of *wireless* has been generally used (or abused?) to refer to cellular services provided by major wireless service providers such as Cingular Wireless, T-Mobile, or Sprint.

1.1.1.3 Smart Phone

The term *smart phone* was initially coined by unknown marketing strategists to refer to a then-new class of cell phones that could facilitate data access and processing with significant computing power. In addition to traditional voice communication and messaging functionality, a smart phone usually provides personal information management (PIM) applications and some wireless communication capability. Roughly speaking, a smart phone is like a small, networked computer in the form of a cell phone. The very first generation of cell phones, despite their large size, could barely offer anything other than making phone calls. Later on, because of astounding advances in semiconductor technology, cell phones were generally equipped with far more powerful processors, larger storage, and a liquid-crystal display (LCD) screen that made it possible to perform some computing tasks locally. Common cell phone applications, collectively referred to as PIM applications, include calendar, address book, organizer, and calculator functions. Data networking capabilities are generally very limited on these cell phones. In essence, this generation of cell phones can be considered a combination of voice-centric cell phone and PDA. As the need for mobile data access on a cell phone became evident, the next generation of cell phones — smart phones — emerged. A smart phone usually supports one or more short-range wireless technologies such as Bluetooth and infrared, making it possible to transfer data via these wireless connections in addition to cellular data connections. It is the smart phone that can provide computer mobility, ubiquitous data access, and pervasive intelligence for almost every aspect of business processes and people's daily lives. Aside from traditional cell phone PIM applications, other typical smart phone applications include simple games, built-in camera, audio/video playback and recording, instant messaging, e-mail, and wireless Internet, among others. In addition, smart phones can be used as a mobile terminal for e-commerce, enterprise applications, and value-added, location-based services. In a word, the smart phone is the future of today's cell phone, as it offers dramatically enhanced wireless capability, computing power, and on-board storage. This book uses the terms *cell phone* and *mobile phone* interchangeably when referring to a voice-centric cellular wireless device.

Today, the general public perceives of *smart phones* as high-end, multifunctional, business-centric cell phones with high-resolution color displays and fast mobile processors which are unaffordable to ordinary consumers due to the cost of the phone device (hundreds of dollars) and wireless data services ($30 to $60 per month). Like many other popular computing devices, however, smart phones will certainly follow the same track and be embraced by the mass market as a result of ever-dropping hardware and service prices and emerging killer applications.

The vision of "anytime, anywhere, from any device" for mobile computing naturally leads to the issue of building a universal mobile platform for reliable and high-performance computing with heterogeneous, seamless wireless access via limited computing resources. Smart phones are generally considered to be one of the promising candidate for such purposes.

1.1.2 Smart Phone: The Universal Mobile Terminal

Today's smart phone packages a potpourri of wireless technologies and applications, and many more are being considered to become integral parts of a smart phone. Indeed, usage of a smart phone is actually driven by the convergence of the different network infrastructures of computing and communication. In addition, we will see functional components of some consumer electronics devices on a smart phone, as well as mobile applications that bridge these otherwise largely independent components: computing, communication, and consumer electronics (3 C's). Moreover, the computing and communication environment is moving to interact with the physical environment or even become a part of it. Looking several years ahead, we can envision the convergence of computing, communication, and consumer electronics, dubbed *digital convergence* (Figure 1.1), in a pervasive way, with the smart phone as the front end — the universal mobile terminal. No doubt this vision of next-generation mobile computing poses a challenge to the design of a smart phone and its application, as well as raising networking infrastructure issues.

1.1.2.1 Convergence of Computing and Communication

The future of communications, as many industry professionals envisage, will involve the convergence of computing and

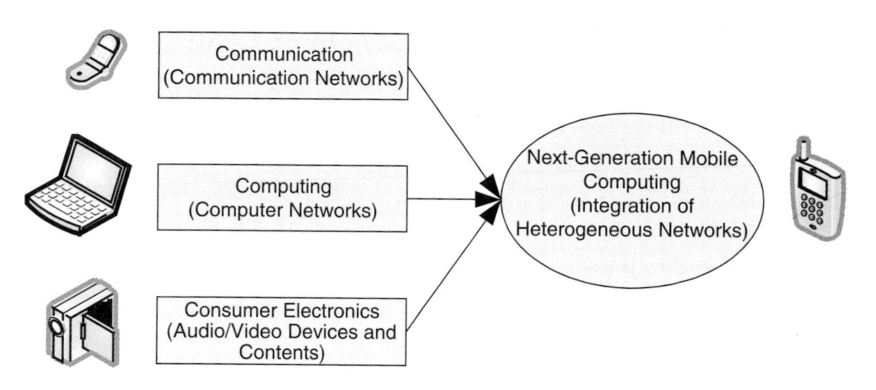

Figure 1.1 Digital convergence.

communications in almost every aspect of information technology, thus allowing information access anywhere, anytime, from any device. This trend will have a huge impact on everyday life and how enterprises, organizations, and government operate. Tomorrow's unwired world will encompass advances in emerging mobile wireless communication networks (the infrastructure), intelligent mobile devices (the terminal), and novel valuable services and applications (the applications).

With regard to the network infrastructure, the remarkable advances in the link bandwidths of wired networks over the last 10 years are an excellent example of how technologies can evolve beyond our expectation. Recall that the early 10-Mbps Ethernet utilized coaxial cable and twisted pair cable, and then came the 100-Mbps fast Ethernet, the gigabit Ethernet, and today's OC192 10-Gbps fiber in the Internet backbone. Network access evolved from 14.4-kbp or 33.6-kbp modems to 64-kbp integrated services digital network (ISDN) channels to ADSL and cable modems and, soon, to wireless LAN and WiMax. The wired network not only supplies an increasingly higher link capacity but has also managed to reach almost everywhere in the world. People are connected via computer networks more than ever before. Similarly, the wireless networks, despite the current poor bandwidth of data transmission, are likely to follow the same journey of wired networks with respect to link data rate and scale. In fact, wireless networks are poised to have a greater impact on our society than wired networks. The traditional voice-centric, circuit-switching, cellular networks are being replaced by data-centric, packet-switching 2.5 G and 3 G networks. Wireless LANs are being deployed in office buildings, residences, hotels, coffee shops, restaurants, and airports. WiMax, arguably a better "last-mile" solution for wireless broadband network access compared to cable modem or DSL, is likely to be rolled out on a large scale very soon; satellite digital television and radio are reaching an increasing number of subscribers. All of these wireless networks, coupled with the underlying wired networks, offer mobility and portability with sufficiently large bandwidth and high-quality wireless data transmission.

With regard to the mobile terminal aspect, it is clear that a mobile device will no longer be considered merely as a communication handset; rather, by providing more powerful computing features such as high-performance mobile processors and large flash memory, a mobile device will be able to accomplish more intensive and intelligent computing tasks such as multimedia processing and data logistics. In particular, the cell phone and PDA are beginning to converge into a single device — a smart phone — that offers a full set of applications, including the necessary optimized wireless communication facilities and computing components, seamlessly integrated for an entirely new collection of business and consumer applications. For example, a smart phone may be able to utilize voice over wireless LAN to allow long-distance calls and roaming between wireless LAN and cellular networks. Also, a user could play a location-based, multiplayer mobile game via Bluetooth during a boring conference session; the locations of the gamers could be determined using the location-sensing capability of the smart phone. For example, the location sensing system can utilize radio-frequency identification (RFID) readers, and RFIDs embedded in the conference badges. In fact, some of these applications have already been demonstrated by researchers using customized phone devices. Indeed, the converged universal mobile terminal provides an open and powerful platform for ubiquitous computing and communication to facilitate a variety of applications that otherwise could not be realized.

No matter how the network infrastructure is built and how the converged device is designed and manufactured, the technology must include desired applications and services to reach users. To this end, mobile network operators, service providers, and software providers must design applications and services with what people really want in mind. Moreover, the convergence of communication and computing makes the applications and services more complex in terms of interfacing heterogeneous networks and dealing with interoperation among different computing units. Nevertheless, many believe that as the enabling wireless technologies continue to mature, "killer apps" offered by the next-generation mobile wireless networks will eventually surface to enjoy widespread acceptance.

Aside from the convergence of computing and communication, a broader vision of convergence, dubbed *digital convergence,* paints an ambitious picture of a future wireless world. In this vision, not only are the computing (computer, data network, and software vendors) and communication (wireless network and service providers) industries involved but also the consumer-electronics industry. Digital convergence involves the same opportunities and challenges of mobile computing but creates many more that could ultimately alter these three industries. As an example, imagine a home network where the computers, televisions, DVD players, game devices, mobile phones, and cordless phones are all connected to a wireless network for communication.

1.1.2.2 Pervasive Computing

Existing mobile and wireless technologies have been, and continue to be, the core building blocks of mobile applications and services; however, an entirely new array of wireless technologies has already emerged from pioneering academia and research laboratories. As a consequence, the new vision of the future of computing blurs the borders of human environment and the computer to produce *pervasive* or *ubiquitous computing*. Some researchers have considered the "anytime, anywhere" goal of mobile computing to be a reactive approach to the need for information access, whereas pervasive computing is perceived as a proactive approach with the vision of "all the time everywhere" [1]. The fundamental distinction between pervasive computing and any other computing framework is the way the user interacts with the computing facility. The traditional computing framework allows users to utilize computers and networks to perform tasks by running programs (a piece of software) in a virtual logic environment that has nothing to do with the users' physical environment. In contrast, pervasive computing promises to build software, devices, and networks that are deeply embedded in the user's physical environment. In fact, users may not even notice that they are dealing with a computing environment because the facilities are seamlessly and naturally integrated with the physical environment.

In addition to those technologies in the domain of mobile computing, the development of new enabling technologies for pervasive computing is gaining the attention of researchers and practitioners. Some examples include wireless sensors, actuators, wireless sensor networks, wearable computers, and large with flexible displays. A unique characteristic of pervasive computing is that its applications and services should disappear into the physical environment, as Mark Weiser envisaged more than 20 years ago [2]. Whether accomplishing this is a Holy Grail or an achievable reality may remain the subject of debate over the long time required for pervasive computing to mature.

In the context of pervasive computing, smart phones may assume even more tightly integrated functionalities than in the context of mobile computing. Augmented by the ubiquitous, heterogeneous, and always-on wireless communications and intelligent sensing of both the device and the environment, smart phones, as a universal mobile terminal, will finally become an integral part of the physical environment. This book discusses smart-phone-related mobile wireless technologies, issues, challenges, and innovative applications and services in the context of mobile computing and pervasive computing.

A conceptual diagram of some notions related to mobile computing is shown in Figure 1.2. As it shows, nomadic computing can be considered a subset of mobile computing, which is in turn a subset of pervasive computing. Applications and services within

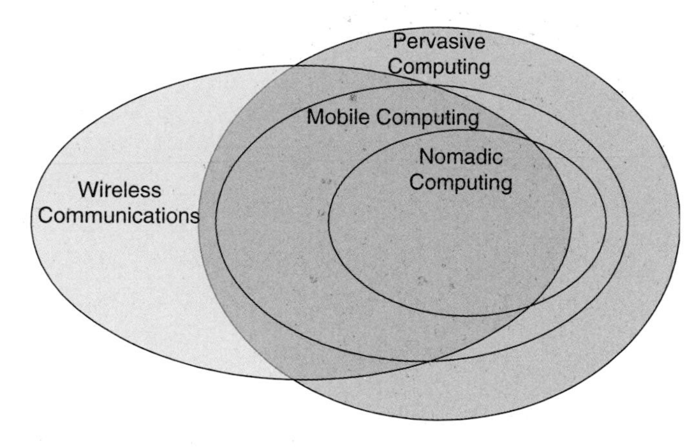

Figure 1.2 Some notions of mobile computing.

these three computing domains, for the most part, leverage wireless communications to provide mobility support on various scales. It is worth noting that the notion of *wireless* often encompasses both fixed and mobile wireless technologies and systems.

1.2 Emerging Mobile Technologies and Applications

Despite economic conditions in the first few years of the 21st century and slowdowns in the computer and communication industries, new mobile and wireless technologies have never stopped evolving and have paved the way for an intelligent and pervasive mobile environment. The following is a brief survey of those emerging technologies.

1.2.1 Cellular Networks

The second generation (2G) of digital cellular networks and services is an unprecedented success worldwide. Voice-centric mobile wireless communication services successfully capitalized on the mobile network operators' (MNOs) huge investment in the purchasing of spectrum frequency and the wireless network infrastructure. For example,

2.5 G wireless data services such as GPRS and EDGE are being used by a large number of subscribers in Asia and Europe. UMTS/WCDMA and WCDMA2000, the long-awaited 3 G networks and services, are expected to deliver high bandwidth and reliable and secure data access with a data rate of several megabits per second. 3 G services available in the United States include cdma2000 1x-EVDO (Verizon) and WCDMA/UMTS (Cingular/AT&T Wireless), among others. In Europe, where UMTS/WCDMA dominates the market, Vodafone and Orange began to provide 3 G data services in selected cities in 2004. In Asia, Japan's NTT DoCoMo (WCDMA) and Korea's SK Telecom (cdma2000 1x-EVDO) were the first to launch 3 G mobile services, in 2001 and 2002. (For a complete list of 3 G services available worldwide, see http://www.3gtoday.com/operators/index.html.)

Furthermore, some European and Asian MNOs are considering building 4 G networks, which have demonstrated a data rate around 300 Mbit/sec in field trail. No matter which mobile technology will win in the future, one thing is for sure: The cellular network will continue to be a core component of the service package on smart phones, for both technological and economical reasons.

1.2.2 802.11 Wireless LAN

The Institute of Electrical and Electronic Engineers (IEEE) 802.11 wireless LAN technology is a perfect example of a technology that initially maintains a low profile yet experiences explosive growth in the residential network market. A survey conducted by IDC in 2003 found that 34% of those who responded connect to a wireless LAN network at home, compared with 27% at work, and network operators are expected to install more than 55,000 new hot spots in the United States over the next 5 years.

Such a huge success is largely due to the flexibility in setting up a wireless LAN, and the low cost of the wireless LAN devices. In the United States, the IEEE 802.11 b and 802.11 g standards operate on the unlicensed 2.4-GHz industrial, science, and media (ISM) band and can provide data rates of 11 Mbit/s and 54 Mbit/s, respectively, whereas the IEEE 802.11a standard uses the 5-GHz unlicensed national information infrastructure (U-NII) band and can deliver a data rate up to 54 Mbit/sec. Despite its growth in the residential market and in some enterprises, many government departments and large companies have banned the use of this technology due to security concerns; both the initial wired equivalent privacy (WEP) and interim Wi-Fi protected access (WPA) protocols have been proven to have serious security flaws [3,4]. IEEE has formed several work groups to address these security issues, as well as other open issues including quality of service and interference with other wireless technologies operating at the same band. For example, in 2004, IEEE ratified the 802.11i security standard for wireless LANs.

1.2.3 Wireless Mesh Network

Because of the remarkable popularity of wireless LAN technology and the considerably higher bandwidth it can provide for wireless data access as compared to cellular networks, a few upstarts are building citywide and even nationwide wireless LAN networks that allow seamless roaming between access points on a large scale, effectively creating a *wireless mesh network*. A key characteristic of a wireless mesh network is that it is a peer-to-peer, self-configuring system in which each mesh node can relay messages on behalf of others, thus increasing the coverage, available bandwidth, and system reliability. Numerous wireless mesh network routers are currently in the market. IEEE has formed an ESS Mesh Networking Task Group (802.11s) in an effort to standardize future wireless mesh network protocols and interoperability.

1.2.4 WiMax

Another example of wireless LAN is WiMax (Worldwide Interoperability for Microwave Access). WiMax is an industry forum formed to promote point-to-point broadband wireless access to home and mobile users. The WiMax technology is based on IEEE 802.16 and European Telecommunications Standards Institute (ETSI) HIPERLAN.

It has two forms of wireless service: line of sight (between backhaul towers) and non-line of sight (to residences and business users). It has the potential to achieve a range of 31 miles and can provide a bandwidth of up to 70 Mbit/sec, which is equivalent to dozens of T1 lines and Internet access of well over a few hundred using DSL or cable model. Intel is the strongest supporter of WiMax. As of this writing, Internet service providers in some major cities are in the process of conducting field tests. The large-scale rollout of WiMax services to end users may begin within the next 1 to 2 years.

1.2.5 Wireless Sensor Network

A sensor is a tiny, low-cost, low-power device that can sense fields and forces in the physical world and translate stimuli, such as motion, heat, light, sound, or pressure, into numerical data. With radio capability, a wireless sensor node can transmit data regarding a state change to other nodes. Tens of thousands of inexpensive wireless sensors can essentially form a wireless sensor network that, once deployed to urban environments or inhospitable terrains, will allow coordination of information gathering and processing in a fully self-organized fashion over a long period of time. The wireless sensor network technology is a key component of pervasive computing because it allows computing facilities to deeply immerse themselves in the physical environment. Some applications of wireless sensor networks include environmental monitoring, habitat monitoring, and traffic control. Ongoing research on the marriage of sensors and mobile devices such as PDAs and smart phones has revealed many interesting potential applications. The challenges of wireless sensor networks include query-informed routing, in-network query processing, security and privacy issues, and sensor system design issues.

1.2.6 RFID

A radio-frequency identification (RFID) tag is a small integrated circuit that can respond to an interrogating signal with some simple data using an on-board antenna. The RF data transmission does not require line of sight. Passive RFID tags do not require batteries for data transmission, whereas active RFID tags with more complex circuits do requires batteries to operate. In a typical application scenario, an RF reader is able to collect data from many RFID tags wirelessly and even update data on those RF tags. This makes the RFID technology extremely compelling for asset and people tracking, inventory monitoring, access control and ticketing, etc. As a sign of the ultimate widespread use of RFID tags, Wal-Mart asked its top 100 suppliers to initiate use of RFID tags on cases and pallets of consumer goods by January 2005. Most of these top 100 suppliers have developed RFID solutions in order to comply with this mandate. Security concerns, privacy issues, and the lack of standards, however, have effectively slowed down the deployment of RFID tags.

1.2.7 WPAN

The wireless personal area network (WPAN) is a collection of short-range, low-power, low-cost wireless technologies, including infrared (IR), Bluetooth, and ultra-wideband (UWB). Infrared is a point-to-point cable replacement technology commonly used on laptop computers and portal devices. Its operating range is quite limited at less than 6 feet, and the effective bandwidth is not suitable for large file transfers. Bluetooth is already widely used in cell phones and PDAs for wireless synchronization within a range of 30 feet using the unlicensed 2.4-GHz spectrum. It also has the potential to build an *ad hoc* piconet for group communication. The latest addition to the WPAN family, and probably the most revolutionary one, is UWB. UWB offers very high bandwidth data transmission: up to 480 Mbit/sec within a range of up of 6 feet and 110 Mbit/sec within 30 feet. The beauty of UWB is that it can coexist with technologies on the same spectrum, thanks to its unique radio characteristics. UWB is still in its infancy; however, it may eventually replace Bluetooth as the prevalent WPAN solution on mobile devices in the next few years.

1.3 Issues and Challenges

The relentless evolution of mobile wireless technologies has given rise to novel applications and services to satisfy the needs of businesses for cost-effective, highly efficient, and robust operations and to provide a rich experience for consumers. These developments in turn raise research issues and challenges yet to be addressed.

1.3.1 Mobile Localization and Location-Based Services

Not until the era of mobile computing did physical location become a significant issue for mobile applications and services. The questions pertinent to a user's location include: Where am I? What information is available in the vicinity? What is happening in those places I am interested in? How do I get there? Where are the other people I know or I want to know? In some cases, the network operator may be asked to report the location of a user with a mobile device. These challenging questions effectively spur the convergence of mobile positioning and tracking techniques, either indoor or outdoor, with geographical information databases to create an adaptive virtual world for each user.

1.3.2 *Ad Hoc* Networks

An *ad hoc* network is a self-organized group of nodes without a fixed infrastructure. Each node in an *ad hoc* network relies on its neighboring nodes to communicate and cooperate with anyone else in a fully distributed way. The topology of the network may constantly change if the nodes are free to move around, thus creating a mobile *ad hoc* network. Potential applications of *ad hoc* communication and collaboration among mobile devices include battlefields, conference, sports stadiums, disaster relief efforts, and even classrooms. The absence of a fixed infrastructure in an *ad hoc* network introduces some challenging issues, such as dynamic network configuration, service discovery, and multihop routing. Because nodes in the network may constantly move, both the moving nodes and the affected stable nodes must adapt to this change in terms of addressing and topology updates. Furthermore, a node must conduct a reliable and fast discovery procedure to determine which resources or services are available on which nodes while moving around. To access a resource and service shared by some nodes outside of its transmission range, the node will rely on an efficient multihop routing mechanism to reach the destination node.

1.3.3 Integration of Heterogeneous Wireless Networks

The aforementioned trend of convergence of computing and communication faces a few challenges in the technological and business domains. Heterogeneous network infrastructures have to coexist and operate with each other to support internetwork roaming and handoff and must be transparent to users. Some trial tests of wireless LAN and cellular integration by wireless operators are already underway. Unified mobility support and seamless handoff in hybrid wired and wireless networks and among multiple wireless networks have been explored by academic researchers.

1.3.4 Security and Privacy

As mobile computing works its way into every aspect of daily life, security and privacy concerns are becoming more widespread. Without a well-designed security mechanism, reliable applications and services are not possible. Specifically, in the wireless world, the theme of security is authentication, authorization, and accounting; we want the technologies to assist us in our business and in daily lives, not expose our privacy to others. It is of paramount importance to design a highly secure mobile system that does not carry the risk of exposing sensitive privacy data. No doubt these mobile computing security and privacy issues will spur further research into their resolution.

1.3.5 Multimedia on Mobile Devices

Multimedia applications for mobile devices are surfacing due to improvements in the data transmission rates of 2.5 G and

3G services. MNOs are seeking "killer apps" for these data services, and multimedia applications such as image/video messaging and mobile television are being considered. In contrast to text-based communication, mobile audio and video communication targeting mobile devices (such as smart phones) must maintain both adaptability and high performance while coping with fundamental design limitations such as power consumption, form factor, wireless connection bandwidth, and mobility. Some of these problems may be addressed by applying peer-to-peer technology.

1.3.6 Smart Devices and Smart Space

Given the diversity of wireless technologies, a mobile device such as a smart phone must be able to integrate multiple types of radio access. In particular, as the universal mobile terminal, a smart phone must be smart enough to discover which radio access is most suitable for the underlying application and then switch to it on the fly without the user's intervention. Each type of radio access must be optimized on a smart phone to mitigate overall interference. To this end, software radio, a technology that employs the same radio for various wireless protocols, may be the answer. Aside from being adaptive to various radio connections, a smart phone will also adjust its power consumption, display light, and wireless link bandwidth in favor of a long operation time when the battery is about to drain out. Smart space further improves the power of mobile computing by extending the adaptability of a smart device to everywhere in the physical environment. Embedded computers, information appliance, sensors, and people wearing mobile computers or smart phones constitute the smart space, which is able to systematically organize and coordinate those components. The vision of smart devices and smart space requires breakthroughs in a number of fields, including chip design, communication protocol design, middleware, and information retrieval.

1.3.7 Context-Aware Computing

In the mobile computing paradigm, *context* is defined as a collection of varying conditions and situations that exist or occur in the computing, physical, and social environment of a user, such as user location, user movement, environmental noise, temperature, user activity, and any sensing data from accessible devices or other users. Context-aware computing is an effort to taking advantages of this contextual information to automatically adapt the operations of computing services for the user. A classic example of context-aware applications is using a cell phone in a movie theater. The cell phone is equipped with wireless sensors that are able to detect the dimmed light, high ambient noise, and slight body movement of the user, and it silently switches itself into vibration mode. Challenges in the domain of context aware computing include context discovery and sensing, context data modeling and interpretation, contextual adaptation, and augmented context cognition.

1.3.8 HCI and Middleware

A key issue in mobile computing is human–computer interaction (HCI) on a mobile device, which requires research into such areas as speech recognition and processing, advanced computer vision, graphic user interface design, and interaction modeling. Middleware is another key component in the mobile computing framework, as it mediates interactions with the networking kernel on the user's behalf and will keep users immersed in the pervasive computing space [1].

Indeed, the challenging issues of next-generation mobile computing mentioned briefly here do not include every aspect of this extremely broad field. Certainly some issues have been left out. For instance, quality of service (QoS) in wireless networks has been actively researched in the networking community for years. In the mobile world, QoS implies building a networking infrastructure, mobile terminals, and software that as a whole offer reliable and adaptive connectivity in accordance with a set of performance metrics. The big picture of next-generation mobile computing calls for dramatic advances in the research and development of a number of research fields, including mobile hardware, wireless sensing, networking, software design, middleware, and HCI. Needless to say, it is literally impossible to cover all of the challenges in these fast changing areas. This book focuses on networking-related issues, as well as service and application design issues, by using the smart phone as a vehicle for discussion and analysis of terminal design and service delivery.

1.4 Summary

As explained in this chapter, mobile computing is an enormously broad area, encompassing enabling wireless technologies, evolving wireless networks, novel service and applications, standardization, and of course business perspectives and opportunities. This book approaches this fast-paced, diverse domain by concentrating on state-of-the-art advancements in mobile computing research and development in both industry and academia, aiming at bridging the gap between mobile computing research and business-centric commercial mobile application and service development. We have included a brief summary of industry practice and recent commercial developments to provide some insight for industry professionals who have grasped the key ideas and want to see how they work in the real world. A solid understanding of the technical foundations (such as protocols and algorithms) of current technologies, systems, and networks is essential for everyone working in this field; therefore, for each topic covered in the remaining chapters, an in-depth discussion of these foundations is presented in such a way that the reader with an average background in computing technologies will gain from it.

References

[1] D. Saha and A. Mukherjee. Pervasive computing: a paradigm for the 21st century, *IEEE Computer*, 36:25–31, 2004.

[2] M. Weiser. The computer of the 21st century, *Scientific American*, 265(3):94–104, 1991.

[3] S. Fluhrer, I. Mantin, and A. Shamir. Weaknesses in the key schedule algorithm of RC4, in *Proceedings of the 4th Annual Workshop on Selected Areas of Cryptography*, 2001.

[4] V. Moen, H. Raddum, and K. J. Hole. Weaknesses in the temporal key hash of WPA, *ACM SIGMOBILE Mobile Computing and Communication Review*, 8(2):76–83, 2004.

2 The Next Wave of Computing

The next wave of computing, many believe, will be mobile computing. Mobile computing does not simply imply using a cell phone on the move and checking schedules, e-mails, weather, and news occasionally with a laptop computer or a PDA. In fact, future mobile computing reflects a far broader and deeper vision than our current perception of using mobile devices to access cellular wireless networks and services, which represents just one of the components of the mobile computing paradigm. The rapid evolution of enabling mobile wireless technologies and the need for information access to be available "anytime, anywhere, from any device" and, even more aggressively, "all the time, everywhere, from all devices" have naturally raised a few key questions regarding technical trends in the mobile arena, enabling technologies, and the role of cell phones in the future. This chapter aims to answer these questions by looking into the evolution of cellular technologies, PDAs, and other mobile devices. The chapter discusses the rise of smart phones, the increasingly popular converged mobile device, and then moves the discussion a step further to identify the trend of convergence within the broader context of general computing, as well as research and development directions in both industry and academia.

2.1 The Evolution of Cellular Networks and Cell Phones

Cellular communication is by far the most widely used wireless communication technology worldwide. A study by EMC (Hopkinton, MA) in 2004 showed there were about 1.5 billion cell phone users worldwide, and the number is expected to surpass

2.45 billion by 2009. Today, cell phones have become so popular around the world that in some places such as Singapore it has reportedly replaced traditional landline Public Switched Telephone Network (PSTN) telephones as the first choice of communication. The U.S. mobile wireless market will hit $158.6 billion in 2005 and $212.5 billion by 2008, according to a Telecommunications Industry Association (TIA) study. Cellular networks and cell phones are by all means a tremendous success. The evolution of cellular networks, services, standards, and cell phones will reveal future directions in the mobile wireless world.

2.1.1 The History of Cellular Systems

The first commercial cellular network system, the Advanced Mobile Phone System (AMPS), was designed by researchers at Bell Laboratories in 1982; however, the key technology of the first analog cellular systems and current digital systems has its origins back in the 1940s, as shown in Figure 1.1. The cellular concept was first proposed by D. H. Ring and his colleagues in an internal Bell Labs memorandum in 1947. The notion of *cell* comes from the concept of frequency reuse that divides a geographical area into small hexagonal cells. A relatively low-powered base station transceiver in each cell uses radio frequencies unused by its neighboring cells to service cell phones within its own range. Those base stations, as well as other network components, constitute a cellular network infrastructure. Another key concept in cellular systems is *handoff*, which enables mobile users to move across different cells without experiencing communication interruption. Some sophisticated handoff techniques are employed in a cellular system to allow almost unnoticed switching of base stations and frequencies. To increase the capacity of a cellular system, a cell can be further split into smaller areas.

Cellular communication between a mobile station (in this book, we use the terms *mobile station*, *cell phone*, and *mobile phone* interchangeably) and its associated base stations is in the form of two-way radio, as the mobile station transmits and receives signals in two different frequencies (channels) to achieve full duplex communication.

As a result, the capacity of a system, or the maximum number of cell phone users a system can accommodate, is largely determined by the allocated frequencies for the cells. More details of the cellular technology are presented in the next chapter.

The frequency spectrum is generally considered a public asset; thus, the government has to control the allocation of nongovernmental uses of frequency bands and signal power to avoid interference. For cellular mobile network operators or mobile network operators (sometimes also called *carriers*) to build cellular networks and provide mobile communication service to the public, they must obtain approval of the use of a licensed frequency band from a regulatory agency of the government, such as the Federal Communications Commission (FCC) in the United States. Because of scarce frequencies available for business operations, the FCC holds auctions where mobile network operators bid for a license, resulting in high license fees. Therefore, even before mobile network operators begin to build a cellular network infrastructure, they have already made a considerable investment in their business plan. Whether or not it is a good strategy for the government to hold bids for the use of available frequencies is highly debatable. While the auction essentially leads to the efficient use of frequencies because only companies with deep pockets can bid, it does place a huge burden onto mobile network operators because of the gigantic initial investment that is required, and it may hamper such companies from moving forward to provide mobile services. For that reason, some counties are very cautious about allocating frequencies to third-generation systems (explained below). A timeline of cellular technologies and commercial developments is shown in Figure 2.1.

2.1.1.1 First-Generation (1G) Systems

In the 1960s, the Improved Mobile Telephone System (IMTS), the first-generation analog cellular network using only 23 channels, was approved by the FCC. It was implemented in several places but never received much attention due to its low capacity. In 1982, AMPS went into operation. It utilized frequency-division multiplexing access (FDMA) over as many as 832 channels in the 800-MHz

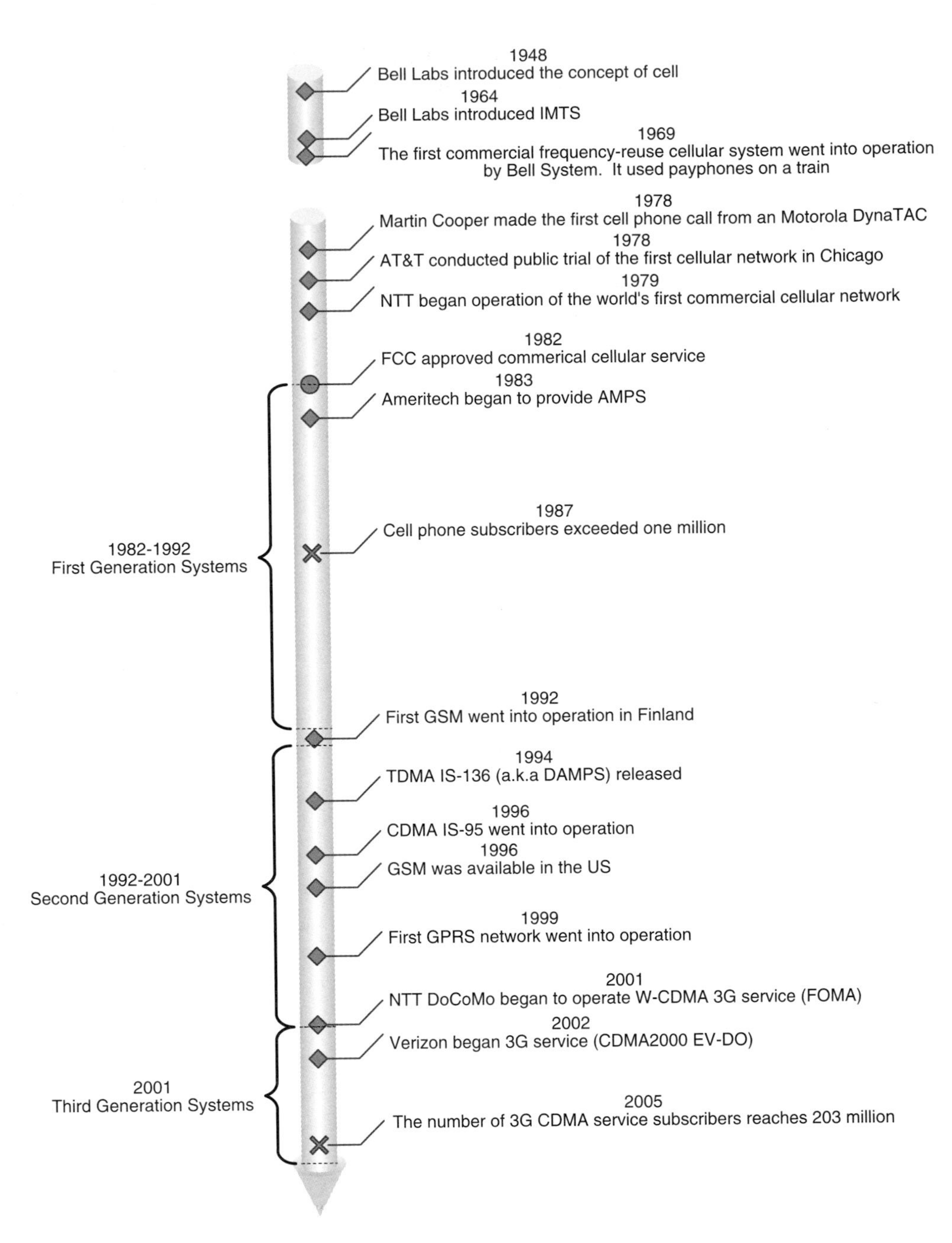

Figure 2.1 Evolution of cellular networks.

band, thus greatly improving the capacity. The first-generation systems also included the Total Access Communication System (TACS) in the United Kingdom, and the Nordic Mobile Telephony (NMT) in Scandinavia nations. Cell phone roaming among different first-generation systems was not possible, and a lack of standardization substantially impeded development of these systems.

2.1.1.2 Second-Generation (2G) Systems

To accommodate the ever-growing number of subscribers, mobile network operators of the first-generation systems soon realized that a more efficient use of the allocated frequency bands was needed, as the FDMA-based first-generation systems could not fulfill the need for greater capacity. Also, the first-generation cellular systems did not offer any security mechanism, effectively making the signal wide open to interception and eavesdropping. To address these issues, a new set of cellular systems was standardized and rolled out in the early 1990s. The 2G cellular systems are digital rather than analog. Two major types of 2G communication networks are global system for mobile (GSM) communication and code-division multiple access (CDMA). The GSM system was originally proposed by European nations as an effort to design a single pan-European mobile communication system across the continent. The multiple-access method in GSM systems is time-division multiple access (TDMA), which enables the same frequency to be used by multiple users simultaneously but each sequentially over a very short interval only in their own time slots. Such a scheme allows eight users to share a single 200-KHz channel in a time-division manner. GSM operates on the 900-MHz and 1800-MHz frequency bands in Europe and Asia.

The CDMA standard was developed and promoted primarily by Qualcomm (San Diego, CA), based on the IS-95 standard (now commonly referred to as cdmaOne). The major difference between GSM and CDMA lies in the spectrum-spreading technique in the radio layer of the wireless network. In short, GSM uses FDM to divide the spectrum into channels, and then uses TDM to further divide channels into time slots, whereas in CDMA, the entire spectrum could possibly be used by an individual node, thanks to a sophisticated spread-spectrum theory. In effect, high-spectrum efficiency and dynamic allocation of bandwidth are considered two major advantages of CDMA over GSM. At the same time, aside from GSM and CDMA, AMPS evolved to D-AMPS, which is essentially a digitized AMPS system.

It was not until the second-generation cellular systems that worldwide roaming of mobile services became a reality, although it was only possible within the same type of systems. To cope with different frequency bands used in different countries within the same system, such as 900-MHz for GSM in Europe and Asia but 1900-MHz for GSM in the United States, *dual-band* cell phones became available. A *dual-mode* cell phone may operate in both analog (AMPS) and digital (GSM or CDMA) networks. Some cell phones even support both GSM and CDMA. The second-generation cellular systems and business operations turned out to be a huge success and exceeded everyone's expectations.

The next chapter will introduce some technological concepts and general ideas of GSM and CDMA within the context of mobile computing. The technical details of CDMA and GSM are beyond the scope of the book. Interested readers may refer to other books on wireless communications for more in-depth discussions; for example, more details about CDMA and GSM are discussed in Glisic and Leppanen [1].

The first- and second-generation cellular networks were predominantly designed for mobile telephony utilizing circuit switching. In addition, paging services were generally provided as a one-way communication service. These services seemed sufficient to satisfy people's needs for mobile person-to-person communication until the late 1990s, when wired networks and the Internet exploded onto the scene. The revolutionary Internet significantly extended people's information access capability by bringing the World Wide Web as well as other feature-rich applications to nearly every desktop computer via a large array of networking technologies. In response, the wireless world saw the opportunity to move beyond voice and paging services by offering data access in cellular networks, mainly to business professionals, in the hopes of riding another wave of rapid user growth. To achieve this goal, new data-centric applications had to be introduced, and the cellular network itself had to be upgraded to provide a higher data transmission rate. To this end, GSM systems offered short message service (SMS), a pilot data service that allows the exchange of short messages of less than 160 characters. Interestingly, SMS turned out to be a phenomenal success in Asia and Europe, partly due to the billing models and for some cultural reasons. Still, GSM suffers from a low data rate of around 14.4 Kbps imposed by the underlying circuit-switching transmission scheme.

On the road toward the third-generation, high-data-rate, digital, cellular networks, some interim systems were developed to fulfill the need for mobile data service without deploying an entirely new system and to fully leverage the legacy systems. These so-called 2.5G systems, such as GPRS, EDGE, and CDMA 2000 1x, for the most part are compatible with their 2G base systems because operators wanted to make the most out of the existing networks. GPRS overlays GSM networks, providing a data rate that is equivalent to a landline dial-up connection. One of the most important features of GPRS is its "always-on" service; that is, the connection stays open without further dial-ups. Both D-AMPS and GSM can be updated to GPRS without a major overhaul. EDGE is a further improvement over GSM with respect to data rate. It is sometimes categorized as a third-generation mobile communication system. On the cdmaOne side, the 2.5G systems are called cdam2000 1x and offer a packet data rate up to 144 kbps. With regard to the D-AMPS system, mobile network operators chose to move toward EDGE and to the 3G systems. Here, our focus is on presenting a big picture of 2G and 3G systems. The evolutionary relationships between these systems are further illustrated in Chapter 3 (Supporting Wireless Technologies).

The applications offered by 2.5G systems include wireless access protocol (WAP)-based wireless Internet, text messaging, voice and picture e-mail services, and instant messaging, among others. As a result, the cell phones used for 2.5G services must be designed to support these applications. The evolution of cell phones is discussed later in this chapter.

2.1.1.3 Third- and Fourth-Generation Systems

Along the winding way toward third-generation (3G) cellular networks, standards bodies are playing increasingly important roles. The International Telecommunication Union (ITU), known as CCITT prior to 1993, is a United Nations organization responsible for coordinating global telecommunication standardization. In response to the advent of 3G cellular systems, in 1999 the ITU approved an industry standard known as International Mobile Telecommunications-2000 (IMT-2000) [2]. This standard defined the general data rate requirements for 3G services, including:

- 2 Mbps in fixed or in-building environments
- 384 kbps in pedestrian or urban environments
- 144 kbps in wide area mobile environments

It officially recognized five industry standards as 3G services: WCDMA, cdma2000, TD-SCDMA, UWC-136, and DECT+. Among these, the dominant standards are UMTS/WCDMA (wideband CDMA) and cdma2000. UMTS/WCDMA is the evolution of GPRS/EDGE, whereas cdma2000 is the next generation of cdmaOne. cdma2000 can be further divided into such systems as cdma2000 1x (a 2.5G system), cdma2000 1xEV-DO (Evolution-Data Optimized), cdma2000 1xEV-DV (Evolution-Data Voice), and cdma2000 3x. Except for cdam2000 1x, the other types of cdma2000 are able to offer data rates higher than 2 Mbps.

As of this writing, 3G cellular systems have been commercialized in many counties, using either WCDMA or cdma2000, but the dominant systems worldwide are still 2G systems. Some mobile network operators, as well as standards bodies, foresee a relentless demand for higher data rates for such emerging applications as real-time multimedia communications and therefore are considering the next generation of cellular networks (4G), which are expected to provide a bandwidth of up to 100 Mbps. Unlike 3G networks, the 4G systems are completely packet switched. As 3G systems are being rolled out, 4G research is already underway.

2.1.1.4 The Theme of Technological Advances in the Wireless Industry

The three generations of cellular systems are excellent representations of the theme of technological advances in the wireless industry during the 1980s and 1990s. As shown in Figure 2.2 the shift between the generations took place in three important areas: voice signal sampling, transmission or switching, and major payload in the system, all closely intertwined at each shift. Value-added applications and services such as wireless e-mailing and SMS subscription gradually began to account for more traffic in the system and a larger portion of the total revenue. The penetration of voice-centric cell phones in most developed countries is so great that growth has been rather

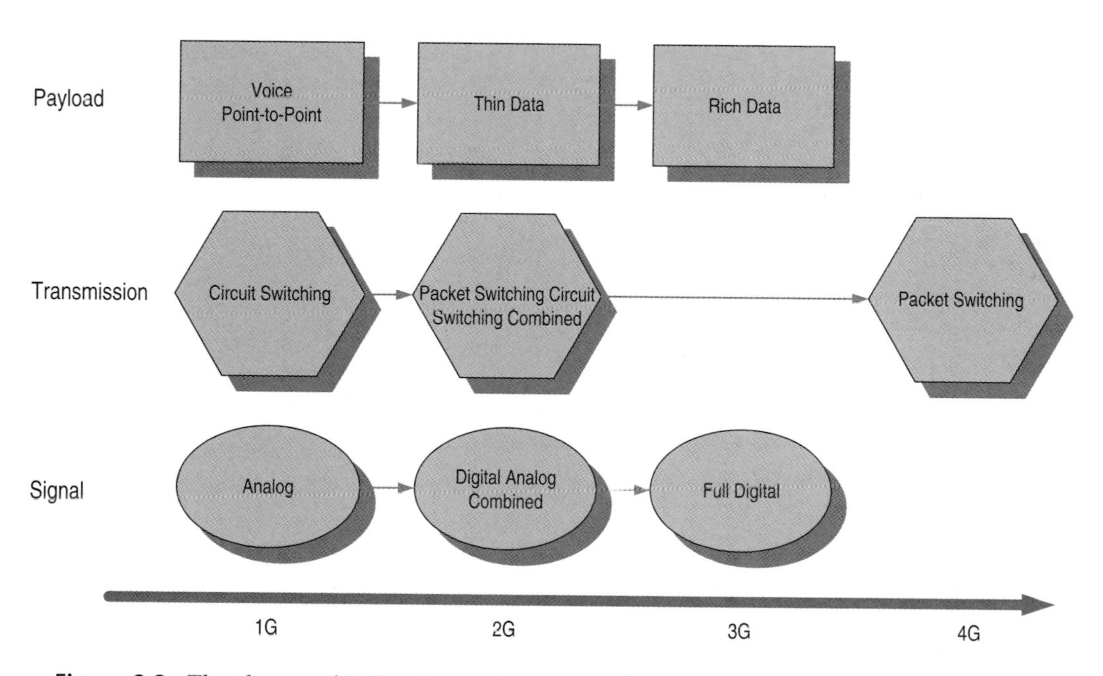

Figure 2.2 The theme of technology advances in the cellular network domain.

stagnant and largely, if not completely, dependent on the upgrade of cell phone devices. In some sense, these value-added applications and services fall into the domain of computing rather than communication, thus calling for an entirely new design model of the cell phones to facilitate computing services.

2.1.2 The Evolution of Cell Phones

The terms *cell phone*, *cellular phone*, *mobile phone*, and *wireless phone* all refer to the same type of voice-centric mobile device that has become the essential personal communication tool everywhere. Sometimes the term *handset* is used instead of *cell phone*, which usually has a monochrome or color screen display, a built-in antenna, and a keypad. The inventor of cell phone is Dr. Martin Cooper, then a Motorola engineer, who set up the first base station in New York and used the world's first cell phone, the Motorola DynaTac, to place the first cell phone call on April 13, 1978, while walking on a street of New York City. The DynaTac was typical of the first-generation analog cell phones in that it resembled a brick in terms of shape, size, and weight (Figure 2.3). Power consumption of these cell phones was fairly high and they had to be recharged frequently. Accordingly, the first-generation cell phones were indeed mobile or, more precisely movable, but they were not very convenient to use, which limited the number of users of first-generation systems.

On the heels of second-generation digital cellular systems such as GSM, cdmaOne, and TDMA came the digital cell phones in 1990s. Thanks to quickly advancing integrated circuit technologies, digital cell phones are very small and lightweight, around 200 g (about 7 oz.) or less and offer longer talk time and standby time. These cell phones also have a small monochrome LCD screen display to show phone numbers, dialing status, and signal strength, as well as some simple information regarding advanced applications such as phone book, call log, voice messages, and so forth. In addition to the plain palm-like cell phones, some "clamshell" cell phones began to appear; this type of phones typically put the screen display on the back of the cover pane to save space. Not long after that, 2.5G cellular systems

Figure 2.3 The first cell phone: Motorola's DynaTAC 8000x (http://www.i4u.com/article421.html. © Copyright 1994–2005 Motorola, Inc. Courtesy of Motorola, Inc.).

went into operations, which in turn brought about a dramatic change in cell phone design with regard to a better user interface and support for short-range network connectivity. Among their features are a brightly colored screen, sharp images and text display, ease of use text input, synchronization with computers, and Personal Area Network (PAN) capability such as IrDA or Bluetooth. These cell phones account for most of the cell phone units used in the world today.

Smart phones were developed by mobile network operators and cell phone manufacturers in late 1990s as a response to the need for a converged personal communication and assistant tool. Powered by the latest mobile processors such as Intel XScale and featuring comparatively large and high-resolution color liquid-crystal display (LCD) screen displays and large memory capacities, smart phones are capable of performing more computing tasks than any other cell phones. One way to leverage the computing capability of a smart phone is to use it as a mobile terminal interacting with wireless-enabled devices, equipment, computers, and even a physical environment embedded with sensors. Imagine a smart phone that communicates with Bluetooth-enabled automated teller machines (ATMs), radiofrequency identification (RFID)-tagged products in a retail store, or an infrared-enabled parking lot gate. These applications transform the smart phone into a general computing and communication device for data collection and processing, in addition to enhancing the user's interactions with other people and other computing systems. Moreover, a smart phone can also work as a consumer electronics device such as a music player or a digital camera. Despite the small 4% share of the entire cell phone market in terms of number of shipped phones in 2003 (The Yankee Group, 2004), smart phones are widely expected to become popular in the mass market over the next several years.

Table 2.1 provides a general comparison of cell phones along with the three generation of cellular systems. Notice that, while cell phones are becoming smaller and lighter, they are more powerful and sophisticated and can perform a wide variety of operations. Not surprisingly, this trend is in line with the evolution of personal computers. Smart phones are not specifically for 2.5G or 3G networks; in fact, many cell phone manufacturer identify their products as smart phones because these cell phones use an advanced operating system on which customized mobile computer applications and services are based. At the time of this writing, the major providers of mobile operating systems for advanced cell phones and smart phones are Symbian OS®, Palm OS®, and Microsoft Windows Mobile® for Smartphone. The market leader is clearly Symbian OS developed by the Symbian Company, a privately held, independent company founded by Ericsson, Nokia, Motorola, and Psion in 1998. Palm OS was developed by Palm, Inc., the company that invented the first successful personal digital assistant (PDA), the Palm Pilot®. Palm split into two companies in 2002: Palm Source, Inc., which continues to make Palm OS, and Palm One, Inc. (later renamed as Palm, Inc.),

Table 2.1 Evolution of Cell Phones

Category	Analog	Digital	Smart Phone
Size	"Brick"	"Palm", "clamshell", or "candy bar"	"palm", "clamshell", or "candy bar"
Weight	1 to 2 lb	6 to 8 oz.	< 5 oz.
Display	N/A	Monochrome or color, small, 172 × 120 pixels	Color, 320 × 240 pixels
Processor	For very basic communication tasks	For preliminary tasks	For some advanced tasks such as multimedia playback
Memory	Only to store phone numbers	Several megabytes	64 MB or more, plus flash memory
Noncellular interfaces	N/A	Sync with computers	Bluetooth, WiFi, GPS, etc.
Battery	Short talk time and standby time	Longer talk time and standby time	Longer talk time and standby time
Price	Several thousand dollars	A wide range, from free to several hundred dollars	Several hundred dollars or less

which focuses on PDAs and smart phone devices such as the Palm V™, Treo™, Tungsten, and Zire™ series. The Windows CE-based Microsoft Windows Mobile for Smartphone, or Microsoft Smartphone, is new to the smart phone OS market. Aside from these general-purpose operating systems, another strong competitor that deserves mention is RIM's BlackBerry™. BlackBerry devices were traditionally wireless data devices that allowed business people to stay connected with wireless e-mails. Along with the rise of smart phones, RIM began to add voice and web access to newer BlackBerry devices. Chapter 4 (Mobile Terminal Platforms) provides a detailed discussion of these software platforms.

2.2 The Evolution of PDAs

The personal digital assistant, sometimes referred to as a *handheld* in the computer industry, commonly refers to a palm-size, lightweight, battery-powered electronic device for personal information management (PIM). Despite the rise of smart phones among business users,

Figure 2.4 Apple Newton MessagePad. (Courtesy of Apple Computer, Inc., Sunnyvale, CA. © Copyright 2005 Apple Computers, Inc.)

PDAs still hold a strong position in the market. The term *personal digital assistant* was coined by John Sculley, then CEO of Apple Computers, at the Winter Consumer Electronics Show in 1992, when the computer company decided to develop a product for the consumer electronics market after the success of Apple's Mac II. In 1993, Apple released the world's first PDA, the Newton MessagePad, which was manufactured by Sharp (Figure 2.4). The Newton MessagePad had an ARM610 20-MHz processor, 4 MB of ROM, 640 k of SRAM, and a 336 × 240 monochrome screen, all compacked into a 7.25 × 4.5 × 0.75 case running the Newton operating system.

2.2.1 Apple Newton

Apple Newton essentially defined a set of core functions that a PDA should offer, including a personal organizer, an address book, quick notes, infrared beaming capability, and some utility tools. It can also send e-mail and faxes with optional extension cards. All these applications are based on handwriting-recognition technology, a key technology in the domain of pen-based computing. Users write with a stylus on the screen using "digital ink" technology, as opposed to using keypads or a keyboard to input English letters. Almost all PDAs today rely on the same design philosophy of pen-based computing. The recent tablet PC is a new platform of portable computers that employ pen-based handwriting recognition technology. Another feature of Newton, as well as most its siblings, is that it does not use any persistent block-based storage such as a hard disk. Instead, it offers some amount of flash memory for user storage, which is much faster than traditional hard disk or tape devices in terms of data access rate. Such a choice was based partly on the design objective that a PDA should be highly responsive while interacting with users. Power consumption was another factor because hard disks or tapes consume far more power than flash memory.

2.2.2 The History of PDAs

Figure 2.5 depicts some of the milestones of PDA history. The entire PDA industry could be traced back to late 1980s, when pen-based computing was considered to be the next big thing. In 1987, a company named Go created an operating system for pen-based computers. The idea of using a pen rather than a keyboard on a computer was apparently innovative to investors as well as large companies such as Microsoft, AT&T, and IBM. Although Go went bankrupt in 1994 due to financial problems, its competitors carried on. First, Apple (Newton), then Casio (Zoom), IBM (Simon), and Microsoft (WinPad) all announced the release of their PDA products within a few years, but none of them became very popular. In 1996, Palm released Palm Pilot, which became a really big hit. Unlike other PDAs in the market at that time, Palm Pilot did not focus on the problem of free-form handwriting recognition and wireless communication; rather, it identified a key feature that had been largely ignored by other PDA vendors: easy synchronization with a desktop computer.

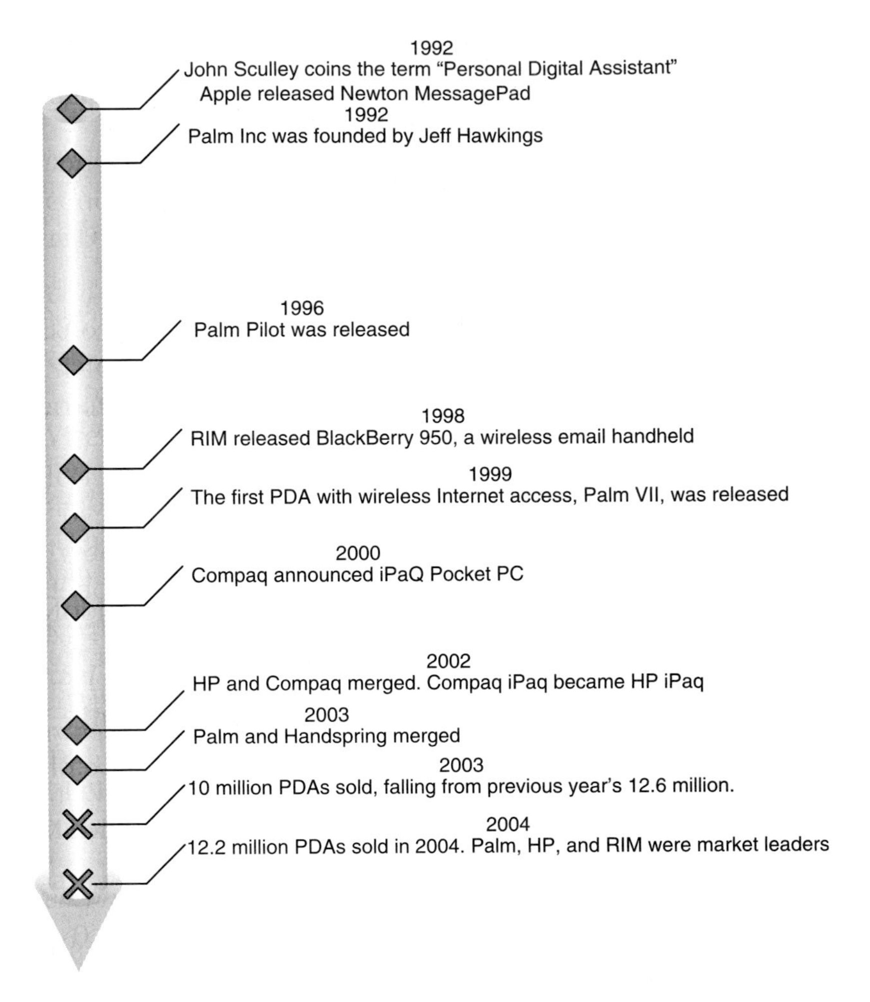

Figure 2.5 Evolution of PDAs.

Palm Pilot allows a user to synchronize with a computer by pressing a single button on the cradle, as shown in Figure 2.6 Some researchers noted that the correct problem framing of the PDA resulted in the success of the Palm Pilot [3]. The Palm Pilot was designed with the vision that a PDA should not be a compact portable version of a desktop computer, thus it should never attempt to provide a complete clone of the set of applications available on a desktop computer.

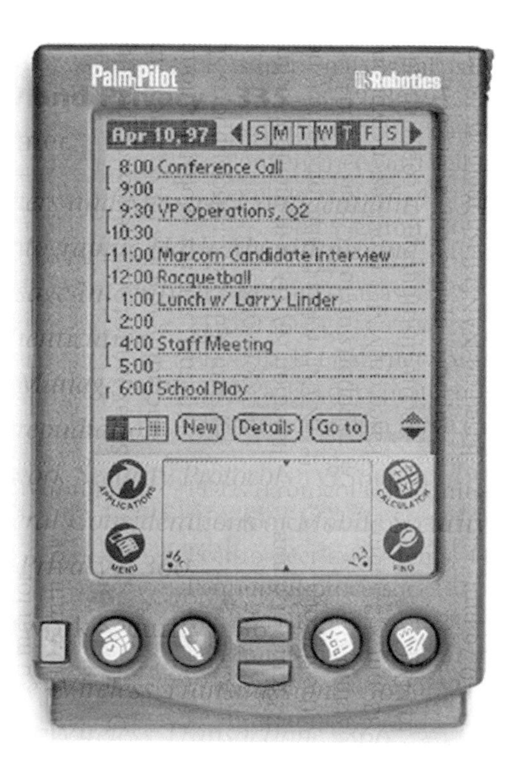

Figure 2.6 Palm Pilot. (Courtesy of Palm, Inc. © Copyright 2005 Palm, Inc.)

The applications running on the Palm Pilot have been carefully selected and tailored to take advantage of portability, the feature that sets PDAs apart from desktop computers. Because computing requirements were significantly reduced, the initial price of the Palm Pilot was very competitive compared to other PDAs. After the release of this pioneering and successful model, several PDA products from other vendors came into play. Almost all of them followed the same design model defined by Palm Pilot.

Along the road to today's PDAs, a handful of computing technologies substantially facilitated the widespread acceptance of these small mobile devices. Below is a summary of key PDA technologies.

2.2.2.1 Handwriting Recognition

The first handwriting recognition engine, dubbed "Calligrapher," did not help the Newton MessagePad at all. It allowed too much freestyle writing anywhere on the screen, and recognition was rather poor. As a result, the general public, not surprisingly, showed little confidence and interest in early versions of Newtons. Even a character in the well-known TV show "The Simpsons," a school bully called Nelson, was disappointed by the ineffectiveness of the handwriting recognition of Newton. In a show first aired in November 1994, the boy told his buddy to write "Beat up Martin" on the screen of a Newton, but the handwriting recognition engine came up with "Eat up Martha." Irritated, Nelson chose to use the device as a weapon when he threw the Newton at Martin's head.

Palm Pilot introduced a much better handwriting engine, Graffiti, which only allows a user to write within a small rectangular area, and freeform writing is not acceptable; instead, users have to learn to write letters and punctuation according to a set of defined shapes for the characters. It turned out that these restrictions have effectively eased the otherwise cumbersome process of handwriting on a Palm Pilot.

Most PDAs are equipped with a stylus for handwriting recognition. Only a small portion of PDAs have a tiny keyboard for data entry, and they are primarily targeted for business applications, such as point-of-sale (POS). An exception is BlackBerry's wireless e-mail handset, which uses an integrated PC-style 33-key QWERTY keyboard to facilitate datacentric applications such as e-mail, PIM, and wireless web. Compared with PDAs, cell phones and smart phones are inherently phone based, meaning that a phone keypad (1 through 9 number keys, each representing three English letters) is most frequently used.

2.2.2.2 Communication Interfaces

The first model in the Newton MessagePad series released in 1993 does not include any wireless communication interface. An optional 9.6-Kbps modem was the only way to transfer data to and from this device. It also had a Personal Computer Memory Card International Association (PCMCIA) slot for extra storage, but the Newton MessagePad was not designed for frequent data migration. Later models in the same series, beginning with MessagePad 120, were equipped with an infrared beam capability and a serial port. Infrared beaming allows a Newton MessagePad to communicate conveniently with infrared enabled printers, laptop computers, or another Newton MessagePad within a short range of several feet. The infrared technology later has become a *de facto* standard component for mobile computers and PDAs.

The next big leap in terms of the communication capability of PDAs has been synchronization with a computer via a serial port or a USB port, first available on Palm Pilot PDAs. Simply push the "sync" button on the device cradle, all data on the PDA will be synchronized with the hosting computer. After establishing a companion relationship between the PDA and the host computer, a user is able to conduct more frequent data updates to the PDA, thereby extending the use of the device, especially in a business environment. Serial communication was first introduced as the synchronization interface. It provides a data rate of only a few hundred bits per second, which is obviously not fast enough for large data transfer. USB was designed to replace the serial interface on computers with two practical advantages: high data rate up to 12 Mbps and ease of communication setup. Two standards, USB 1.x and 2.0 (up to 480 Mbps), have subsequently been developed and are widely used in computing devices, including PDAs, music players, and personal computers.

With easy-to-use synchronization and point-to-point infrared beaming, the capabilities of PDAs have been greatly expanded. The host computer acts as a proxy for the PDA for information access over a network. The information that can be pushed to a PDA over synchronization includes news, stock quotes, flight schedules, restaurant reviews, etc. An example of this kind of proxying application is AvantGo. Initially, the PDA did not provide direct communication capability for data access to a network. This was not a big problem in mid-1990s, as the Internet and wired networking technologies were still it their infancy and were not available to general public. When the Internet exploded in late 1990s and most computers were online, the need to have direct network access, be it the Internet or a local area network (LAN), became imperative. As a consequence, wireless technologies such as wireless cellular access, wireless LAN access, and

Bluetooth, started to become a necessary communication interface on PDAs. To date the most frequently included wireless interfaces on a PDA are Bluetooth and wireless LAN.

The evolution of PDAs has followed a course similar to that of cell phones in many respects, from monochrome screens to color screens; from low computing capabilities to state-of-the-art, high-performance mobile processors; from IrDA to Bluetooth and Wireless LAN. Because a PDA remains by and large a standalone computing device without direct network access, it has evolved primarily in the hardware components and supported functions. On the other hand, a cell phone relies heavily on the cellular network to operate, thus the evolution of cell phones has paralleled that of cellular networks. Nonetheless, the fundamental difference between PDAs and cell phones is that a PDA is a pure computing device and a cell phone is primarily a communication device.

Worldwide PDA shipments experienced a sharp decline in 2002 and later after more than 10 years of growth. Aside from worldwide economic downturn, another major factor contributing to the decline of PDA sales was probably competition from cell phones that have built-in wireless Internet access and common PDA functions - in other words, smart phones.

2.3 PDAs Versus Cell Phones

The success of Palm in the past 10 years sheds some light on the future of PDAs. As always, the key issue for any business is to identify what users really want and how much they will pay for it. Back in the late 1980s and early 1990s, wireless communication was still in the research-and-development stage in research laboratories and academia, and references to the wired Internet were not heard very often. It was becoming clear that a small handheld portable device with easy-to-use applications and a simple communication interface could gain some ground with people who were tired of the increasingly time-consuming personal data management and would like to get rid of paper organizers and calendars. It was also becoming evident that portability or mobility was an important issue with regard to the design of mobile computing devices.

Cell phones, on the other hand, are voice communication devices with mobility support enabled by the cellular networks. Due to the inherent resource constraint nature of cell phones, many sophisticated communication tasks are performed on the backend network infrastructure. The functionality of a cell phone is diverse but none is computing intensive or requires large memory. On the other hand, radio communication consumes far much power than common computing tasks. Table 2.2 provides a brief head-to-head comparison of PDAs and cell phones and some low-end smart phones.

The distinction between a PDA and a voice-centric cell phone is evident in almost all respects; however, both PDAs and cell phones are undergoing drastic changes in terms of offered functionality. A multifunction mobile device that can do both and more is a natural direction; consequently, the boundaries between these types of mobile devices continues to blur and may eventually melt away.

Table 2.2 PDA Versus Cell Phone

Category	PDAs	Cell Phone (Voice Only)
Major use	Personal information management	Mobile telephony and messaging
Size	Larger, postcard	Smaller, palm size
Weight	6 to 9 oz.	6 to 8 oz.
Display	Monochrome or color, large, 240 × 320 or higher	Monochrome or Color, small, 172 × 120
Processor	ARM, StrongARM, DragonBall, several hundred MHz	Proprietary processors of several hundred MHz
Memory	64 Mb or more	4- to 8-Mb SRAM with an 8- or 16-Mb NOR Flash
Interfaces	USB, serial, IrDA, Bluetooth, Wireless LAN	Bluetooth
Battery	Very long working time	A few hours of talk time
Price	Ranging from less than $100 to several hundred dollars	Comparatively inexpensive or even free

2.4 The Convergence of Mobile Devices

When it comes to mobile, the need for convergence is evident. Nobody wants to carry around a handful of mobile devices. Converged mobile devices stand a good chance of becoming popular, because in a mobile scenario, size, weight, functions, wireless communication capability, and of course price are often equally important to end users. If two mobile devices can merge into one and work well in both application scenarios, it would clearly be unnecessary to use both. Also, a converged mobile device may offer the functionality of data integration and interoperation, which are otherwise difficult to achieve with two separate devices.

2.4.1 The Convergence of PDAs and Cell Phones

Only a few years ago, a PDA was usually considered a small, easy-to-use electronic organizer that did not offer too much beyond replacing paper organizers. Communication capabilities of a PDA were limited to synchronization with a host computer. Once disconnected from the host computer, operations of a PDA are totally dependent on its locally maintained snapshot of data until the next synchronization. Real-time data access and processing are rather difficult, if not impossible.

When cell phones become ubiquitous, voice communication via the cellular networks with mobility support became not only a means of personal communication but also an element of lifestyle, bearing social and cultural implications. In contrast to PDAs, cell phones are real-time communication tools. The high penetration of base stations in urban areas operated by various wireless carriers effectively makes it possible for cell phone users to access the backend cellular networks anytime they want.

Both of these two types of mobile devices are indispensable for many people who have a strong need for communication and personal information management. Both of them can comfortably fit into a pocket, but it is undoubtedly cumbersome to try to cram both in one pocket. On the other hand, some of the functions provided by

PDAs and cell phones actually overlap, such as the phone book and calendar. Thus, a question has been raised by many business professionals: Can we have the best of both worlds with a single device? Or, even more intriguing, can we use a single device to do the things that are otherwise impossible on either of them?

The concept of smart phones is the answer to these questions. In essence, a smart phone is a converged mobile device that supplies a rich set of datacentric computing and communication applications and services in addition to conventional voice communication. Smart phones are generally designed to seamlessly integrate PDAs and cell phones into a single set of hardware components. An example is the Motorola MPX 200 smart phone. As many other smart phones enter the market, the Motorola MPX 200 will allow users to check e-mails, manage address books and calendars, play music and video clips, and surf the web. It should be noted that PDAs and cell phones possess some conflicting design goals. PDA functionality generally requires a fairly big screen display and a stylus-based method of operation, which has resulted in PDA-centric smart phone such as Microsoft's Pocket PC Phone edition, which is bigger than regular cell phones and has poor phone functionality. Very often smart phone designers have to sacrifice some PDA functionality in exchange for a usable smart phone. This situation may change as display technology, human-computer interface technology, and semiconductor technology evolve to provide better solutions to these conflicting requirements.

The first generation of smart phones is simply a multifunction cell phone with PIM support. All smart phones at the time of this writing fall into this category. The second generation of smart phones is likely to support many other wireless data networks such as wireless LAN, ultra-wideband (UWB), and wireless sensing, thereby allowing extensive data access and cooperation between the smart phone and various wireless networks. In addition to voice communication and personal information management, an entirely new set of applications and services will be available for the next generation of smart phones. The design philosophy is that rather than cramming multiple separated functions into a single device, a smart phone should leverage the surrounding wireless environment as much as possible to provide a new dimension of applications for convenient information access, personal and group communication, and human-machine and human-environment interactions.

Microsoft's use of another term, "Smartphone," represents something totally different from a smart phone (even if the only difference is the space between the two words). Smartphone is Microsoft's software platform for cell phones.

2.4.2 The Convergence of Cell Phones and Mobile Entertainment Devices

Mobile entertainment devices refer to handheld audio, video, and graphics entertainment systems such as music players, video players, handheld televisions, and gaming devices. Mobile entertainment devices are extremely popular among teenagers. For example, worldwide sales of mp3 players doubled in units and dollars in 2003 to more than 24 million units and $3 billion, as reported by research firm In-Stat/MDR. The firm also predicted that, by 2009, mobile gaming services in the United States will generate $1.8 billion in sales annually. Most of the mobile entertainment devices do not require wireless data connections to operate. Multimedia contents are downloaded into the device's storage via USB or are stored in a removable media Memory card. For example, portable mp3 player users usually collect music on a computer, and transfer their selections to their mp3 players through a USB connection. An example of such a device is the popular Apple iPod, shown in Figure 2.7 which has been closely bundled with Apple's iTunes online music store. Similar needs are also evident in mobile gaming. In addition to allowing gamers to obtain games from somewhere online, more compelling is to enable multiplayer mobile gaming via wireless networks. Moreover, gaming experiences can be enriched dramatically by a wide array of services and applications, such as location-based gaming and *ad hoc* gaming, supplied by underlying wireless networks. At the same time, the mobile gaming device can also work as a cell phone or PDA. Nokia's N-Gage QD Game Deck, shown in Figure 2.8 is such a device.

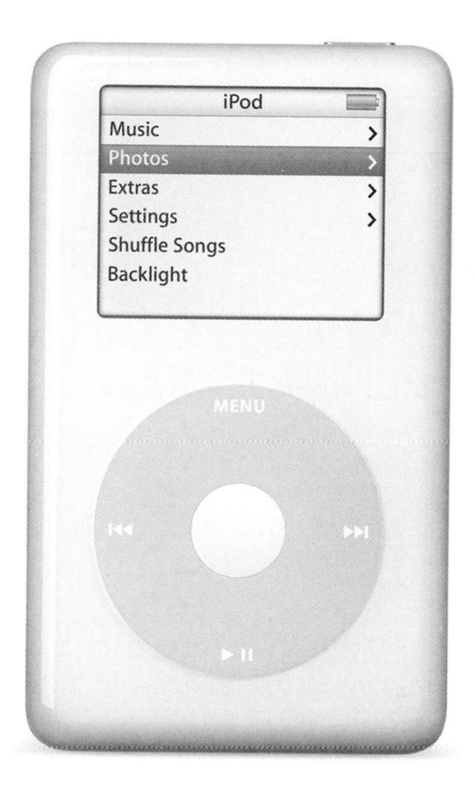

Figure 2.7 Apple iPod. (Courtesy of Apple Computer, Inc., Sunnyvale, CA. © Copyright 2005 Apple Computer, Inc. http://www.apple.com/ipod/.)

Personal digital assistants and cell phones also offer an assortment of entertainment applications as part of the software package. Needless to say the quality of these applications and the entertaining experiences they can provide are nowhere close to those dedicated mobile entertainment devices, primarily due to their limited computing capabilities and the form factor. However, PDA and cell phone users want to be able to enjoy simple forms of entertainment that are optimized for such devices.

Like cell phone services, mobile entertainment services have the potential to become another subscription-based multibillion dollar market. While communication among cell phones is primarily comprised of voice and text messaging, mobile entertainment users will

Figure 2.8 Nokia's N-Gage QD Game Deck. (Courtesy of Nokia Corporation, Espoo, Finland. © Copyright 2005 Nokia Corp.)

be supplied with a variety of data communication methods, ranging from infrastructure-based music download, to peer-to-peer multiplayer gaming, to multicast multimedia streaming, etc. A number of parties, including consumer electronics manufacturers, PDA and cell phone or smart phone manufacturers, mobile network operators, software providers, and traditional game device providers, have demonstrated their interest in this phenomenal opportunity. No matter which industry segment will take the lead in the new frontier, it must first address two fundamental issues in the mobile entertainment domain: (1) how to leverage emerging mobile wireless technologies for online/offline entertainment with a converged mobile device, and (2) identifying what mobile entertainment applications and services best fit into this market. In the near term, we probably will see different parties cooperate with each other to leverage the stronghold of each one. For example, cell phones may be able to access online music store such as Apple iTunes via wireless data connections and play high-quality music through its earphones. In the long run, companies from one sector may encroach on each other's strongholds. Imagine an iPod-like smart phone (the iPhone?) or a Palm-based online multiplayer mobile gaming device.

Aside from convergence among cell phones, PDAs, and mobile entertainment devices, some other applications such as consumer electronics, industrial handheld equipment, and home automation are expected to converge to a universal mobile device as well. Consider remote controls for televisions, digital videodisc (DVD) players, and stereo systems. These small wireless devices only differ in the frequency used. Moreover, other electronic devices may also be remotely controlled, either from within a small room using direct wireless communication or from a remote site via a network. For example, a digital surveillance camera with a built-in motion detection sensor or a temperature sensor can possibly send images, data, or full motion video back to a remote site or a cell phone via cellular networks whenever a predefined condition of sensing data has been met. Users may use a cell phone to control the camera and the sensors or directly control some wireless-enabled devices in the monitored environment.

2.4.3 Smart Phone: The Universal Mobile Terminal

It is clear that cell phones and PDAs will converge to a single mobile device, a smart phone. It is also evident that a smart phone will incorporate some functions of mobile entertainment devices and other consumer electronic wireless devices. In essence, a smart phone will be a universal mobile terminal in the future mobile computing realm. It is not only a communication device, but also a computing device, powered by the next generation of wireless networks ranging from wireless PAN to worldwide cellular networks. In addition, a smart phone may be used as a wireless control device to allow unified monitoring, positioning, and control over consumer electronics devices. These function sets are not merely crammed into a smart phone using separate hardware and software components; rather, they are systematically integrated to allow data sharing and interoperability.

From the mobile network operator's point of view, the success of smart phones hinges on the degree to which smart phones can replace these single-function mobile devices. In order to be universal, the following challenging issues have to be resolved first:

- Power consumption of a smart phone will increase as a result of more hardware components used and more wireless transmission. Battery life becomes a critical problem. To date, there are three types of cell phone batteries — NiMH (nickel-metal hydride), Li-ion (lithium-ion), and Li-polymer, all of which support a few hours of talk time and a week of standby time. New battery technologies for mobile devices such as fuel cell batteries have emerged but are not in mass production.
- The form factor of a smart phone makes it impractical to watch television or video on the small screen even if wireless link bandwidth is sufficiently large. Breakthroughs in display technology are highly anticipated. Both clamshell-like cell phones and regular "candy bar" cell phones now provide 2-inch TFT thin-film transistor (TFT) displays with more than 64 k colors with 128 × 160 or 176 × 220 resolution. Some have even better QVGA display (320 × 240).
- The common use of flash memory on mobile devices may eventually give way to large-capacity, low-cost persistent storage such as micro hard disks so as to accommodate increasing demand for greater storage space. Offering very large flash memory is not economically feasible due to its high cost. Many cell phones now provide a secure digital (SD) interface that allows stamp-size SD memory cards to be used for extensive storage. Another standard, multimedia card (MMC), is also supported by many cell phones. These memory extensions can support up to 1 Gb or larger add-on memory. The SD interface may also be used for communication, such as SDIO Bluetooth cards. Details of mobile memory technologies are introduced in Chapter 4.
- Wireless interference is a critical issue, as a smart phone will deal with heterogeneous wireless networks in different ranges, especially when many of these wireless networks operate on the

unlicensed spectrum. Bluetooth, some types of wireless LANs, and most microwave ovens use the same unlicensed 2.4-GHz spectrum band. Although the underlying radio frequency schemes are different, interference may occur in some cases. For example, it has been reported that a Bluetooth-enabled cell phone may turn on an electronic shaver. On the other hand, wireless LANs are considerably faster than cellular access, thus Internet access from a cell phone or smart phone could go through a wireless LAN hotspot if available, rather than always connecting to cellular networks. In the long run, with mobile access becoming more ubiquitous, a systematic approach to heterogeneous wireless networks is needed.

2.5 Smart Phone Applications and Services

The trend of convergence in the mobile device domain calls for a combination of mobile applications and services that used to be offered by cell phones, PDAs, or mobile entertainment devices separately. New applications and services will also surface as a result of the ever-increasing wireless link bandwidth, ubiquitous wireless access, or simply a novel business model.

2.5.1 First-Generation Smart Phone Applications

First-generation smart phone applications are mostly natural extensions of PDA and cell phone functionality. Most of these applications are available as of 2005. Below is a list of first-generation smart phone applications:

- *Mobile telephony* is the traditional cellular telephony service including phone call and voice mail services. Some phones support voice dialing. Mobile telephony is, and will likely remain, the most important application in the mobile wireless realm.

- *Short message service* (SMS) enables sending and receiving text messages to and from a phone number. According to the GSM Association, about 30 billion SMS messages are sent globally every month, primarily in Europe and Asia.
- *Enhanced message service* (EMS) and multimedia message service (MMS) — EMS allows a user to send formatted text, animations, images, and simple melodies in a message, whereas MMS supports polyphonic melodies, large images, and audio and video clips.
- *Cell phone positioning* complies with government regulations. In the United States, the FCC requires mobile network operators to implement E911 service to position a cell phone in case of emergency. For accurate outdoor positioning, the global positioning system (GPS) is the best choice. Some GPS receivers can be plugged into a smart phone's extension slot (compact flash), and some smart phones even have a built-in GPS. Chapter 3 will discuss GPS technology in more detail.
- *Navigation systems with traffic information and geographic information system* (GIS) — Location services are facilitated by GPS, cellular networks, or other wireless technologies such as WiFi hotspots. Driving directions are provided with respect to real-time traffic information.
- *Instant messaging* (IM) is real-time text messaging service via a directory server. Unlike SMS, users can engage in one-to-one or group chat sessions. Some well-known desktop IM services have been made accessible from cell phones, such as AOL instant messaging and MSN messaging.
- *E-mail, calendar, organizer, and notepad* are typical PIM applications. Web-based PIM services such as Yahoo! allow users to synchronize personal data, including e-mail, personal calendars, organizers, address books, etc., from the web to PDAs or cell phones.
- *Address book* is another common PIM application that can interoperate with mobile telephony applications.

- *Wireless Internet browsing* — Web access is a core function of a smart phone. The website a user visits from a smart phone could be a set of specially designed wireless web pages or regular web pages.
- *Data synchronization with a computer or another mobile device* — Interfaces of this application can include USB, Bluetooth, IrDA, or Wireless LAN.
- *Information push service* — News, weather, movie show time, games, blogs, and financial data can be downloaded on a timely basis to a cell phone and are viewed via specially designed programs.
- *Audio/video/television streaming service* — Such services are enabled by 2.5G or 3G cellular networks and wireless broadcasting networks. Online music subscription services are also available for cell phone users in the United States.

2.5.2 Second-Generation Smart Phone Applications

The rapid advancements in wireless technologies have resulted in a broader spectrum of data-centric applications and services. Some of them have been commercialized in a few countries; many more remain in the research or trial stages. Many second-generation smart phone applications require significantly higher data rates than first-generation smart phone applications. Below is a list of these smart phone applications and services.

- *Voice over IP (VoIP) or voice over wireless IP (VoWIP)* — VoIP refers to the packet switching telephony technology used on the Internet. Compared with conventional landline PSTN phone service, VoIP is more cost effective and has larger capacity. VoIP has already been used extensively in today's telephone networks. For mobile wireless communications, instead of using cellular networks (e.g., base stations, mobile switching centers, etc.) to transmit digital voice, VoWIP utilizes wireless LANs and wireless Metropolitan Area Network MANs for the same purpose. Such services mainly target enterprises that are interested in using an existing wireless IP infrastructure to reduce the cost of phone calls.

- *Mobile commerce* includes online shopping, stock transaction, location-based advertisement, location-based in-store shopping assistance, location-based travel assistance, and so on. Mobile commerce can be transactional or merely adaptive information provision, where the adaptability lies in location information, user's business profile, and related business logic. A number of service vendors are already in the increasingly hot wireless advertising market, such as Vindigo, Skygo, and AdvantGo.

- *Mobile enterprise* — Companies with a large portion of mobile workers can use cell phones or smart phones to improve productivity. Mobile workers generally refer to people frequently on the go during work hours, such as employees of shipping and delivery companies, police forces, pet control workers, and whoever needs to hit the road often. Usually mobile workers can only update assignments at the office. The only means of communication on the road is via phone calls back to the office. It would be more efficient if mobile workers could have real-time access to backend business information so they can take advantage of dynamic scheduling. Moreover, an intelligent application running on a smart handheld used by mobile workers can automatically determine the most suitable assignment for a worker, thus lowering the cost of transportation and labor. For example, a taxi company in Singapore uses a location-based wireless system to dispatch cab drivers to the nearest customers in real time [4]. This is likely to occur in a variety of business sectors that are concerned with dynamic scheduling of workers as an effort to reduce operational cost.

- *Mobile gaming* — The mobile multiplayer gaming service may have a good chance to mature when 3G is widely deployed, or *ad hoc* gaming utilizing *ad hoc* wireless networks is technically and economically possible. In spite of the fact that a few critical issues remain to be resolved, such as the small cell phone display, the mobile gaming market is poised to grow rapidly. According to an estimate by Frost & Sullivan in 2003, the European mobile gaming market is expected to increase from a little more than $800 million to a massive $7 billion by 2006. Despite the unsuccessful launch of Nokia's N-Gage, other newer models in this series have managed to gain substantial ground among mobile gamers. In addition, the two leading game device vendors, Nintendo and Sony, have announced the release of portal game devices, both with some sort of built-in wireless network capability.

- *Mobile music* — Motivated by the enormous success of online music subscription services, some mobile network operators, cell phone manufacturers, and Internet content providers have teamed up to offer music download services for cell phone users. The quality of audio playback on cell phones will be improved as well.
- *Remote access* provides secure access to a remote network, such as enterprise networks, or allows the user to watch and control a remote site through a virtual private network (VPN) or other secure transports.
- *Remote monitoring* — Sensors embedded in the environment at a remote site will be able to respond to wireless signals. An industry forum, ZigBee [5], has been formed for the standardization of such applications.
- *Mobile wallet and mobile ticketing* — In this application scenario, a smart phone serves as a payment device. It can be used to order electronic tickets by sending an SMS to a service number, or the user can simply wave it at a wireless enabled POS for small purchases. For example, in Helsinki, Finland, people commuting by the city transport system can use SMS to buy tickets from a GSM cell phone. A special code is sent back to the cell phone and used for validation at the entrance of the transport system. Similar systems are used in Japan, where the electronic ticket is a dot-matrix bar code. SK Telecom, the largest mobile network operator in South Korea, offers some cell phones that can be used to purchase beverages from vending machines and pay for gasoline.

- *Mobile social networking* applications may leverage the mobile wireless capability of cell phones to augment person-to-person communication or group communication and collaboration within a community. For example, location-based social networking applications can find physically proximate people sharing similar interests.

- *Mobile ID/key* is a cell phone that can be used as a personal electronic ID, such as student ID, company ID, conference badge, driver license, door key, or car key; to gain access to a building or parking lot; to automatically log in to a computer system; to borrow books from a library; and so on. User's identity data can be stored in a smart card or in the phone's secure storage. As for identify authentication, in addition to the traditional PIN/password approach, some biometric schemes such as fingerprint matching and iris matching have become available on some PDAs. Some software vendors offer voiceprint applications for mobile devices.

The applications and services listed above suggest that smart phones can play a significant role in realizing those otherwise impossible tasks or next generation mobile computing. With quite a few fledging wireless technologies on the horizon, more applications and services are destined to appear in the mobile world and bestow freedom of communication and computing to the next unprecedented degree.

2.6 The Vision of Next-Generation Mobile Computing

The discussion on smart phones and the trend of convergence in the preceding section focuses on the mobile device side in association with a range of applications and services regarding first-generation (multifunction) and second-generation (cellular plus PAN access) smart phones. How will mobile wireless networks evolve in the next 5 to 10 years? What is the impact of ubiquitous wireless networks on our daily life? To answer these questions, researchers and practitioners are looking into new-generation mobile computing systems, communication networks, and applications and services.

2.6.1 Pervasive Computing

The concept of pervasive computing has been briefly introduced in Chapter 1. Pervasive computing is the future of mobile computing. The fundamental difference that separates pervasive computing from other visions of computing is that it emphasizes the relationships between three basic entities: people, computers, and environments. People are working or living in environments with large quantities of connected computers, visible or invisible. The use of these mobile or fixed computers is so natural and unobtrusive that we do not even notice their existence. Not only do environments assist us in accomplishing our tasks in an automatic and well-coordinated way, but we also influence the characteristics and dynamics of our environments. The following is a list of key conceptual components of pervasive computing:

- *Embedded computing* — Our physical environment is embedded with many small intelligent computing devices, spreading to nearly every corner of our daily life. Here the notion of "embedded" implies a more general meaning of functional integration and interoperation as opposed to the common understanding of embedded systems. In the realm of pervasive computing, we do not adapt to the space surrounding us; we do not need to know which application to use, where they are, what they can do, and how they can do what they do. Instead, the space adapts to us.
- *Ubiquitous data access* — Wireless and wired connectivity are always available anywhere from any mobile devices. Because knowledge of embedded devices is not required in order for us to take advantage of them, the focus is shifted from smart devices to smart objects, which embody data models that encapsulate attributes, patterns, and any other relevant information of various operations a user may perform.

- *Mobility and adaptability* — Both smart devices and their users are mobile; the geographical locations may change dynamically at varying speeds and in diverse scopes. In addition, other factors to pervasive computing may also change, such as environmental context and user's physical and psychological conditions. The computing systems embedded in the user and environments have to facilitate undisruptive mobility and adapt to internal and external physical and computing dynamics expeditiously and inconspicuously.
- *Extensibility and scalability* — Open interfaces between smart devices and networks allow the computing environment to be extended and intelligently cooperate with newly deployed devices and services. Data objects are frequently transferred and converted automatically to allow information sharing on a very large scale enabled by widely used open data service platforms such as Web Services.

No doubt this brave vision of pervasive computing poses a number of challenges to researchers in both the wireless industry and academia. Some of the interesting issues pertaining to the aforementioned conceptual components are listed below:

- Embedding a huge number of smart computing devices in environments requires rethinking of system design with regard to power consumption, user interface, network capability, storage, and software architecture.
- Ubiquitous data access calls for integration of various wireless and wired networks and open data access interfaces. Taking advantage with multiple wireless links from different data sources at the same time, one will be able to obtain rich data from different circumstances, enriching the user's experience.
- The objective of context aware should be maintained throughout the design of mobile wireless networks, systems, and smart devices to support mobility and adaptability.

- The software infrastructure of pervasive computing must be open and extensible to allow convenient and effective interpersonal and inter-device communication, as well as scalable data collection, processing, and dissemination.

Pervasive computing is a very broad area that subsumes many disciplines of computer science. In the context of smart-phone-based mobile computing, the focus of the book is on the role of a smart phone in the future mobile world and how it will engage in a large set of applications and services with emerging wireless technologies and networks. Among those smart computing devices that will exist and cooperate with each other in the pervasive computing paradigm, smart phones are most likely to be able to become a universal computing and communication platform for potential mobile wireless services and applications, thereby gaining enormous popularity in people's daily lives. The following discussion is centered on the technological trend of smart phones and related issues in the context of next-generation mobile computing.

A smart phone, or a universal mobile terminal, relies on its surrounding wireless networks, such as cellular networks, wireless LANs, Bluetooth PAN, wireless MANs, and wireless sensor networks, to operate. Thus, building reliable, high-performance, and secure wireless networks becomes an issue equally important as the issue of mobile terminal design. Additionally, both wireless networks and mobile terminals would be completely useless if no viable applications and services were offered to the user or subscriber. The three components of network infrastructure, mobile terminals, and applications and services comprise a tripod of mobile computing. As shown in the preceding sections, each component of mobile computing itself is undergoing drastic changes. Nonetheless, the vision of next-generation mobile computing is quite clear and can be summarized as follows (Figure 2.9):

- *Convergence of mobile access* — We will see a blend of wireless technologies being used in the business world and people's daily lives. More significantly, system components and network elements

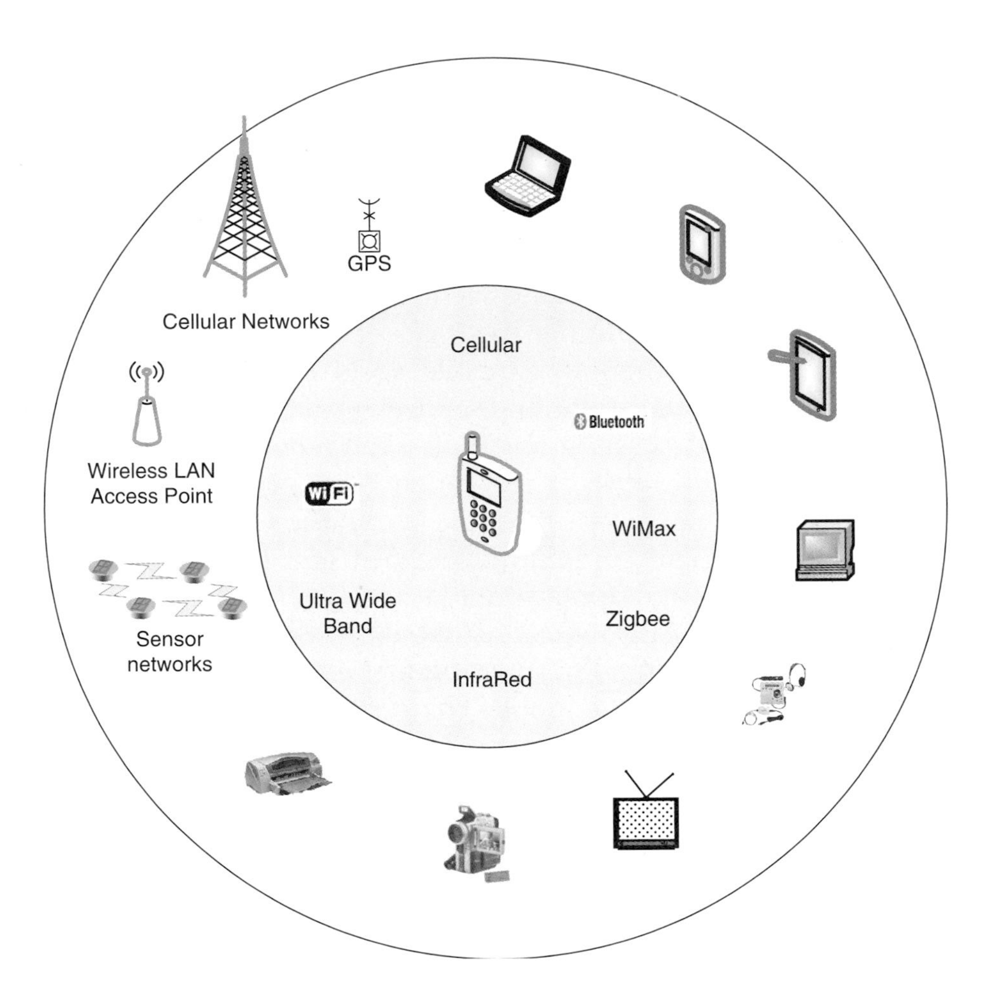

Figure 2.9 The vision of mobile computing.

utilizing those technologies will be well coordinated and interoperable, prompting the use of some converged mobile terminals that are able to provide applications and services with a set of systematically integrated hardware and software components.

- *Pervasiveness of mobile intelligence* — Because wireless is everywhere, access to backend systems and networks will become ubiquitous. In addition to providing always-on, always-available network connectivity, a key challenge in this regard is that the services and applications must be unobtrusive, meaning that the system of well-coordinated wireless components must be smart enough to choose the best way to accommodate a user's needs, and this process must be completely transparent to the user. In other words, the user will not need to explicitly use the system any more; instead, the system will help the user achieve a task automatically and intelligently.

The inner circle in Figure 2.9 depicts a few wireless technologies that may interface to the center of the picture, a smart phone. The outer circle shows surrounding environments of the smart phone, whereby mobile devices, consumer electronics, sensory devices, wired network devices, and wireless infrastructures interconnect with each other through a range of wireless networks. Devices in this big picture can be categorized into a number of classes as follows:

- *Mobile devices*, including smart phones, PDAs, cell phones, laptop computers, handheld PCs, table PCs, and specialized mobile devices used in various industries
- *Instrumental devices*, including wireless sensors, actuators, RFID tags, in-car sensors, and so on
- *Wired network devices*, including networked desktop computers, network switches, routers, printers, scanners, wireless access points, wireless switches, interface adaptors, and specialized computers used in various industries
- *Consumer electronics*, including televisions, DVD players, camcorders, digital cameras, VCRs, stereo systems, set-top boxes, home theaters, audio recorders, portable music players, portable game devices, mobile game devices, and watches, among others
- *Home appliance*, including microwave ovens, refrigerators, cookers, washing machines, dishwashers, coffee makers, heaters, air conditioners, vacuum cleaners, clocks, lamps, security systems, landline telephones, cordless phones, and so on.

The trend of convergence of mobile access suggests unified access from a smart phone to surrounding wireless devices, whereas the pervasiveness of mobile intelligence represents how data are collected, processed, and disseminated among all the components in a mobile environment.

2.6.2 Convergence of Mobile Access

Mobile communication and computing are usually considered two different arenas. In the business world, they are two industries. The mobile communication industry provides voice communication using cell phones via the backend cellular networks. Parties in the mobile communication industry are mobile network operators, mobile device manufacturers, and mobile software vendors. The mobile computing industry is part of the general computer industry, offering a portable computing platform and data access using handheld devices, laptop computers, and tablet computers that may or may not need to access wired and wireless data networks. Internet protocol (IP) is the dominant network protocol over almost all physical networks. The mobile computing industry consists of nearly all sectors in the computer industry, including computer hardware manufacturers, software vendors, IP network device manufacturers, and IP network access providers.

In the more general technological terminology, mobile communication and computing are still two distinct areas. Mobile communication or, strictly speaking, mobile wireless communication concentrates on the communication aspect, primarily on the physical and data link layers, and enabling hardware technologies in chip design. Mobile wireless networks range from cellular networks to wireless LANs to Bluetooth to any other networks utilizing radio signals. Conversely, mobile computing is primarily concerned with issues at the network and transport layers and with mobile applications and services. It emphasizes building computing systems to support mobility at a variety of scales.

The rapid advancements of communication and computing technologies eventually led to the convergence of communication and computing, which is currently taking place across the entire mobile computing paradigm. Communication networks and IP data networks are converging into a logically unified mobile network that allows vertical handoff and tight data integration. For example, integration of cellular networks and wireless LANs essentially makes it possible for wireless phone calls to be handled more flexibly and cost effectively via wireless LAN hotspots. Cellular data services are necessary to ensure wide area wireless Internet access when a wireless LAN hotspot is not available close by. The coexistence of Bluetooth, wireless LAN, and other WPAN networks is technically conceivable. Research on challenging issues in this domain such as mobility support, distributed resource management, service discovery, signaling protocols, QoS provisions, and billing and accounting is underway. A general mobile access abstraction layer, arguably mobile IP, is needed to enable mobility across different wireless networks.

Moreover, handheld mobile devices will converge to leverage mobility. Smart phones will become the universal mobile terminal to interface with both communication and computing networks on-the go. The functionality supplied by a smart phone will be far richer than a simple combination of phone calls and PDA organizers. On the hardware side, rather than simply laying out many chips for different radios on the handset, a smart phone will be able to engage in a number of wireless networks with the very same radio chip set. On the software side, smart phone software will be designed to take full advantage of the convergence of communication and computing to make the smart phone more adaptable to changing environments.

Applications and services in mobile computing are also set to converge in many respects. Voice communication will be augmented by instant messaging, e-mail, videoconferencing, landline telephony services, and mobile gaming. Mobile payment services can be an essential element of music download applications, online television, sports broadcasting services, and digital art and document sale services. A mobile application will be able to post text and photographs taken by a built-in camera to web blogs and wiki web sites. (Wiki is a web application server technology that allows web surfers to directly edit hosted web pages using a web browser and can be used as a web-based collaboration tool among web users; the most famous wiki site is wikipedia [http://www.wikipedia.org/].) The very same application can also pull in the latest news, weather, stock quotes, enterprise calendar, and location-based business information via really simple syndication (RSS), a lightweight XML-based content syndication protocol. Interoperability of applications and services on multiple mobile terminals is likely to be enabled by XML-based web services.

The convergence of mobile wireless networks, mobile terminals, and mobile applications and services has begun to have a tremendous impact on the way we perceive mobile wireless environments. Regardless of the underlying wireless networks, mobile terminals, or applications and services, optimized mobile access for a particular operation is always needed and should always be available. Wired networks will also be involved in the converged mobile computing paradigm, serving as a substantial backhaul of the IP networks.

2.6.3 Limitations of Convergence

Convergence of mobile access is not a cure for everything. It has its inherent technical and nontechnical limitations. On the mobile terminal side, simply juxtaposing quite a few function sets into the same device may not always result in sizable business success. In many cases, simple and easy-to-use single-function models are preferred by users over multifunction combos. Likewise, converged networks and applications may cause inflexibility and reliability issues, particularly in the mobile wireless domain. In addition, the needs of people from different cultures and social background are sometimes too diverse to allow the use of converged systems, making it practically impossible to build converged systems and networks in a cost-effective way. The key to overcoming these limitations is to strike a balance between diversity and convergence based on relevant technical factors and user analyses. To achieve this goal, design tradeoffs are inevitable, as has been the case in the evolution of many other technological segments. Chapter 7 (Mobile Application Challenges) addresses these issues in more detail.

2.6.4 Pervasiveness of Mobile Intelligence

Another trend in mobile computing is mobile intelligence deeply immersed in the computing environment. Mobile intelligence refers to the schemes and techniques incorporated into the mobile application and service architecture that utilize a wide range of datasets in surrounding wireless networks to facilitate context awareness, decision making, location awareness, and user profile awareness. Convergence in mobile access makes it easy for an application or a service to connect to various wireless networks, whereas pervasive mobile intelligence harnesses ubiquitous mobile access to assist users in adapting to the surrounding environment, interacting with others, and enriching their personal experiences. In some sense, the trend of pervasiveness of mobile intelligence can be attributed to the evolution of pervasive computing, which, as explained in Chapter 1, is to extend mobile computing into a broader vision dealing with the physical environments of numerous wireless computing devices and providing ubiquitous mobile access. Because of the massive, highly varying data available in different wireless networks, it is not feasible to capture every aspect of each mobile computing operation; the data has to be collected, interpreted, and exchanged in a way that adapts to the user's context; on the other hand, such a heterogeneous computing environment, along with varying physical environments, may provide a valuable opportunity for an application or a service to leverage immersive computing and communication embedded in physical environments, an apparent Holy Grail that could nevertheless have a phenomenal impact on people's lives.

Pervasive mobile intelligence can be applied to many scenarios, including enabling ubiquitous mobile access, context-aware information retrieval and processing, and large-scale data collection and dissemination. In each case, the mobile software infrastructure on both the network side and terminal side embodies the intelligence that is invisible to the end user.

2.6.4.1 Enabling Ubiquitous Mobile Access

The coexistence and integration of heterogeneous wireless networks require a reliable and efficient system to support ubiquitous mobile access, as well as mobility and authentication, authorization, and accounting (AAA). Moreover, such a system should be unobtrusive to the user. Imagine a saleswoman with a smart phone walking into a technical conference session while text-messaging her boss back at the company's headquarters. The smart phone will be able to quickly find all conference attendees who are current customers of the company (assuming they agree to expose this information to other conference attendees). At the same time, the saleswomen may use the smart phone to pull in other useful information, such as which conference sessions they have attended and what questions they have raised during those sessions. With a little help from the customer relation management (CRM) software at the company's headquarters, the saleswoman could come up with a business plan that incorporates issues raised during the conference sessions and products of the company. She may then send messages or voice mails to selected customers in the room asking for a brief meeting sometime later. This application scenario may take place in a location where many wireless technologies such as cellular networks, wireless LAN, and Bluetooth are being used in a more systematic way. The software on the mobile terminal scans the data flowing among these wireless networks to find the most likely sales targets and presents to the user accordingly. Challenging issues in this regard are privacy and possible spamming; there must be a well-defined mechanism that allows users to decide whether or not to expose specific personal or sensitive business data to others or to receive unsolicited messages from others. (Mobile privacy issues are discussed in Chapter 5.) Moreover, this decision-making process itself must be unobtrusive.

2.6.4.2 Context-Aware Information Retrieval and Processing

The notion of context awareness was briefly discussed in Chapter 1. Context-aware information retrieval and processing deals primarily with determining the most appropriate context profile based on a collection of knowledge and data in the computing, communication, and physical environment. A list of suggested actions is made available to the user afterwards. The need for context-aware information retrieval and processing originates from the desire to make use of vast quantities of pervasive information and knowledge for effective decision-making. For example, when a user with a smart phone is driving down a highway, the smart phone will be able to detect that the user is moving at a speed of more than 60 miles per hour. The context will be identified as "driving" in this case. For safety reasons, whenever there is an incoming phone call, the smart phone will automatically turn on the car's microphone and speaker and will use them as input and output components (they are connected to the smart phone over some wireless connections), instead of using the smart phone's built-in microphone and speaker. In this example, the context is determined quite easily, but in the real world it is often difficult to accurately and quickly determine the context based on uncertain factors.

2.6.4.3 Large-Scale Data Collection and Dissemination

Large-scale data collection and dissemination are very common in a pervasive computing environment, where many small wireless devices such as sensors, actuators, and RFID tags, as well as smart phones and computers form a very large, heterogeneous distributed database. Numerous wireless networks are interconnected with each other to allow query processing over the large distributed database. To make this procedure more efficient and scalable, in-network query processing with coordinated nodes in the networks is necessary. These interconnected wireless networks are also referred to as a wireless grid [6]. Data can be collected locally or remotely from a wide range of wireless devices in different settings as diverse as wireless sensors in a terrestrial area, *ad hoc* smart phones within an urban area, RFIDs of products and palettes in a huge warehouse, consumer information appliances in a home data network, and even from people with wearable computers.

Figure 2.10 highlights the relationship between the two directions of next-generation mobile computing. Many aspects, including those of pervasive computing, are still in the very early stages of research

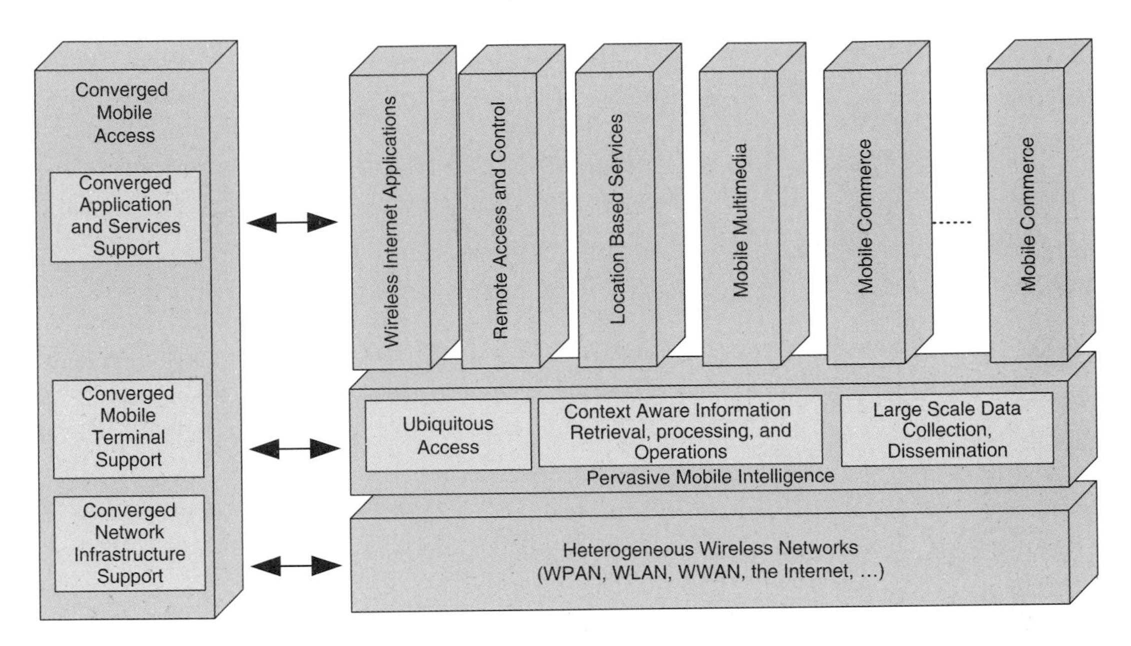

Figure 2.10 Conceptual diagram of the vision of mobile computing.

and development and are far from maturity. While mobile computing promises to offer robust, unobtrusive, and open access to any data from any devices, everywhere, all the time, the first practical step, particularly for the mobile wireless industry in the near term, is to combine existing wireless networks and mobile devices into an interoperable computing and communication environment to allow open access to relevant data and to offer feature-rich applications and services on converged smart phones that satisfy the need for high productivity of mobile professionals.

2.7 Mobile Computing Challenges

The research and development of next-generation mobile computing is at a crucial stage. The answer to the question of whether the vision is solely hype or a potential step toward augmented user experience and improved productivity remains unclear as many challenging issues in this domain have yet to be addressed. To this end, the vision of next-generation mobile computing essentially encompasses cutting-edge research and development in a number of directions. Some of these efforts have demonstrated dramatic achievements and that are being commercialized by industry practitioners. The last section of this chapter presents a summary of these research directions, as well as some emerging mobile wireless applications that utilize smart phones to operate in a variety of circumstances. The emphasis here is on the need and motivation of the technological trend, rather than the technical details of these research issues, which are thoroughly discussed in the next chapter. The remaining chapters will further discuss these issues.

2.7.1 Integration of Wireless and Wired Networks

The notion of networks in the context of computing technology has a very broad meaning beyond common IP-based data networks and cellular networks. Let's first take a look at the networks currently available. On a very large scale, the Internet, of course, is the most widely used network. The Internet is actually a network of networks around the globe. Other networks include the cellular networks, such as the 2G cdmaOne, GSM, DAMPS, 2.5G GPRS and EDGE, and early 3G networks, which have managed to attract billions of subscribers collectively. There are numerous civilian radio networks such as ham (amateur) radio and the two-way radios used by law enforcement personnel. There are also many commercial and home wireless LANs that allow people to connect to the Internet. Wireless sensor networks have been deployed to islands, forests, and even glaciers to monitor the environment and inhabitants. Security surveillance video networks are commonly used to monitor streets, buildings, stores, and other critical areas. Television cable networks send analog and digital programs to home users. PSTN provides the very basic means of voice communication via landline telephones. Communication satellites are used to transmit voice, data, radio, and video to receivers throughout the world, and are used for militant purposes as well. The vision of next-generation mobile computing foresees that not only will all these networks be interconnected but also those devices that currently do not have any network interfaces will be connected and possibly remotely controlled. Though this scenario seems overwhelmingly far from reality, some portion of it has begun to take place in our daily lives. As an example, an obvious trend in the mobile wireless industry is the integration of two of the most popular networks: cellular networks and wireless LANs hotspots.

There has been some confusion with regard to WiFi and wireless LAN among industry professionals and the general public. In fact, many people think WiFi and Wireless LAN are the same, but they are not. The term *WiFi* refers to the WiFi Alliance, which is a nonprofit international association formed in 1999 to certify the interoperability of wireless LAN products based on IEEE 802.11 specifications [7]. The year 2003 witnessed the great success of the free-spectrum, high-bandwidth WiFi technology. As of 2005, WiFi has become the predominant wireless home networking technology, thanks to the dropping prices of 802.11b and 802.11a wireless access points and WiFi interface cards. Many small businesses consider WiFi as an alternative to wired enterprise networking solutions.

Wireless LAN technology is inherently a local area network technology because the 802.11 protocols are essentially designed to clone Ethernet protocols in a wireless environment. The typical operational mode of a wireless LAN network is infrastructure based, where every mobile terminal communicates with a wireless access point that in turn connects to the Internet. The WiFi network used by the general public commonly works in this operational mode. In the United States, many such small-scale WiFi networks (also known as WiFi hotspots) have been established in airports, hotels, coffee shops, hospitals, schools, shopping malls, office buildings, and residences. For example, Chicago's Sears Tower, the highest building in the United States, will soon have WiFi access on all floors within the 110-story skyscraper. Another operational model of a wireless LAN is *ad hoc* based in that two mobile terminals communicate with each other directly without a central access point. Moreover, a number of mobile terminals in *ad hoc* mode can coordinate with each other and self-organize into multihop communities (namely, a wireless mesh), in which mobile terminals must rely on others to relay messages.

WiFi has become the first choice of mobile data access solution for people on the go. As a consequence, many laptop computers now have built-in WiFi support. It is also not unusual to see PDAs shipped with a built-in WiFi interface. Smart phones will definitely follow the same trend (already some models provide built-in WiFi capability). Internet access can be achieved using either WiFi hotspot access or cellular data access. For voice communications, besides cellular network access, voice over wireless LAN technology has emerged to replace traditional PBX within some enterprise networks. To complement the service offerings, mobile network operators have begun to consider integrating cellular networks with WiFi hotspots to allow cell phones users to roam freely between these two types of networks. This is a strategy of mobile network operators striving to maintain a good position in competition with WiFi data access providers (i.e., Wireless Internet Service Providers, or WISPs) that may offer voice communication in the future. The benefit of integrating cellular networks and WiFi hotspots are summarized below:

- The bandwidth of WiFi Internet access (11 Mbps for 802.11b and 54 Mbps for 802.11a and g) is much higher than what 2.5G and 3G cellular networks can provide. As long as a user stays within the transmission range of an access point, WiFi is apparently a better choice for Internet access than cellular data access. Taking into account the volume-based billing model of cellular data access versus the simple flat rate of the WiFi hotspot model, users are more likely to prefer WiFi data access that resembles wired Internet access in many ways. Of course, this is a subtle issue largely affected by a number of nontechnical factors such as WiFi infrastructure costs and billing plans.
- The security concerns of WiFi can be mitigated by combining cellular network security mechanism with WiFi. Wireless LAN standards are well known to have security vulnerabilities in their key generation and distribution mechanisms, which has considerably

hindered the use of wireless LANs in critical environments. (The IEEE 802.11i security standard for 802.11 wireless LAN addresses wireless LAN security.) Cellular networks, on the other hand, can provide a high level of encryption and authentication. The ubiquity of cellular networks will further help WiFi to be accepted by more companies and organizations.

- VoWIP has the potential to complement or compete with cellular communications, and the cost will be significantly lower than cellular communications for an enterprise or for consumers making long distance calls via a wireless mesh network. Roaming among WiFi hotspots on a very large scale poses a significant challenge to VoWIP providers, though.

Open issues with regard to the integration of cellular networks and WiFi hotspots are, to name a few, seamless voice and data access roaming between WiFi hotspots and cellular base stations, cell phone dual connection management, and AAA integration. As most WiFi hotspots connect to a backend wired network, some dedicated gateway servers are required at the border of cellular networks and wired networks to manage mobility-related data of a mobile terminal. In addition, voice communication services of wireless mesh networks can be fused with cellular services as well. These interesting issues surrounding the integration of cellular networks and WiFi are discussed further in Chapter 5 (Mobile Networking Challenges).

Aside from the integration of cellular and WiFi hotspots, some researchers have proposed integrating cellular- and *ad hoc*-based wireless LANs to eliminate any wireless LAN access points. Such an idea can be applied to a range of application scenarios for data access, such as sports stadiums, battlefields, and remote areas where it is more or less impractical to establish a centralized network infrastructure and the majority of communication is performed among mobile terminals within the *ad hoc* network. Instead of establishing an outbound Internet connection from each mobile terminal, a more efficient scheme is to choose some mobile terminals to serve as the designated proxy for Internet access and let others use this proxy for Internet access. Research issues regarding this scenario include how to choose mobile terminals as proxies and how to manage them to achieve system optimality that could be quantified using some metrics for in terms of high performance, fairness, and fault tolerance.

Other wireless networks may also be integrated to allow cross-network communication and cooperation. For example, the coexistence of Bluetooth and 802.11b wireless LANs is very common in a WPAN. Because they both use the unlicensed 2.4-GHz spectrum band, interference may occur in some cases when simultaneous radio transmissions are underway.

A galaxy of new applications and services will emerge to leverage these heterogeneous networks to deliver ubiquitous mobile access, monitoring, control, and analysis for consumers. Those applications that are impossible prior to this wave of mobile computing for either technical or economical reasons may finally become a reality. Among these, mobile multimedia and mobile commerce have already begun to reach the general public; nevertheless, many challenging issues still exist with regard at the network layer and application layer. We will revisit these issues in the next chapter.

2.7.2 Mobile Security and Privacy

Security in mobile computing is always a hot issue. 2G and 3G cellular systems are designed to be secure to some extent for voice communication, due to the encryption that makes eavesdropping quite difficult. Because data service is about to take off in the mobile wireless world with the highly anticipated forthcoming 3G deployments, security problems that we have experienced on the Internet will appear in the mobile world. An example is the malicious e-mail virus attack on 110 million cell phones operated by NTT DoCoMo as early as 2001. In this incident, cell phones that were infected by opening an e-mail continuously dialed national emergency numbers or froze up. In August 2004, Bradoor became the first Windows Mobile worm to infect ARM-based Pocket PCs. Once infected, the device is fully under the control of the attacker, who can issue commands remotely, display messages on the screen, or download and upload files. Worms targeting other operating systems have also been found, such as Cabir against some Symbian cell phones and LibertyCrack against Palm OS. SMS spam has reportedly grown significantly in some countries where mobile users rely more heavily on SMS than voice calls. Other wireless technologies also have security problems. Wireless LANs have long been criticized for the incredibly weak security mechanism, which ironically boosted its wide acceptance. Many Bluetooth-enabled cell phones are reported to be vulnerable to certain short-range network attacks.

Security issues in mobile computing can be divided into four major categories:

- *Message interception and falsification* — Using sniffer software to monitor and analyze wireless traffic and inject falsified packets into a stream to gain network access or affect legitimate communication
- *Impersonation, identity theft and fraud* — Using technical or social engineering techniques to obtain authentication credentials of legitimate users
- *Mobile virus and device hijacking* - Surreptitiously installing viruses, worms, or Trojans into a mobile device such that the device can be fully exposed to the attacker and controlled remotely
- *Mobile network attack* — Utilizing a number of compromised mobile devices to launch a DOS attack against the wireless network, effectively making the network unable to provide service
- *Spamming* — Sending a large volume of unsolicited SMS messages or IM messages to users
- *Phishing* — An attack against e-mail users in order to steal sensitive information; phishers usually send an e-mail to a user in which they falsely claim to be an established, legitimate company such as a credit card company or a bank, and; the embedded links in the e-mail direct the user to a bogus website that looks very similar to the legitimate company's website

The heterogeneous mobile wireless networks in place today make it extremely difficult to establish a common security framework to address all the potential problems. In reality, security issues are approached in a number of ways. Wireless network operators should take the responsibility to secure communication channels for message transfer and to offer strong authentication schemes for mobile terminals and users. Mobile devices such as smart phones and PDAs, along with their system software, should incorporate system-level security to defeat viruses and network attacks. Services and applications offered by mobile network operators should be considered as performing in a very likely insecure mobile environment and be subject to strong authorization, authentication, and accounting procedures. In many cases, just as in the wired networks, addressing security problems in the mobile world also involves nontechnical issues, such as business strategies, political and culture concerns, and social engineering.

Cryptography can help to ensure message confidentiality by applying encryption to the message. It can be done end-to-end, such as in IPSec, or in the network. The problem of end-to-end encryption is the overhead on mobile terminals. Because mobile devices do not have computing power comparable to that of a desktop computer, it would be impractical to apply the same end-to-end encryption used on desktop computers to mobile devices.

Digital signatures can be used to ensure data integrity and authentication. Usually the digital signature is generated by first applying a one-way hashing functions to the message body. If public-key cryptography is used, the result (namely the hash code or hash) will be signed with the private key of the sender. The digital signature will then be appended to the message and sent to the recipient. To check the integrity of the message, the recipient must first use the public key of sender to decrypt the digital signature. Any modification to the message body will cause a mismatch between the decrypted digital signature and newly computed hash code. If secret-key cryptography is used, then the sender will use a secret key to sign the hash code, and the recipient will use the same secret key to regenerate the hash code. Message digest works in a similar but simpler way, without providing encryption to the hash code. Nevertheless, the one-way hash algorithms serve as the core of the approaches widely used in the wired and wireless networks. Unfortunately, MD4, MD5, SHA-0, and even SHA-1 have been reported to occasionally collide; that is, two messages are hashed into the same code [8]. Although it is still theoretically impractical to find an alternative message for an arbitrary hash code, the recent advancements of cryptography have demonstrated that it might become possible someday.

Another significant issue in the domain of mobile security is authentication, authorization, and accounting (AAA). Multiple service providers might be involved in the process of dealing with these issues for a mobile terminal. To date, a SIM card is widely used on GSM cell phones as a unique ID that carries AAA information of the user. Handling the AAA of a cell phone user is not a critical issue, as the SIM card is always in the phone; however, the problem becomes more difficult to solve within the context of integrated heterogeneous networks operated by different parties. Moreover, as mobile *ad hoc* networks emerge to become a practical service in the future, issues of AAA transition and transaction among mobile terminals will also have to be addressed.

Mobile privacy is not the same as mobile security. A mobile system may be secure at the expense of the user's privacy. In essence, mobile privacy refers to the total control of users over their personal data that may be exposed to, collected, and used by other persons or parties for a variety of reasons. Mobile privacy has become a major concern in mobile computing, even more so when mobile wireless networks and systems begin to be more pervasive in people's daily life. To this end, mobile privacy has been regarded as the Achilles heel of pervasive computing [9]. Interestingly, some of the promising mobile services and applications tend to utilize more personal information such as user's location, history data, and user's actions. Mobile privacy could be a key barrier for the adoption of such services if users are aware of these privacy threats (although they should be informed anyway). The fundamental question, then, is how to draw a line between full exposure of a user's personal data and interactive informed data sharing such that both mobile privacy and system design objectives can be fulfilled. Mobile security and privacy issues are discussed in detail in Chapter 6.

2.7.3 Location-Awareness Mobile Computing

The freedom of mobility offered by mobile computing presents enormous opportunities for innovative business services and applications. One of the most promising opportunities is location-aware mobile computing, which uses the mobile terminal or user's geographic location to provide value-added location-based services utilizing a wide variety of wireless and wired technologies. For example, the worldwide location-based services market is expected to rise to 3.6 billion by the end of 2009, according to the research firm ABI.

The heart of location-aware mobile computing lies in mobile positioning, tracking, and analysis schemes based on various wireless technologies. The term *positioning* in the context of location-aware mobile computing refers to techniques for determining the geographical location of a mobile device or a user, indoors or outdoors. The past, present, and estimated future positions of a mobile device or a user are monitored by location-based services that combine position data with other location-related information to enable context-aware operations. The following is a list of location based mobile services that are available or being deployed very soon.

- Emergency, such as E911 in the United States or E411 in Europe
- Authorized mobile user positioning, such as tracking the location of a child
- Navigation, in association with GIS (Geographic Information System)
- Real-time traffic information updates, coupled with maps and navigation
- Mobile workforce monitoring (i.e., logistical telematics) and real-time scheduling, such as delivery service tracking, scheduling, routing, and road assistance services

- Asset management, such as product tracking and positioning
- Location-based business searches in conjunction with mobile advertisement
- Location-based message services, such as SMS within proximity
- Location-based computing service discovery and automatic configuration, such as finding the nearest WiFi hotspot or *ad hoc* gaming network and making use of them.

When mobile users actively initiate a location-based service from a mobile device, it is very likely that these users want to obtain location-based information within their proximity. On the other hand, mobile users could be passively tracked by a wireless network that would indicate their position to someone. In both cases, the mobile device must be location aware, and the network must have an efficient way to collect, compute, and analyze the locations of a large number of mobile devices. For example, a smart phone with a built-in GPS (such as the Motorola i730) will extend its use to another level, especially for trip planning and driving assistance. With a built-in WiFi interface on a smart phone, a user may be able to conduct indoor mobile device positioning and find WiFi-enabled printers in a building.

Enabling positioning technologies for location-aware mobile computing include cellular networks for outdoor positioning, GPS for outdoor positioning, and a few indoor positioning approaches utilizing various sensing technologies such as radiofrequency tags, ultrasonics, wireless LAN, Bluetooth, and wireless sensor networks. It should be noted that the positioning of a network device or a user is also a complex issue in fixed wired networks. Because IP addresses normally serve as unique identifiers of computers in a wired network, location-dependent IP address allocation information can be used to locate the computer holding a specific IP address. Such a scheme is also applicable in the mobile world, where IP plays a key role in identifying mobile devices. Some of these positioning and tracking schemes are discussed further in Chapter 7 (Mobile Application Challenges).

Depending on the requirements of a location-aware application and wireless networks infrastructure in place, one or more positioning techniques may be combined to allow both indoor and outdoor positioning with desirable accuracy and response time at modest cost. Furthermore, because a mobile user's location is also context data, location-based services can be fused with other context-aware operations, services, and applications to improve adaptability over context changes. For example, a smart phone roaming into the range of a wireless Internet web proxy server may silently switch to it for better web access performance.

Location-aware mobile computing usually utilizes GIS to enrich its applications and services. A GIS essentially consists of a multidimensional map with numerous information overlays for a geographic area. Like a telephone directory, GIS is a location-based information database for mobile users. Any location-dependent information can be compiled into a GIS for special purposes as diverse as business strategies, political voting estimates, investment risk analysis, and demography, among others.

2.7.4 Human-Computer Interface of Mobile Applications

Mobile computing imposes another profound level of challenges on human-computer interfaces (HCIs). Because applications and services are predominantly used on the move, the interaction between the user and the mobile terminal and between the user and a backend service system must be sufficiently simple, convenient, intuitive, flexible, and efficient. In addition, the user interface has to be designed to save power if at all possible. The human-computer interface incorporates far more than GUI design. Unlike desktop computers that use a standard keyboard and a mouse as the predominant input devices and a monitor for display, user interfaces of mobile terminals may support more input and display methods such as keypad input or voice recognition, and other means of feedback, including LEDs, sound, and handset vibrations, in addition to visual images and texts. More importantly, for desktop computers HCI has been isolated from the variances of input and output devices such that a word processor application on a desktop computer is designed to work well with a monitor regardless of the monitor type or size. This is not the case for mobile terminals. For example, a smart phone may have a small keyboard as the input device. The keyboard could have one letter per key or could have two or three letters shared by one key. Or, the device may use a stylus for handwriting recognition, so users can write anywhere on the screen or only within a text input area on the screen. Voice recognition is also possible for device-based and network-based backend service interactions. A mobile terminal may also have the capability to use the built-in speaker or LED flashlights to notify the user of state changes.

Beyond conventional means of HCI, recent research has addressed multimodal interfaces, which combine two or more input modes of body languages, such as body gestures, facial expression, and eyeball movements, to facilitate effective communication between a user and the computer. A wide range of recognition technologies are incorporated into multimodal interfaces, including speech recognition, vision recognition, and biometrics. Furthermore, researchers in the HCI area envisage futuristic multibiometric-multimodal-multisensor (M3) systems that could interpret and respond to natural language and user behaviors [10]. As multimodal interfaces continue to mature and computation overhead keeps being reduced, applying these approaches to mobile HCI is just a matter of time.

Key issues in the context of HCI are as follows:

- *User interface design on a mobile device* — A number of interleaved factors that are intrinsic to mobile wireless networks and mobile devices will affect the design of a user interface for a mobile application. These factors are size, color depth, and resolution of the display, battery time, weight of the device, wireless connection bandwidth, link reliability, processor, capacity, etc.
- *Coherent user interface design across multiple mobile devices* — Due to the lack of standardized representation layer of input and output devices, a user interface has to be designed and implemented on each type of targeting device. The layout of the GUI

components and the operational logic must be consistent to allow easy navigation and intuitive operations.

- *Adaptive user interface design* — The application and supporting system software on a mobile device must be able to dynamically optimize the user interface on the device, based on the hardware configuration of the mobile device and context of the running application.

In contrast to the well-established design philosophy of desktop application interfaces, in the mobile world no such guidelines exist. Depending on the nature of an application or a service, different aspects of the requirements are likely to be prioritized in favor of design goals. As a universal mobile terminal, future smart phones may simplify the user interface design by supplying a converged hardware platform with a flexible software infrastructure that allows adaptive configuration to meet the needs of disparate applications and services. The software infrastructure is able to coalesce knowledge and information obtained from diverse operations and represent them as fine-grained smart data objects to applications and services in a coherent manner.

The traditional taxonomy of mobile devices treats a smart phone as simply a multifunction cell phone with scarce data integration. Such a conceptual understanding is too rigid to be applied to the user interface design for future smart phones, as it fails to facilitate systematic and efficient use of data gathering from users, other devices, networks, and the surrounding environment. For example, a smart phone that can download and play mp3 songs is not just a combination of a cell phone and an mp3 player with distinct user interfaces; rather, it may allow the user to find people in adjacent areas who are likely to share the same interest in a musical artist by looking at a music download database. Then, the user may use the integrated user interface to send instant messages, e-mails, or whatever type of communication SMS to these people. In this sense, a virtual mobile community can be built with data objects from various sources.

2.7.5 Context-Aware Software Design

The notion of *context* has a very broad meaning across many circumstances. It is generally defined as any interrelated information, data, knowledge, or event that represents a condition with respect to temporal, spatial, and state change. Context is an extremely important factor in mobile computing, as it well denotes the dynamic characteristics of mobile computing, such as a mobile user's location, status of an operation, noise of the physical environment, etc. The idea of context-aware mobile computing is to design systems, applications, and services that take advantage of detected and inferred context data to meet users' needs in a more intelligent and unobtrusive fashion. The following is a list of categories of context information generally exposed in the domain of mobile computing:

- The *environmental context* includes such variables as temperature, humidity, height, light, weather, noise, surrounding buildings and objects, traffic conditions, demographic information, and history.
- The *computing context* includes static and dynamic computing contexts. Static computing context includes computer identity, CPU speed, capacity, storage space, display size, speaker volume, device weight, remaining battery power, program size, wireless connections, etc. Dynamic context includes free memory and storage, network bandwidth and delay, available mobile access points and their signal strength, adjacent mobile neighbors, computational operation history, audio/video compression ratios and transmission rate, CPU load, I/O rates, etc.
- The *user context* includes the user's identity, geographic location, body temperature, body and face profile, gestures, social status, and relationships, among others.
- The *service context* includes service level agreement, billing information, user account balance, usage patterns, and statistics, etc.

Context data from different categories may pertain to each other. For example, the amount of transferred wireless data in the computing category of context directly relates to the wireless data quota defined as part of the service contract, and the environmental temperature, either indoor or outdoor, will obviously affect the user's body temperature, the power consumption of a mobile device, and even wireless signal strength.

Context-aware mobile applications and service have been intensively researched for many years. A number of prototypes have been developed by researchers and industry practitioners. Earlier experiments showed that context-aware mobile applications and services may have a strong potential to be widely accepted by the masses. Below are some examples of context-aware mobile applications and services:

- *Smart homes* (context-aware home networks and systems) - The network or system is able to adapt to the needs of the residents for computing, communication, and entertainment by constantly monitoring changes in their identities, behaviors, and computing operations. Perhaps the best example of a smart home is Bill Gates' "home of the future," which reportedly can sense a person entering a room and adjust the light, temperature, and even the wall-mounted entertainment system for that person.

- *Smart phones* — A smart phone is able to leverage sensing technologies to detect a user's surrounding environment, such as in a movie theater or on the street, and to change its operations accordingly. In addition, services and applications on a smart phone can make use of a user's computing profile to provide personalized information or user interfaces.

- *Smart conferences* (including smart display, smart presentation, and smart meeting) — A sensory badge or a smart phone can be used to identify a conference attendee. When an attendee walks close to a display, preauthorized personal information regarding the attendee will be shown on the screen, as well as information regarding others he or she knows. Before a presentation, when the presenter enters the room, the projector will display the presenter's information. Information about attendees who are asking the presenter questions can be displayed as well. People chatting casually in the venue will be able to find out more about each other through the use of their smart phones. Some of these applications were demonstrated at the 2004 Pervasive Computing conference.

- *Location-based services* — The location of a mobile device or a mobile user is definitely a key context. Location-based services incorporate such context data to provide personalized, geographically dependent services.

The quest of context-aware mobile computing lies primarily in dealing with pertinent context data according to the needs of an application or a service. Because of the vast amount of context data in a mobile computing environment, context-aware systems, applications, and services must be able to address the following fundamental issues:

- Accurately identify which context data sets are needed.
- Collect desired context data efficiently.
- Process context data against some patterns or rules.
- Adapt to context change or present options to the user.

To solve these problems, a good model of context data for a specific application or a service is needed. The model should determine what context datasets are needed and by what means they should be collected, as well as the underlying logic to determine the most appropriate context based on predefined or self-learned rules. Like HCI in mobile computing, context-aware applications and services also rely heavily on the software infrastructure of a mobile device to operate. Smart phones are expected to be among the first mobile devices to have context-aware applications and services implemented, partly because it is by far the most pervasive mobile computing device in our daily life.

2.7.6 Low-Power Mobile Computing

The success of mobile computing hinges on battery and power consumption technologies to a large extent. For a smart phone, the talk time and standby time are always two of the most important specs users are concerned with. After all, people choose to use mobile devices because they can be used on the move, thus a power charge may be possible only intermittently. The dominant power source for mobile devices is a rechargeable battery. The types of mobile battery include Ni–Cd, Li–ion, Li–polymer, and the emerging fuel cell. Unfortunately, throughout the history of computing, even though many segments have demonstrated an amazing increase in computing power, bandwidth, or capacity, battery technology has not improved at the same pace. Within the box of a desktop computer or a PDA are hardware components that are far more powerful than they were only a few years ago. CPU speed, hard-disk capacity, and network bandwidth are the most evident examples. For example, it takes about 5 years to double the capacity of a hard disk and 3 years to double CPU processor speed. In contrast, battery technology does not change that much over time: it has taken 30 years for Ni–Cd battery to double its power. However, we may be on the verge of a breakthrough in battery technology; some chemical, electrical engineering, and interdisciplinary scholars have demonstrated the use of some innovative materials and techniques to improve battery life, such as using nuclear microbatteries that harness an incredible amount of energy released naturally from tiny bits of radioactive material [11].

Table 2.3 provides a list of batteries commonly used for portable devices. Lithium-ion (Li-ion) batteries are the most widely used batteries in cell phones and PDAs. A Li-ion battery can usually supply a talk time of about 3 to 5 hours. Fuel cell batteries have shown the highest gravimetric and volumetric density in laboratories, but commercial use is still at the very early stage.

Research shows on a laptop computer, about one third of the power is consumed by the display, and another third by the CPU, and 12% by the hard disks. On a cell phone or a smart phone,

Table 2.3 Battery Technologies.

Battery Type	Used Since	Energy Density (Wh/kg)	Volume Density (Wh/L)	Recharge Time (hours)	Charge and Discharge Cycle Life
Alkaline (AA)	1949	80–150	350	N/A	N/A
Rechargeable Alkaline	1992	80	200–300	2–3	25–100
Sealed lead acid	1970	30	50–100	8–16	200–500
NiMH	1990	50–80	100	2–4	500
Ni–Cd	1950	40–60	50–100	1–1.5	1500
Li–ion	1991	100	250–300	3–4	500–1000
Li–polymer	1999	150–200	300–450	8–15	100–150
Fuel cell	2005	300–1000	600–1500	Simply use a new cartridge	N/A

Source: Data from Buchmann [12] and Wikipedia [13].

wireless radio becomes a major factor. To tackle the power constraints from the perspective of mobile computing, researchers have been concentrating in two directions: low power computing and power management.

2.7.6.1 Low-Power Mobile Computing

A significant amount of effort has been devoted to reducing the power consumptions of individual hardware components and applications by incorporating low-power techniques into the design, development, and implementation of mobile devices and applications. The mobile device, constrained by the limited battery time, must be power aware; it should be able to dynamically adjust the power consumption of hardware components to extend operation time. Also, mobile applications are designed with low power in mind and should be power aware too. Applications should automatically change, pause, or terminate power-intensive operations in order to extend operation time. For example, when the battery of a smart phone is quickly being drained by continuous wireless Internet access via cellular networks, the wireless Internet application should switch to low-power mode, if any, to download text only, and start to use more aggressive caching scheme for the mobile browser.

2.7.6.2 Power Management in Mobile Computing

Power management in mobile computing is concerned with, given a limited power supply, making better use of limited battery power by orchestrating the power usage of various hardware components and corresponding applications. The goal of power management is the overall systematic efficient use of battery power, which can be performed at the system and application levels. In both cases, a good model that accurately profiles each component is necessary. Many wireless network protocols in mobile computing have been designed with power management in mind such as power-aware mobile ad hoc routing protocols (Chapter 5 will talk about mobile ad hoc routing in detail); in addition, a device can be put into power-save mode or sleep mode after some idle time.

2.8 Notable Mobile Computing Projects

For years researchers and industry practitioners have actively been working to develop pioneering prototype mobile computing systems and networks in a number of interesting projects. Although some of them are not directly related to smart phones and most are far from being mature, they all invariably make use of emerging mobile wireless technologies to enhance the user's experience. Moreover, these projects represent important research-and-development efforts to identify the enormous potentials and fundamental limitations of wireless technologies in the foreseeable future.

2.8.1 Oxygen

MIT Oxygen (http://oxygen.lcs.mit.edu/) is an effort geared toward a human-centered, pervasive computing vision. Oxygen aims at using a spectrum of technologies for the following themes:

- Distribution and mobility of user, data, and services
- Semantic content and related services
- Adaptation and change in response to whatever dynamics in a network or a physical environment
- Information personalities in terms of privacy, security, and customization

Oxygen identifies three classes of devices in a mobile environment: embedded devices, called Enviro21s (E21s), which are embedded in our homes, offices, and cars and are able to sense and affect our immediate environment; handheld devices, called Handy21s (H21s), which allow anywhere, anytime communication by various means; and dynamic, self-configuring networks (N21s) that effectively and reliably connect those devices and other networks, as well as people, services, and resources. Oxygen software that adapts to changes in the physical environment and computing environment or in user requirements (O2S) provides control and planning of abstraction of mechanisms in response to these changes as well as leveraging changes. Oxygen's perceptual technologies are based on speech and vision for interaction in a multimodal fashion.

A prototype Oxygen handheld device (H21) is equipped with a microphone, a speaker, a camera, an accelerometer, and a display for use with perceptual interfaces. It utilizes several hardware and compiler technologies to achieve low power computation. For example, StreamIt is a language and high-performance compiler designed to facilitate data streaming applications in Oxygen.

Oxygen's network technologies include the Cricket indoor location sensing system, Intentional Naming System (INS), Self-Certifying System (SFS), and Cooperative File System (CFS). Cricket uses ultrasound sensing devices to position moving objects with very high accuracy. Details of Cricket will be discussed in Chapter 7. INS focuses on on-demand resource discovery and management; SFS and

CFS provide secure access to a remote site over insecure networks without introducing centralized control servers.

Similar projects focusing on building intelligent environment with a variety of wireless technologies and perceptual/display technologies include Stanford iWork (http://iwork.stanford.edu/) and Microsoft EasyLiving (http://research.microsoft.com/easyliving/). A key function in such systems is localization and tracking of object in an indoor environment. Aside from ultrasonic sensors used in Cricket, 802.11 wireless LANs, Bluetooth, and RFID tags can also be used in different settings under various constraints. It is also conceivable that combining multiple wireless sensing technologies may yield improve accuracy. Examples of such systems include Active Badge (infrared based; http://www.uk.research.att.com/ab.html), Microsoft RADAR (802.11 signal fingerprint based) [14], and LANDMARC (RFID based) [15]. Chapter 7 discusses these location sensing systems in more detail.

2.8.2 Smart Dust

A major research direction of pervasive computing is wireless sensor networks that potentially could be integrated into a general wireless network infrastructure. The goal of the Smart Dust project at the University of California, Berkeley, is to create massively distributed wireless sensor networks consisting of hundreds to many thousands of tiny sensor nodes (motes) and some number of interrogators to query the network and retrieve in-network processed sensor data of interest [16,17]. The sensor nodes will be completely autonomous in the sense that they coordinate with each other to form an intelligent distributed network. Each mote will contain an microelectromechanical system (MEMS) sensor, optical transceiver, signal processing and control circuitry, and a power supply based on thick-film batteries and solar cells, all in a package the size of a cubic millimeter. Because of their extremely small size and light weight, some motes can even float in the air, buoyed by air currents. The motes can directly communicate with a base station or work in a peer-to-peer fashion. The base station can be fixed or attached to a handheld device. The challenges of building Smart Dust systems lies in two categories: power consumption and free space networking. To achieve extremely low-power operation on a mote, intelligent control of hardware subsystems and applications are needed. In a centralized single-hop routing scheme, every mote communicates directly with a base station. A base station uses a corner-cube retroreflector (CCD) image sensory array or complementary metal oxide semiconductor (CMOS) camera to accommodate sensor data transmission from many motes at the same time. For those motes that do not have line of sight to the base station, multihop routing can be used.

A large number of motes can be deployed in an environment randomly to monitor environmental changes, object movements, and any other classes of information that can be sensed. Examples of Smart Dust applications include instrumentation of semiconductor processing chambers, wind tunnels, air pollution monitoring, and unmanned spying in a hostile environment. It is also possible to make use of motes for localization and tracking of objects in an instrumented environment, provided a Smart Dust system is able to offer fine-grained node localization in the first place. On the other hand, location information can also help to direct queries to those motes geographically located in a region of interest.

2.8.3 AURA

Microsoft's Advanced User Resource Annotation (AURA) system (http://aura.research.microsoft.com/) is a wireless system for digitally identifying and annotating physical objects and sharing these annotations using a cell phone or a PDA equipped with an ID reader device to capture and annotate user interactions with the physical world. The ID reader can be a widely used bar code reader, an RFID reader, or even a fingerprint scanner. With AURA, a user can associate text, threaded conversations, audio, images, video, or other data with specific tags (such as a bar code of a CD) and submit them to a database. This is the same idea of blogging, but it uses bar codes or any other ID code to uniquely identify an item of interest. In addition, AURA is designed to collect and disseminate users' annotations. Other users who encounter an object may use the mobile device to scan the bar code, feed it into online Aura applications, and obtain annotations of the object in real time. For example, AURA can be used to retrieve detailed annotations (most likely readers' comments) of books while browsing a bookstore.

The goal of the AURA project is to "annotate the planet." Not only objects with a bar code but also anything that can be uniquely identified could be potentially annotated. On one hand, the aggressive vision of AURA has the potential to bring tagged information sharing to an unprecedented level; on the other hand, this vision naturally leads to humans being annotated, as it is feasible to use biometric features to identify a person. To this end, privacy protection will certainly present a big challenge.

2.8.4 Wireless Grid

Wireless Grid (http://wirelessgrids.net/) is a research project supported by National Science Foundation NSF aimed at investigating issues surrounding wireless grids and opportunities utilizing wireless grids. A wireless grid is a network of *ad hoc* networks in nature. Like a power grid, the goal of Wireless Grid is to enable flexible, secure, and distributed resource sharing and service provision among heterogeneous wired, fixed wireless, or *ad hoc* networks and mobile, nomadic, and fixed devices. The challenges of wireless grids encompass three related computing paradigms that all undergo rapid development: grid computing, peer-to-peer computing, and web services [6]. Grid computing essentially allows wired networks of computers to share resources with others in the grid. Peer-to-peer networks provide a way to offer scalable and efficient resource sharing on an overlay network. Web services enable machine-to-machine communication in a self-contained manner, utilizing XML message passing and flexible Web Service Definition Language (WSDL). The Wireless Grid system thus is able to apply the three computing frameworks to a heterogeneous network environment.

As a proof-of-concept wireless grid application, the Wireless Grid research team at Syracuse University developed a distributed audio recording and sharing system call DARC (Distributed Ad hoc Resource

Coordination). Individual mobile devices in DARC can act as mixers or recorders. An audio stream is sent from a recorder to a mixer service on a selected peer. By harnessing collective capabilities of many recorders and mixers in a specific environment, a group of people can engage into distributed music authoring and sharing. The wireless grid project team has also designed applications used in medical and warehousing monitoring with an integrated wireless and wired grid [18]. In both cases, wireless sensor networks at distinct locations are seamlessly integrated with other wireless networks and the wired grid such that data collection, filtering, aggregation, and storage can be accomplished across diverse networks of sensors, mobile devices, and computers. For example, in a medical wireless grid, grid nodes may include patient monitoring sensors (blood pressure, heart beat rate, level of oxygen saturation in the blood, etc.), doctors' or nurses' mobile devices (with sensor readers), and emergency medical technician (EMT) mobile devices. In a warehouse monitoring wireless grid, pallet sensors, Smart Dusts, and a range of mobile or fixed sensor readers can be used.

2.9 Summary

The next wave of mobile computing embodies a far broader spectrum than traditional voice-centric mobile wireless communication with cell phones. The trend of convergence of communication, computing, and consumer electronics is reflected by the strong demand of converged multifunctional smart phones that incorporate cell phone, PDA functions, and protable audio/video device into a single device, which in turn opens up enormous opportunities with regard to business and consumer applications and services. However, the vision of next-generation mobile computing encompasses more profound changes to people's daily life. First, the future wireless network infrastructure will be a systematic integration of a variety of heterogeneous wireless technologies ranging from personal area networks to global wide area networks. This wireless world will eventually enable ubiquitous mobile access anytime, anywhere, from any devices. Second, a galaxy of "smart" wireless mobile devices will be deeply embedded into environments to enable pervasive mobile intelligence everywhere. In particular, smart phones perhaps will play an important role as the universal mobile terminal in tomorrow's mobile computing paradigm, primarily due to their unparalleled popularity among computing devices. A number of innovative mobile applications and services are being explored to take advantage of this ubiquitous access and pervasive intelligence, such as location-based services, mobile commerce, mobile gaming, and mobile social networking, to name a few.

Challenges within the domain of next-generation mobile computing calls for dramatic advances in almost every discipline of computer science and engineering, including mobile processors, smart displays, mobile operating systems, and communications protocols, context-aware mobile computing, and human interface design. Moreover, power consumption and security issues are particularly important in the mobile computing realm.

This chapter was designated to give the reader a vision of next-generation mobile computing, a big picture of today and tomorrow's mobile world, and a brief survey on promising applications and challenging issues. Rather than elaborate on the details of research and development in this broad area, this chapter has emphasized the motivation behind and impact of these trends, challenges, and research outcomes. The next two chapters will introduce supporting wireless technologies and hardware and software platforms, respectively, as they are the building blocks for future mobile wireless systems, services, and applications.

Further Reading

The Evolution of Cellular Systems

Mobile Telephony: Wide Area Coverage, Bell Laboratories Technical Memorandum, December 1947.

W. R. Young, Advanced mobile phone service: introduction, background, and objectives, *Bell System Tech. J.,* January, 1979.

Mobile telephone history, http://www.privateline.com/PCS/history.htm.

The Evolution of PDA and Pen Computing

J. Hawkins, creator of Palm Pilot, *On Intelligence,* Time Books, New York, October, 2004.

The Concept of PDA, Apple Museum, http://www.theapplemuseum.com/index.php?id=tam&page=pda.

The Vision of Next-Generation Mobile Computing

S. Baker *et al.*, Big Bang!, *Business Week*, June 21, 2004 (http://www.businessweek.com/magazine/content/04_25/b3888601.htm).

W. Gates, Keynote speech, 2004 Mobile Developers Conference (http://www.microsoft.com/billgates/speeches/2004/03-24-VSLive.asp).

S. Levy, The next frontiers. I. The wireless revolution, *Newsweek*, June 7, 2004 (http://msnbc.msn.com/id/5092820/site/newsweek/).

M. Copeland, How to ride the fifth wave, *Business 2.0*, June 15, 2005 (http://www.business2.com/b2/web/articles/0,17863,1071030,00.html).

R. Berger, Open Spectrum: A Path to Ubiquitous Connectivity, *ACM Queue*, May 2003 (http://www.acmqueue.com/modules.php?name=Content&pa=showpage&pid=37).

B. Zenel and A. Toy, Enterprise-Grade wireless, *ACM Queue*, May 2005 (http://www.acmqueue.com/modules.php?name=Content&pa=showpage&pid=301).

Daily Wireless (http://www.dailywireless.org/): an excellent website for wireless technologies and state-of-the-art industry development. For WiFi and cellular integration, see http://www.dailywireless.org/modules.php?name=News&file=article&sid=225. For the latest development of WiFi based ID tracking, see http://www.dailywireless.org/modules.php?name=News&file=article&sid=4070.

Howard Rheingold's blog (http://www.smartmobs.com/) has many insightful comments on the development of mobile communication technologies, pervasive computing, and their social implications.

G. Abowd and L. Iftode, The smart phone - a first platform for pervasive computing, IEEE Pervasive Computing, April-June 2005 (Vol. 4, No. 2). This issues features several papers on a variety of pervasive computing applications using smart phones, such as "Social Serendipity: Mobilizing Social Software" by N. Eagle and A. Pentland, and "ContextPhone: A Prototyping Platform for Context-Aware Mobile Applications" by M. Raento, A. Oulasvirta, R. Petit, and H. Toivonen.

Tim Berners-Lee, who invented the Web in early '90s, talked about how web page design hampers the usability of wireless Web, March 2005 (http://www.msnbc.msn.com/id/7218432/).

Google mobile services, including web search, location based search, mobile web search, blog for mobile, and Google SMS, http://mobile.google.com/

MSN mobile services including Hotmail, MSN Messager, mobile web, and MSN alerts, http://mobile.msn.com/

Yahoo mobile services, including Yahoo messager, Yahoo mail, SMS from a PC, blog for mobile, mobile games, web search, SMS search, mobile web, and mobile alerts, http://mobile.yahoo.com/

Mapquest mobile services, including mobile maps, GPS locating service, and traffic maps, http://www.mapquest.com/features/main.adp?page=slashmobile

References

[1] S. G. Glisic and P. A. Leppanen, *Wireless Communications: TDMA vs. CDMA*, Kluwer Academic, Dordrecht, 1997.

[2] IMT, *IMT-2000,* International Mobile Communications, Geneva, Switzerland, 2003 (http://www.itu.int/home/imt.html).

[3] J. P. Allen, Who shapes the future? Problem framings and the development of handheld computers, in *Proceedings of the Ethics and Social Impact Component on Shaping Policy in the Information Age*, May 1998, Washington, DC.

[4] Z. Liao, Real-time taxi dispatching using global positioning systems, *Commun. ACM*, 46(5):81–83, 2003.

[5] ZigBee Alliance, San Ramon, CA, 2004 (http://www.zigbee.org). (http://www.zigbee.org/," 2004.

[6] L. W. McKnight and J. Howison, Wireless grids: distributed resource sharing by mobile, nomadic, and fixed devices, *IEEE Internet Computing.*, Vol. 8, No. 4, July/August, 2004.

[7] WiFi Alliance, 2004 (http://www.wi-fi.org/).

[8] X. Wang, D. Feng, X. Lai, and H. Yu, Collisions for hash functions MD4, MD5, HAVAL-128, and RIPEMD, *Cryptology ePrint Archive, Report 2004/199, 2004.*

[9] M. Satyanarayanan, Privacy: the Achilles heel of pervasive computing?, *IEEE Pervasive Computing*, 2:2–3, 2004.

[10] S. Oviatt, T. Darrell, and M. Flickner, Multimodal interfaces that flex, adapt, and persist, *Commun. ACM*, 47(1):30–33, 2004.

[11] A. LaL and J. Blanchard, The daintiest dynamos, *IEEE Spectrum*, September 2004.

[12] I. Buchmann, *Batteries in a Portable World: A Handbook on Rechargeable Batteries for Non-Engineers*, 2nd ed., Cadex Electronics, Richmond, British Columbia, 2001.

[13] Wikipedia, 2004 (http://en.wikipedia.org).

[14] P. Bahl and V. N. Padmanabhan, RADAR: an in-building RF-based user location and tracking system, in *Proceedings of IEEE INFOCOM '00*, March 2000, Tel-Aviv, Israel.

[15] L. M. Ni, Y. Liu, Y. C. Lau, and A. P. Patil, LANDMARC: indoor location sensing using active RFID, in *Proceedings of IEEE International Conference on Pervasive Computing and Communication (PERCOM '03)*, March 2003, Dallas-Fort Worth, Texas.

[16] J. M. Kahn, R. H. Katz, and K. S. J. Pister, Next century challenges: mobile networking for smart dust, in *Proceedings of MobiCom '99*, 1999, August 15–19, 1999, Seattle, Washington.

[17] B. Warneke, M. Last, B. Liebowitz, and K. S. J. Pister, Smart dust: communicating with a cubic-millimeter computer, *IEEE Comput.*, 34(1):44–51, 2001.

[18] M. Gaynor, S. L. Moulton, M. Welsh, E. LaCombe, A. Rowan, and J. Wynne, Integrating wireless sensor networks with the grid, *IEEE Internet Draft*, 8(4):32–39, 2004.

3 Supporting Wireless Technologies

This chapter provides extensive coverage of existing mobile wireless technologies. Much of the emphasis is on the highly anticipated 3G cellular networks and widely deployed wireless local area networks (LANs), as the next-generation smart phones are likely to offer at least these two types of connectivity. Other wireless technologies that either have already been commercialized or are undergoing active research and standardization are introduced as well. Because standardization plays a crucial role in developing a new technology and a market, throughout the discussion standards organizations and industry forums or consortiums of some technologies are introduced. In addition, the last section of this chapter presents a list of standards in the wireless arena.

3.1 The Frequency Spectrum

The fundamental principle of wireless communication is electromagnetic wave transmission between a transmitter and a receiver. Signals are characterized by their frequencies in use. Multiple signals or noises of the same frequency will cause interference at the receiver. To avoid interference, various wireless technologies use distinct frequency bands with well-controlled signal power which are portions of the so-called frequency spectrum. As a scarce public resource, the frequency spectrum is strictly regulated by governments of countries around the world. In the United States, the Federal

Communications Commission (FCC) has the responsibility of regulating civil broadcast and electronic communications, including the use of the frequency spectrum, and the National Telecommunications and Information Administration (NITA) administers the frequency use of the federal government. In Europe, the frequency spectrum is managed on a national basis, and European Union (EU) members coordinate via the European Conference of Post and Telecommunications Administrations (ECPT) and the Electronic Communications Committee (ECC). Worldwide unified regulation of wireless communication is understandably difficult to achieve for various technological, economic, and political reasons. To this end, the International Telecommunications Union (ITU) has been formed as an international organization of the United Nations. The ITU allows governments and private sectors to coordinate development of telecommunication systems, services, and standards. In almost all countries, portions of the frequency spectrum have been designated as "unlicensed," meaning that a government license is not required for wireless systems operating at these bands. In effect, wireless system manufacturers and service providers are required to obtain an exclusive license for a frequency band from regulatory bodies or resort to the use of the unlicensed spectrum. In either case, the emitted power of the wireless systems must comply with the power constraints associated with the regulations in question. In addition, frequency allocations of a country may change over time. (For the latest information regarding frequency allocation in the United States, see http://www.ntia.doc.gov/osmhome/allochrt.html.)

A radio signal is characterized by wavelength and frequency. In a vacuum, the product of wavelength and frequency is the speed of light (about 3×10^8 meters per second); in general, a higher frequency means shorter wavelength. For example, visible light is in the frequency band of 4.3×10^{14} to 7.5×10^{14} Hz, with wavelengths ranging from 0.35 to 0.9 μm. FM radio broadcasts operate within the frequency range of 30 to 300 MHz at wavelengths between 10 and 1 m.

The frequency spectrum can be divided into the following categories: very low frequency (VLF), low frequency (LF), medium frequency (MF), high frequency (HF), very high frequency (VHF), ultra-high frequency (UHF), super-high frequency (SHF), extremely high frequency (EHF), infrared, visible light, ultraviolet, x-ray, gamma-ray, and cosmic ray, each of which represents a frequency band. Figure 3.1 shows the frequency spectrum up to the visible light band. Notice that in the context of electronic communication, there are two categories of transmission medium: guided medium (*e.g.*, copper coaxial cable and twisted pair) and unguided medium (for wireless communication in the air). The guided medium carries signals or waves between a transmitter and a receiver, whereas the unguided medium typically carries wireless signals between an antenna and a receiver (which may also be an antenna). Nevertheless, each medium operates at a specific frequency band of various bandwidth determined by its physical characteristics. For example, coaxial cable uses many portions of frequencies between 1 KHz and 1 GHz for different purposes: television channels 2, 3, and 4 operate at frequencies from 54 to 72 MHz; channels 5 and 6, 76 to 88 MHz; and channels 7 to 13, 174 to 216 MHz. The optical fiber uses visible or infrared light as the carrier and operates at frequencies between 100 and 1000 THz.

Wireless communication operates at frequencies in the so-called radio spectrum, which is further divided into VLF, LF, MF, HF, VHF, UHF, SHF, and EHF. In addition, infrared data association (IrDA) is also used for short-range wireless communication. The following text discusses frequency bands at which existing mobile wireless technologies operate; notice that very often the frequency regulations enforce emitted power restrictions to avoid interference among wireless devices operating at the same frequency band.

3.1.1 Public Media Broadcasting

- AM radio: Amplitude modulation (AM) radio stations operate at a frequency band between 520 and 1605.5 KHz.
- FM radio: Frequency modulation (FM) radio uses the frequency band between 87.5 and 108 MHz.

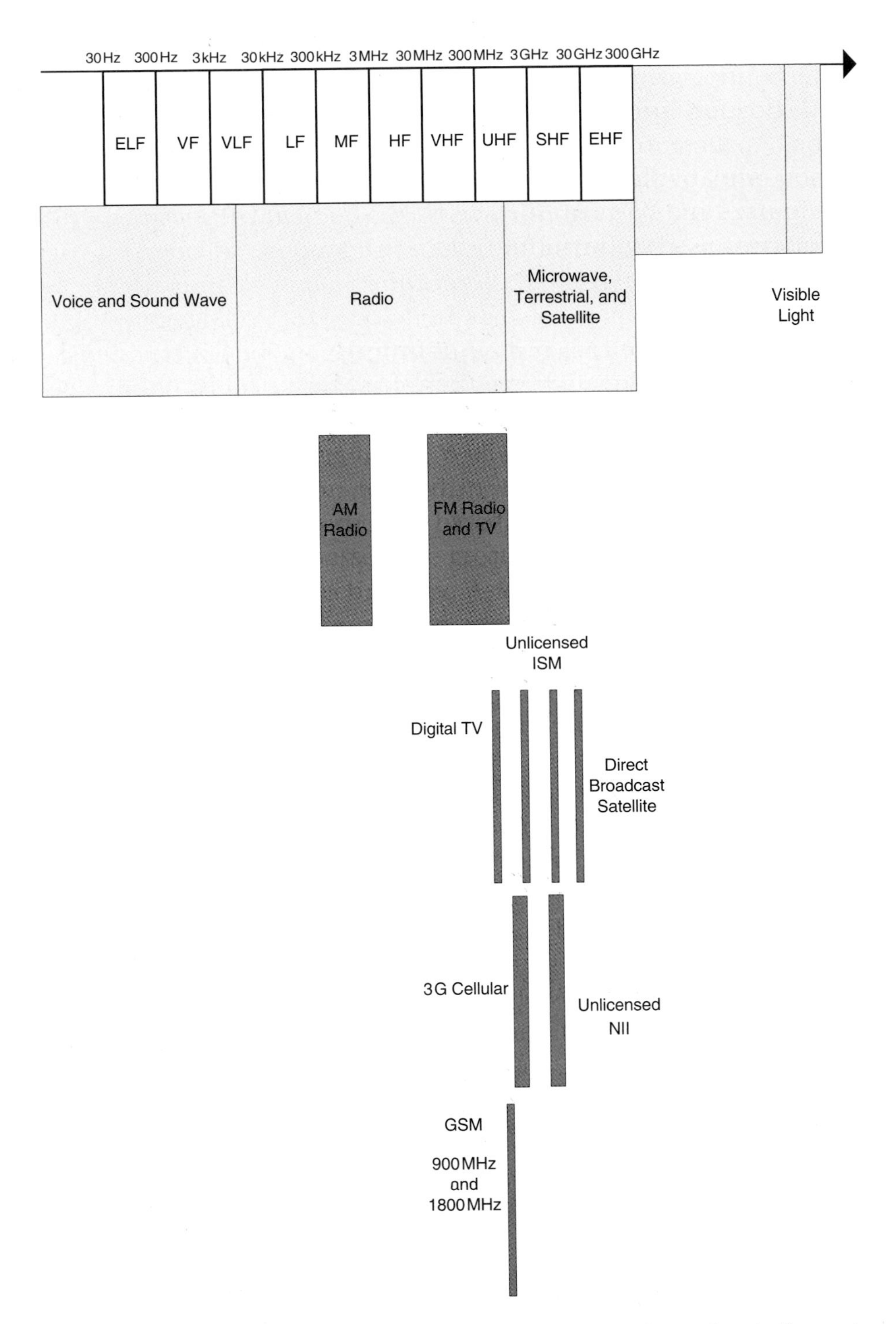

Figure 3.1 The frequency spectrum (refer to the text for the exact frequency band allocated to each system).

- SW radio: Shortwave (SW) radio uses frequencies between 5.9 and 26.1 MHz within the HF band. The transmission of shortwave radio over a long distance is made possible by ionosphere reflection. HAM amateur radio, a popular activity enjoyed by over 3 million fans worldwide, relies on the HF band to communicate across the world.
- Conventional analog television: A quite small slice of VHF (30 to 300 MHz) and UHF (300 to 3000 MHz) has been allocated for analog television broadcasting. In the United States, each channel occupies a 6-MHz band. The first VHF channel, channel 2, operates at 54 to 60 MHz, whereas the last UHF channel, channel 69, operates at 800 to 806 MHz.
- Cable television: The frequency bands of channels 2 to 13 are exactly the same for both conventional television and cable television. Beyond those channels, cable television requires frequencies from 120 to 552 MHz for channels 13 to 78.
- Digital cable television: Channels 79 and above are reserved for digital cable broadcasting at frequencies between 552 and 750 MHz.
- Digital audio broadcasting (DAB): DAB is a standard developed by the European Union for CD-quality audio transmission at frequencies from 174 to 240 MHz and from 1452 to 1492 MHz. In the United States, a technique called in-band on-channel (IBOC) is used to transmit digital audio and analog radio signals simultaneously with the same frequency band. The resulting services are generally marketed as high-definition radio.
- Direct broadcast satellite (DBS): The upper portion of the microwave Ku band (10.9 to 12.75 GHz) is used for direct satellite-to-receiver video and audio broadcasting. See Section 3.13 for more details regarding satellite communication.
- Satellite radio: Frequencies from 2320 to 2345 MHz have been allotted for satellite radio services in the United States. See Section 3.13 for more details regarding satellite communication.

3.1.2 Cellular Communication

- Global system for mobile (GSM): The two frequency bands used by GSM are 890 to 960 MHz and 1710 to 1880 MHz. They are sometimes referred to as the 900-MHz band and the 1800-MHz band.
- Code-division multiple access (CDMA): The IS-95 standard defines the use of the 800- and 1900-MHz bands for CDMA cellular systems.
- 3G wideband CDMA (WCDMA)/universal mobile telecommunications system (UMTS): Three frequency bands are allocated for 3G UMTS services: 1900 to 1980 MHz, 2020 to 2025 MHz, and 2110 to 2190 MHz.
- 3G CDMA 2000: This system reuses existing CDMA frequency bands.

3.1.3 Wireless Data Communication

- Wireless LANs: IEEE 802.11b operates at 902 to 928 MHz and 2400 to 2483 MHz, and the industrial, scientific, and medical (ISM) radio bands operate at 2.4 GHz band in the United States. The IEEE 802.11b operates at 2400 to 2483 MHz in Europe, and at 2400 to 2497 MHz in Japan. IEEE 802.11a and HiperLAN2 use 5150 to 5350 MHz and 5725 to 5825 MHz, and the unlicensed national information infrastructure (U-NII) band operates at 5.8 GHz in the United States. They operate at 5150 to 5350 MHz and 5470 to 5725 MHz in Europe, and at 5150 to 5250 MHz in Japan. Section 3.10 discusses wireless LANs in more detail.
- Bluetooth: Seventy-nine 1-MHz channels are allocated from the unlicensed 2.402 to 2.480 GHz in the United States and Europe for Bluetooth signal transmission. Other countries may have fewer channels but all fall into the 2.4-GHz band. Section 3.11 talks more about the Bluetooth technology.

- WiMax: A wide range from 2 to 11 GHz that includes both licensed and unlicensed bands will be used for 802.116a, and 11 to 66 GHz can possibly be used by 802.116c. Section 3.14 introduces WiMax as part of the wireless MANs section.
- Ultra-wideband (UWB): In the United States, the FCC mandates that UWB can operate from 3.1 to 10.6 GHz. UWB is further discussed in Section 3.12.
- Radiofrequency identification (RFID): RFID tags operate at the frequency bands of LF (120–140 KHz), HF (13.56 MHz), UHF (868–956 Mhz), and microwave (2.4 GHz). Section 3.13 explains RFID technology and its applications.
- Infrared data association (IrDA): IrDA uses frequencies around 100 GHz for short-range data communication.
- Wireless sensors: Sensor motes support tunable frequencies in the range of 300 to 1000 MHz and the 2.4-GHz ISM band. In particular, ZigBee, the remote sensor control technology, operates at the 868-MHz band in Europe, 915-MHz band in the United States and Asia, and 2.4-GHz band worldwide.

3.1.4 Other Fixed or Mobile Wireless Communications

- Digital cordless phone: The Digital Enhanced Cordless Telecommunications (DECT) standard in Europe defines the use of the frequency band 1880 to 1990 MHz for digital cordless phone communication. In the United States, cordless phones use three frequency bands: 900 MHz, 2.4 GHz, and 5.8 GHz, each of which is also intensively used by other short-range wireless communication technologies.
- Global positioning system (GPS): GPS satellites use the frequency bands 1575.42 MHz (referred to as L1) and 1227.60 MHz (L2) to transmit signals.
- Meteorological satellite services: The UHF band from 1530 to 1650 MHz (the L band) is commonly used by meteorological

satellites, as well as some global environmental monitoring satellites. Part of the UHF and SHF bands are used for military satellite communication.

- Radio frequency remote control, such as remote keyless entry systems and garage door openers. These short-range wireless system commonly used for automobiles operates at 27 MHz, 128 MHz, 418 MHz, 433 MHz, and 868 MHz in the United States; 315 MHz and 915 MHz in Europe; and 426 MHz and 868 MHz in Japan.

3.2 Wireless Communication Primer

For our in-depth discussion of the many sophisticated mobile wireless network technologies, a basic understanding of wireless communications is necessary. Here, a primer of concepts within the domain of wireless communication is presented. Readers who are interested in further details are referred to Stallings' book on wireless communications and networks [1].

3.2.1 Signal Propagation

A radio signal can be described in three domains: time domain, frequency domain, and phase domain. In the time domain, the amplitude of the signal varies with time; in the frequency domain, the amplitude of the signal varies with frequency; and, in the phase domain, the amplitude and phase of the signal are shown on polar coordinates. According to Fourier's theorem, any periodic signal is composed of a superposition of a series of pure sine waves and cosine waves whose frequencies are harmonics (multiples) of the fundamental frequency of the signal; therefore, any periodic signal, no matter how it was originally produced, can be reproduced using a sufficient number of pure waves.

Electronic signals for wireless communication must be converted into electromagnetic waves by an antenna for transmission. Conversely, an antenna at the receiver side is responsible for converting electromagnetic waves into electronic signals. An antenna can be omnidirectional or directional, depending on specific usage scenarios. For an antenna to be effective, it must be of a size consistent with the wavelength of the signals being transmitted or received. Antennas used in cell phones are omnidirectional and can be a short rod on the handset or hidden within the handset. A recent advancement in antenna technology is the multiple-in, multiple out (MIMO) antenna, or smart antenna, which combines spatially separated small antennas to provide high bandwidth without consuming more power or spectrum. To take advantage of multipath propagation (explained in the next section), these small antennas must be separated by at least half of the wavelength of the signal being transmitted or received.

A signal emitted by an antenna travels in the air following three types of propagation modes: ground-wave propagation, sky-wave propagation, and line-of-sight (LOS) propagation. AM radio is a kind of ground-wave propagation, where signals follow the contour of the Earth to reach a receiver. SW radio and HAM amateur radio are examples of sky-wave propagation, where radio signals are reflected by ionosphere and the ground along the way. Beyond 30 MHz, line-of-sight propagation dominates, meaning that signal waves propagate on a direct, straight path in the air. It is noteworthy that radio signals of line-of-sight propagation can also penetrate objects, especially signals of large wavelength (and thus low frequency). Satellite links, infrared light, and communication between base stations of a cellular network are examples of line-of-sight propagation.

3.2.1.1 Attenuation

The strength or power of wireless signals decreases when they propagate in the air, just as visible light does. As soon as radio waves leave the transmitter's antenna, some amount of energy will be lost as the electromagnetic field propagates. The effect will become more evident over a long distance as the signal disperses in space; therefore, the received power of the signal is invariably less than the signal power at the transmitting antenna. In the most ideal circumstances (*i.e.*, in vacuum), signal power attenuation is proportional to d^2, where d denotes the distance between the transmitter and the receiver. This effect is sometimes referred to as *free space loss*. In reality, beside free space loss, a number of other factors have to be considered to determine signal attenuation, such as weather conditions, atmospheric absorption, and space rays. In addition, signal attenuation is more severe in high frequencies than in low frequencies, resulting in signal distortion.

When it encounters obstacles along the path, a signal may experience more complex attenuation than power reduction. For example, for visible light we are well aware of the following effects: *shadowing*, *reflection*, and *refraction*. Likewise, for high-frequency wireless signals, such effects also exist. Shadowing and reflection occur when a signal encounters an object that is much larger than its wavelength. Though the reflected signal and the shadowed signal are comparatively weak, they in effect help to propagate the signal to spaces where line-of-sight is impossible. For example, when reflection and shadowing are caused by buildings in an urban area, signals from an antenna of a base station may be able to reach cell phone users within a building in the area, although it might be a good idea for the user to walk close to the window for better signal strength (perceived as a number of "bars" displayed on the cell phone screen). Refraction (bending) occurs when a wave passes across the boundary of two media. Moreover, wireless signals are also subject to *scattering* and *diffraction*. Specifically, when the size of an obstacle is on the order of the signal wavelength or less, the signal will be scattered into a number of weaker pieces. Diffraction occurs when a signal hits the edge of an obstacle and is deflected into a number of directions.

3.2.1.2 Noise

The receiver of a wireless communication system must be able to detect transmitted (most likely attenuated and distorted) signals from unwanted noises. Common types of noise are *thermal noise* (white noise) produced by any electronic circuitry; *intermodulation* noise, which occurs when two frequencies of signals are modulated and transmitted over the same medium; *crosstalk* between two channels; and *impulse* noise generated by instantaneous electromagnetic changes. To cope with noises in received signals, a wireless system has to ensure that the transmitted signals are sufficiently stronger than the noises. Another approach is to employ spread spectrum schemes (explained below) that convert a signal over a wide range of frequencies of low power density as random noise. Wireless signals are subject to various impairments or distortion along the way from the transmitter to the receiver. To quantify these effects, the signal-to-noise ratio (SNR) is used to represent the ratio of the power in a signal to the power of the noise. SNR is usually computed in decibels as the product of 10 and the logarithm of the raw power ratio.

3.2.1.3 Multi-Path Propagation

The receiver of a wireless system is exposed to all radio waves in its surrounding environment; therefore, it may receive indirect signals from different paths, such as reflected signals, shadowed signals, and refracted signals, as well as signals generated by other means of propagation, all carrying the same signal with different levels of attenuation and distortion. These signals may impose some negative effect on the direct signal to a great extent. The most severe effect of multipath propagation is intersymbol interference (ISI). Intersymbol interference is caused by overlapping of delayed multipath pulses (of a primary pulse) and subsequent primary LOS pulses, where one or multiple pulses represent a bit. The degree of attenuation of these pulses may vary from time to time due to path changes or environmental disturbances, making it more difficult to recover the transmitted bits. To prevent intersymbol interference from occurring, the first primary pulse and the second have to be separated by a sufficient time difference such that the delayed multipath pulses of the first can be differentiated from the second LOS pulse. This implies that the symbol rate of the signal and bandwidth of the radio channel are limited by multipath propagation.

3.2.2 Modulation

Signal modulation is a technique used to combine a signal being transmitted with a carrier signal for transmission. The receiver demodulates the transmitted signal and regenerates the original signal. Normally the carrier signal is a sine wave of a high frequency. The input signal could be digital (digital modulation) or analog (analog modulation). In either case, the three basic characteristics of a signal are utilized for modulation. The device that performs this modulation and demodulation is the *modem*. Modulation is often referred to as signal encoding. Analog signals can be modulated by the following methods.

3.2.2.1 Amplitude Modulation

For AM signals, the output signal is a multiplication of the input signal with a carrier wave. The amplitude of the carrier wave is determined by the input analog signal. The frequency of the resulting output signal is centered at the frequency of the carrier. As its name implies, AM radio that operates in the frequency band of 520 to 1605.5 KHz uses amplitude modulation.

3.2.2.2 Frequency Modulation

Rather than vary the amplitude of the carrier wave, frequency modulation alters the transient frequency of the carrier according to the input signal. Again, as its name implies, FM radio that operates within the frequency band of 87.5 to 108 MHz uses frequency modulation.

3.2.2.3 Phase Modulation

In phase modulation (PM), the phase of the carrier signal is used to encode the input signal. Like AM, FM and PM shift the frequency of the input signal to a band centered at the carrier frequency. Both FM and PM require higher bandwidths. Analog modulation is necessary for transmitting a wireless analog signal such as voice over a long distance. Directly transmitting the signal itself to the receiver without applying modulation would require a large antenna to be effective, as the frequency of voice signals falls into the range of 30 to 3000 Hz. For digital data, if the medium only facilitates analog transmission (*e.g.*, air), some digital modulation techniques will have to be employed. A carrier wave is also used to carry binary streams being transmitted, according to some keying schemes in digital modulation. Below is a list of digital modulation schemes:

- Amplitude-shift keying (ASK) —ASK uses presence of a carrier wave to represent a binary one and its absence to indicate a binary zero. While ASK is simple to implement, it is highly susceptible to noise and multipath propagation effects. Because of that, ASK is primarily used in wired networks, especially in optical networks where the bit error rate (BER) is considerably lower than that of wireless environments.

- Frequency-shift keying (FSK) — Similar to frequency modulation, FSK uses two or more frequencies of a carrier wave to represent digital data. Binary FSK (BFSK), which employs two carrier frequencies for 0 and 1, is the most commonly used FSK. The resulting signal can be mathematically defined as the sum of two amplitude-modulated signals of different carrier frequencies. If more than two carrier frequencies are used for modulation, each frequency may represent more than one bit, thereby providing a higher bandwidth than ASK.

- Phase-shift keying (PSK) — PSK uses the phase of a carrier wave to encode digital data. Binary PSK simply reverses phase when the data bits change. Multilevel PSKs use more evenly distributed phases in the phase domain, with each phase representing two or more bits. One of most commonly used PSK schemes is quadrature PSK, in which the four phases of 0, $\pi/2$, π, and $3\pi/2$ are used to encode two digits. PSK can be implemented in two ways. The first is to produce a reference signal at the receiver side and then compare it with the received signal to decide the phase shift. This method somewhat complicates things at the receiver end, as the transmitter and the receiver must be synchronized periodically to ensure that the reference signal is being generated correctly. Another method is differential PSK (DPSK). In DPSK, the reference signal is not a separated signal but is the one preceding the current wave in question. One of the second-generation cellular systems, Digital-Advanced Mobile Phone Service (DAMPS), uses DPSK.

Amplitude-shift keying and PSK can be combined to offer more variations of phase shifts on the phase domain. Quadrature amplitude modulation (QAM) is such a scheme in which multiple levels of amplitudes coupled with several phases provide far more unique symbol shifts over the same bandwidth than used by PSK over the same bandwidth. QAM is widely used in today's modems.

Apart from analog and digital modulation, another category of modulation that should be discussed for wireless communication is analog-to-digital data modulation, a procedure sometimes referred to as *digitization*. Two major digitization schemes are pulse-code modulation (PCM) and delta modulation. PCM samples an input analog signal in short intervals, and each sample is converted into a symbol representing a code. To reconstruct the original input signal from samples, the sampling rate must be higher than twice the highest frequency of the input signal. In other words, given a sample rate of *fs*, a frequency higher than 2*fs* in the input signal will not be recovered in the reconstruction. Delta modulation uses a staircase-like sample function to approximate the input signal. The resulting digital data are comprised of a series of 1's and 0's indicating the ups and downs, respectively, of the staircase function.

In the wireless world, signals transmitted through the air are primarily high-frequency analog signals. In wireless voice communication, the user's voice is digitalized into digital data and then modulated to analog based band signals (digital modulation), which are finally modulated with a carrier wave for transmission. For wireless data transmission, the first step of this procedure is not necessary. In either case, the receiver takes the reverse order of these steps to recover the transmitted data or voice.

3.2.3 Multiplexing

Modulations of analog signals or digital data is concerned with a single input signal to be converted efficiently into other forms. In contrast, multiplexing is a collection of schemes that address the issue of transmitting multiple signals simultaneously in a wireless system in the hopes of maximizing the capacity of the system. The devices for multiplexing and demultiplexing are multiplexers and demultiplexers, respectively. If signals of the same frequency are spatially separated from each other such that no frequency overlapping occurs at any given place, then multiple signals of different frequencies can be transmitted and received without a problem. Radio stations are an excellent example of this *spatial division multiplexing*: AM and FM radio signals only cover the area in which the radio stations are located, and they cannot interfere with other radio signals on the same frequency in adjacent areas. Apart from spatial division multiplexing, three prominent schemes of multiplexing have been devised.

3.2.3.1 Frequency-Division Multiplexing

In frequency-division multiplexing (FDM), signals from a transmitter are modulated to a fixed frequency band centered at a carrier frequency (*i.e.*, a channel). To avoid inference, these channels have to be separated by a sufficiently large gap (*i.e.*, a guard band) in the frequency domain; hence, transmission and reception of signals in multiple channels can be performed simultaneously but independently. Analog cellular systems use FDM; in these systems, calls are separated by frequency.

3.2.3.2 Time-Division Multiplexing

Time-division multiplexing (TDM) allows multiple channels to occupy the same frequency band but in small alternating slices of time following a sequence known to both the transmitter and the receiver. Each channel makes full use of the bandwidth of the medium but only contributes a portion of the overall data rate. Coordination among the transmitters is necessary to prevent conflicting use of the frequency band. When applied to digital signals, TDM can be done on a bit level, byte level, block level, or levels of larger quantities. GSM and D-AMPS both use TDM but in different ways. TDM and FDM can be combined to increase the robustness of the system. In this case, signals from a transmitter are modulated onto different carriers for a certain amount of time and jump to another carrier, effectively creating a "frequency hopping" phenomenon.

3.2.3.3 Code-Division Multiplexing

Code-division multiplexing (CDM) makes better use of a frequency band than FDM and TDM. Signals from different transmitters are transmitted on the same frequency band at the same time but each has a code to uniquely identify itself. The orthogonal codes mathematically ensure that signals cannot interfere with each other at the receiver. CDM effectively converts the problem of limited frequency space into ample code space but adds the overhead of implementation complexity. The transmitter and receiver must be synchronized such that individual signals can be correctly received and decoded. Compared to FDM, CDM provides greater security against signal tapping because transmitted signals appear as noise if the receiver does not know the code. CDM is the underlying multiplexing scheme of orthogonal frequency-division multiplexing (OFDM). CDMA cellular systems use similar CDM schemes to provide multiple wireless communication channels access to the same frequency band. Another multiplexing scheme, wavelength-division multiplexing (WDM), is very common in optical networks using fiber as the transmission medium. It is actually FDM for fiber, which offers an extremely high bandwidth. In WDM, a fiber can be divided into a number of wavelengths (nanometers), each of which can be assigned to a transmission channel. Dense wavelength-division multiplexing (DWDM) systems support eight or more wavelengths. Because of their high data rate, WDM and DWDM are the predominant multiplexing schemes used by optical networks in the wired Internet backbone.

3.3 Spread Spectrum

Wireless systems transmit data over a specific, quite narrow frequency band that allows a transmitter and a receiver to differentiate the intended signal from background noise when the signal quality is sufficient. Narrowband interference can be avoided by filtering out any other frequencies except the designated ones at the receiver. The major advantage of narrowband signal transmission is, as the term implies, its efficient use of frequency due to only a small frequency band being used for one signal transmission. Its drawback is evident, though, as it requires well-coordinated frequency allocation for different signals, and it is quite vulnerable to signal jamming and interception.

Spread spectrum takes another approach. Instead of applying all transmitting power to a narrow frequency channel, spread spectrum converts the narrowband signals to signals of much wider band with comparatively lower power density, according to a specific signal spreading scheme. Because of the low power density (power per frequency), the converted signal appears to be background noise to others who are unaware of the spreading scheme; only the designated receiver is able to reconstruct the original signal. For data transmission with multiple channels using frequency modulation, narrowband signals in each channel can be applied with the spread spectrum technique. To be able to differentiate these channels and reconstruct those signals afterwards, each channel is assigned a code with sufficiently large distance separating it from others in the code space, and the codes are made known to the designated receiver only. The advantages of spread spectrum are as follows:

- Code-division multiplexing has greatly improved channel capacity (*i.e.*, the number of signals that can be transmitted at the same time over a given frequency band) as compared to the narrowband spectrum.
- Spread spectrum offers high resistance against narrowband interference and tolerates narrowband interference because the signal is transmitted over a wide band. Even though some portion of the frequencies is distorted, the original signal can still be recovered with the error detection and correction techniques of the coding mechanism.
- Security against tapping and jamming is greater compared to narrowband spectrum techniques. Signals of spread spectrum are indistinguishable from background noise to anyone who does not know the coding scheme (*language*).

The disadvantage of spread spectrum is its relatively high complexity of the coding mechanism, which results in complex radio hardware designs and higher cost. Nonetheless, because of its remarkable advantages, spread spectrum has been adopted by many wireless technologies, such as CDMA and wireless LANs. Depending on the way the frequency spectrum is used, three types of spread spectrum systems are currently in place: direct-sequence spread spectrum (DSSS), frequency-hopping spread spectrum (FHSS), and orthogonal frequency-division multiplexing (OFDM).

3.3.1 Direct-Sequence Spread Spectrum

The direct-sequence spread spectrum spreads a signal over a much broader frequency band. It employs a "chipping" technique to convert a user signal into a spread signal. Given a user data bit, an XOR computation is performed with the user bit and a special chipping sequence code (digital modulation), which is a series of carefully selected pulses that are shorter than the duration of user bits. The resulting signal (the chipping sequence) is then modulated (radio modulation) with a carrier signal and sent out. On the receiver side, after demodulation, the same chipping sequence code is used to decode the original user bits. The chipping sequence essentially determines how the user signal is spread as pseudorandom noise over a large frequency band. The ratio of the spreading (*i.e.*, the spreading factor) varies for different spread spectrum systems. The longer the chipping sequence, the more likely a user signal can be recovered. A transmitter and a receiver have to stay synchronized during the spreading and despreading.

3.3.2 Frequency-Hopping Spread Spectrum

Frequency-hopping spread spectrum uses a frequency-hopping sequence to spread a user signal that is known to both the transmitter and the receiver. User data is first modulated to narrowband signals, then a second modulation takes place — a signal with a hopping sequence of frequency is used as the radio carrier. The resulting spread signal is then sent to the receiver. On the receiver side, two steps of modulations are required: (1) use the same frequency-hopping sequence to recover the narrowband signal, and (2) demodulate the narrowband signal. In effect, the transmitter and the receiver follow the same pattern of synchronized frequency hopping. As a result, only if the hopping sequence is made known to the receiver can it recover the original user data bits; otherwise, the transmitted signals will appear as background noise. FHSS does not take up the entire allotted frequency band for transmission. Instead, at any given moment, only a portion of it is used for hopping. The two types of frequency hopping systems in use are fast hopping systems and slow hopping systems. Fast hopping systems change frequency several times when transmitting a single bit, whereas in slow hopping systems each hop may transmit multiple bits.

Interestingly, the concept of frequency hopping was invented by a Hollywood actress, Hedy Lamarr, and composer George Antheil during World War II. The idea was to use a piano-roll sequence to hop between 88 channels to make decoding of radio-guided torpedo communications more difficult by enemies. Their idea was not implemented because the U.S. Navy refused to consider developing a military technology based on a musical technique. As George Antheil put it, "The reverend and brass-headed gentlemen in Washington who examined our invention read no further than the words 'player piano.' 'My god,' I can see them saying, 'we shall put a player piano in a torpedo." (source: American Heritage of Invention & Technology, Spring 1997, Volume 12/Number 4)

For both DSSS and FHSS, multiple signals with different sequence codes (either chipping-sequence code or frequency-hopping code) can be multiplexed by CDM. To compare, FHSS is relatively simpler to implement than DSSS, but DSSS makes it much more difficult to recover the signal without knowing the chipping code and is more robust to signal distortion and multipath effects. Both are widely used by a large array of wireless technologies operating on the unlicensed spectrum. For example, the IEEE 802.11b standard for wireless LAN employs DSSS over the 2.4-GHz free spectrum, whereas the Bluetooth standard uses FHSS for simplicity.

3.3.3 Orthogonal Frequency-Division Multiplexing

Orthogonal frequency-division multiplexing is a modulation technique that utilizes multiple subcarriers in parallel to transmit user data. These subcarriers are orthogonal in that they are modulated with their own data independently. OFDM was first conceived of in the 1960s in an effort to minimize interference among adjacent channels in a frequency band. Because of multicarrier parallelism, OFDM offers a much higher data rate than single-carrier modulation techniques. In addition, because subcarriers are orthogonal, multipath interference can be largely reduced. In reality, some OFDM systems are actually code OFDM (COFDM), which combines error-control channel coding schemes with OFDM modulation. COFDM has some nice properties, such as being resistant against phase distortion, signal fading, and burst noise. OFDM is used in asymmetric digital subscriber line (ADSL), IEEE 802.11a/g wireless LANs, and the broadband wireless data access technology WiMax. COFDM is predominantly used in Europe for digital audio broadcast (DAB) and digital video broadcasting (DVB).

3.4 Global System for Mobile and General Packet Radio Service

The preceding chapter gave an overview of the evolution of cellular systems, from first-generation analog systems — the advanced mobile phone system (AMPS) and total access communication system (TACS) — to the second-generation GSM and CDMA to the third-generation WCDMA and cdma2000. Although the world of wireless telecommunication experienced remarkable growth in terms of number of subscribers at the beginning of the 21st century, some mobile wireless operators soon realized that mobile voice service was not enough. The industry has evolved from analog voice to digital voice, and now it is time to leverage high-speed data access provided by 2.5G and 3G

Table 3.1 2G Cellular Systems.

2G Cellular System Category	2G Cellular Systems
Time-division multiple access (TDMA)	Global system for mobile (GSM)
	Digitized advanced mobile phone system (D-AMPS) (IS-136)
	Packet data cellular (PDC)
Code-division multiple access (CDMA)	CDMA (IS-95)

cellular networks for data-centric applications and services. Second-generation cellular systems can be divided into two categories according to the multiplexing access scheme of the radio interface: TDMA or CDMA. The TDMA systems include GSM, D-AMPS (also referred to as IS-136 to supersede IS-54), and packet data cellular (PDC, used in Japan). The CDMA category is fairly straightforward in that the cellular systems utilizing CDMA are also called CDMA systems and they comply with the IS-95 standard. In this section, we first review two second-generation cellular systems, GSM and CDMA (see Table 3.1), then discuss 2.5 G data services and 3 G cellular networks being rolled out worldwide. The introduction to GSM and CDMA focuses on the functional components and key operations. Security issues of these systems are discussed in Chapter 6 (Mobile Security and Privacy).

3.4.1 Global System for Mobile

The first-generation analog cellular systems are based on frequency-division multiple access (FDMA), where each cell supports a number of channels of equal bandwidth and each mobile connection takes two channels (one up and one down) for full-duplex communication. A user with a mobile device takes up the two channels exclusively over the connection time. The capacity of the first-generation cellular systems is largely limited by the static channel allocation scheme. GSM is a solution to this problem.

GSM can be regarded as a TDMA-based, circuit-switching, digital cellular system in that the channels allocated to a cell are shared by several mobile connections in a TDM fashion. A mobile connection will use two such simplex channels, each of which is 200 KHz wide, for a time slot of 577 μsec to transmit a 148-bit data frame (see Figure 3.2). Eight such data frames make up a TDM frame of 1250 bits that is sent every 4.615 msec. A multiframe is comprised of 26 TDM frames, 24 of which are used for data traffic. One of the other two is a control frame and the other is a slot reserved for future use. Accordingly, a shared channel offers a data rate of 270 Kbps, so each user is able to access 1/8 of it, or 33.85 Kbps.

The frequency bands of the first European GSM systems, namely GSM 900, are between 890.2 and 914.8 MHz for cell phone transmissions and between 935.2 and 960.8 MHz for base-station transmissions. Both of these frequency bands are designated for 124 200-KHz channels. Because the frequencies used in adjacent cells cannot overlap in the frequency domain, the total number of channels available in each cell is much smaller than the total number of channels available within the allocated frequency bands. Other GSM systems, such as the digital cellular system (DCS) 1800, personal communication system (PCS) 1900, and GSM 400, have been deployed in different

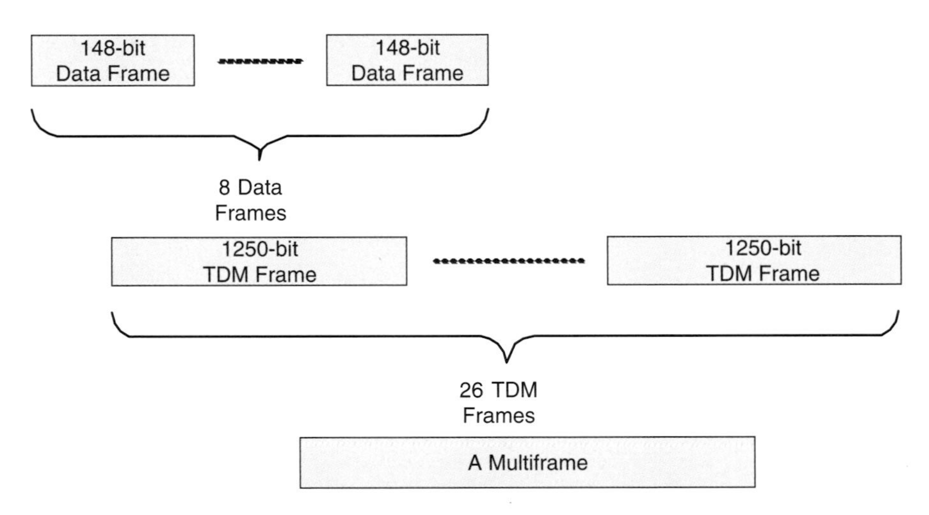

Figure 3.2 GSM data frame structure.

Table 3.2 Frequency Bands of GSM Systems.

GSM	Allocated Frequency Band (MHz)		Deployment
Systems	Uplink	Downlink	
GSM 900	890–915	935–960	Europe, Asia
DCS 1800	1710–1785	1805–1880	United Kingdom
PCS 1900	1850–1910	1930–1990	North America
GSM 400	450–457/478–486	460–467/488–496	Nordic countries

regions. Table 3.2 outlines the frequency bands of these systems and their deployment worldwide.

Speech coding of GSM employs the *regular pulse excited–linear predictive coder* (RPE-LPC), which generates a 260-bit block every 20 msec. The modulation scheme of GSM is Gaussian minimum shift keying (GMSK), in which a symbol represents a single bit.

3.4.1.1 GSM Network Architecture

A GSM network has three components: mobile station, base-station subsystem, and network system (see Figure 3.3). The mobile station is comprised of cell phones or other portal communication devices that can interface with the base-station subsystem. Each GSM cell phone essentially has two functional components: a mobile environment and a subscriber identity module (SIM) card. A mobile environment is a collection of physical elements for radio transmission and digital signal processing. A SIM card is a small chip that stores the identification of a user (subscriber) and service network information, as well as other data necessary for access authentication and authorization. A SIM card is not built into the cell phone handset; instead, it can be taken out and plugged into other GSM cell phones. As for almost all cellular systems, coverage areas are divided into cell sites. A mobile station in a cell directly communicates with a base transceiver station (BTS), which is responsible for handling all mobile stations within the cell site. The base-station subsystem has one or more BTSs and a base station controller (BSC). Handoffs among BTSs in the same base-station subsystem are controlled by the BSC.

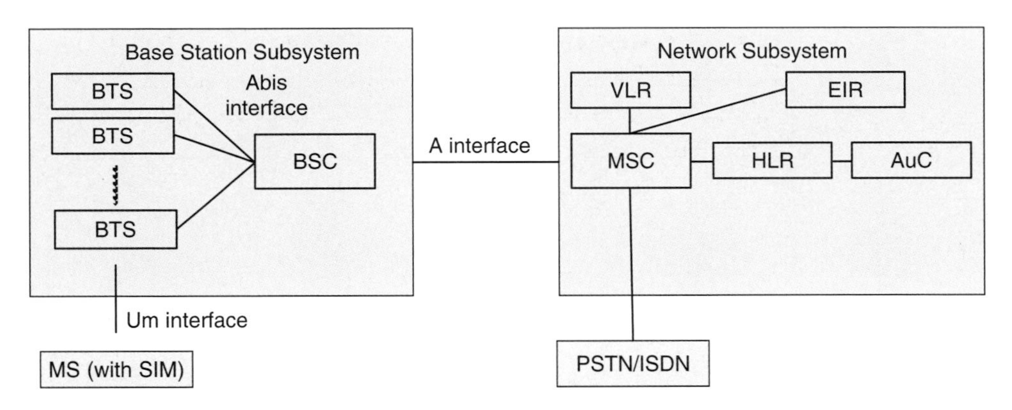

Figure 3.3 GSM architecture.

To cover a large area, a GSM network will build many base-station subsystems, each of which services mobile stations in a smaller area. The network subsystem connects the cellular network to the public switched telecommunication network (PSTN) by means of the mobile switching center (MSC), which also facilitates handoffs of mobile stations among different BSCs. BSCs in an area connect to the responsible MSC, which in turn connects to other MSCs.

The operation of an MSC relies on the home location register (HLR), visitor location register (VLR), authentication center (AuC), and equipment identity register (EIR). The HLR is a central database of mobile stations that maintains updated records of a mobile station's location area and its serving BSC, as well as the mobile subscriber's identification number and subscribed services. The VLR stores a subscriber's current location in terms of a serving MSC. Note that a subscriber always has a home switching center, in which the VLR keeps track of other switching centers (visited switching centers) in the vicinity of the subscriber at that moment. The AuC controls access to user data and encryption keys of subscribers in the HLR and VLR, as data transfer in GSM systems is encrypted. The EIR is a database of mobile station device identifications, particularly for stolen and malfunction devices. In addition to MSCs and these databases, an operation maintenance center (OMC) is set up to monitor and control MSCs in a region. It generates traffic statistics, failure alerts, and security-related reports.

The communication between a MSC and a BSC is defined in the Signaling System No. 7 (SS7). SS7 is a set of protocols of control signaling for circuit switching networks. In a GSM system, many control signaling are based on SS7.

3.4.1.2 Location Area Update

A mobile subscriber is uniquely identified by the phone number, formally known as the mobile station international ISDN number (MSISDN). An MSISDN number is comprised of three portions: a country code, such as 1 for the United States and 44 for the United Kingdom; a national destination code; and a subscriber number. Within a GSM system, an internal number, the internal mobile subscriber identity (IMSI), is used to identify a mobile subscriber. This number is stored in the SIM card.

A mobile station constantly monitors broadcast messages from its serving BTS. Once a change has been noticed, an update request, together with the subscriber's IMSI and 4-byte temporal mobile subscriber identity (TMSI) number, is sent to the new VLR via the new MSC. A temporary mobile station roaming number (MSRN) is allocated and mapped to the mobile's IMSI by the new VLR and sent to the mobile station's HLR. This number contains the new MSC, the subscriber number, and two other codes of the country. Upon receiving the MSRN, the HLR of the mobile station notifies the old VLR to eliminate the previous record of the mobile station. Finally, a new TMSI is allocated and sent to the mobile station to identify it in future paging or call-initiation requests. Throughout this location area update procedure, the MSISDN number of the mobile station is not used at all.

3.4.1.3 Call Routing

Call routing in GSM systems is performed by a procedure similar to that of location area update. There are two cases: mobile terminated call (MTC) and mobile originated call (MOC). An MTC could be initiated by a landline telephone, a cell phone of other cellular networks such as CDMA, or another GSM mobile station. In the first and second cases, a gateway MSC (GMSC) will take the dialed

MSISDN number and determine the HLR of that number. The GMSC then uses the MSISDN number to query the HLR, which returns the current MSRN of the dialed number that identifies the current location. Then, the GMSC is able to route the call to the current MSC of the dialed number. Upon receiving notification of an incoming call to a mobile station currently within its control, the MSC queries its VLR for the TMSI of the mobile station by the MSRN number. After that, a paging message is broadcast among all BSCs of the MSC. The BSC that is serving the desired mobile station will reply to the MSC. Finally, two parties of the call are indirectly located, and the call can proceed. If the call is initiated from one mobile station to another (MOC), a GMSC is no longer necessary; instead, the MSC of the caller will take over, and the remaining procedures are the same. An MOC may target a landline phone number or a cell phone number. In this case, a GMSC that connects two types of networks will link the two parties. In either case, the MSC involved must perform additional tasks so as to validate that the mobile subscribers are allowed to utilize the service.

3.4.1.4 Handoff

The mobility of mobile stations requires that a handoff occur when the subscriber carrying the mobile station moves across cells. Depending on the BTS and MSC arrangement of the fixed cellular network infrastructure, the handoff may occur in the following five scenarios, based on the movement of a mobile station: (1) intracell, (2) intercell but intra-BSC, (3) inter-BSC but intro-MSC, (4) inter-MSC but intranetwork, and (5) internetwork.

Intracell handoff is performed when the channels of a mobile station are changed to prevent narrowband interference. This is managed by the BSC and reported to the MSC. Intra-BSC handoff is the most common type and involves two BTSs in the same BSC. During an intra-MSC handoff, two BSCs within the same MSC manage to allocate channels to the mobile station in the new cell and release old channels used in the previous cell. Intranetwork handoff is carried out when the mobile station moves across the border of cells managed by two MSCs. As long as the mobile station remains in the same network, handoff is transparent to the subscriber, and no communication interruption should occur. For internetwork handoffs, such as roaming across GSM networks of two wireless operators, real-time communication will be interrupted and the connection will have to be reestablished. These five scenarios deal with GSM networks only. A handoff between two entities of the same type of network is referred to as a *horizontal handoff for micromobility*. Conversely, a *vertical handoff for macromobility* refers to handoffs between two different types of networks such as handoff between a GSM network and a CDMA network. Supporting vertical handoff is a critical issue in next-generation mobile computing when heterogeneous networks converge to allow ubiquitous mobile access.

Issues surrounding mobile station handoff within GSM networks can be summarized as follows.

- *When should the handoff be performed?* Typically, some periodic measurement is done on the serving BTS of the mobile station to monitor the signal strength between them. Generally, a gradual reduction of signal strength might indicate the mobile station is moving away from the BTS. A number of handoff strategies have been devised to make this procedure smooth yet fast. Thresholds and handoff margins are used to determine if a handoff is immediately needed and can be done without frequent back-and-forth handoffs. In addition to signal strength, sometimes a set of performance statistics of a cell is also used to decide if a handoff is necessary for better signal quality. In particular, when the call-blocking probability of a cell is considerably high due to heavy traffic load in a cell, the mobile station may choose to handoff to a neighboring cell.
- *How fast is a handoff?* The duration of a handoff is dependent on several factors, including movement speed of the mobile station, signal strength measurement and reporting interval, and handoff thresholds. Handoff may change the frequency and possibly MSRN of the mobile station, yet all necessary operations have to be done without being noticed by the user during a voice call.

The GSM standard requires a maximum duration of 60 msec for a handoff.

3.4.2 General Packet Radio Service

Global system for mobile (GSM) was designed for digital voice communication. As the need for data services such as e-mail, web browsing, and text messaging started to grow substantially, the limited data rate of 9.6 to 14.4 Kbps of GSM systems could not meet the increasing bandwidth demands of these applications. Those readers who have used the very first generation of dial-up modems have some idea of how slow the GSM data rate is. Aimed at leveraging the widely deployed existing GSM systems, general packet radio service (GPRS) has been implemented as a 2.5G cellular system for value-added data services.

3.4.2.1 Packet Switching

Basically, GSM is circuit-switching based, as are landline telephone networks, meaning that when a channel (one of the eight time slots in TDM) is assigned to a user at the beginning of a connection, it will be exclusively used by the user throughput the connection and will not be shared with others, even if there is really nothing going on for some period of time. For voice communications, this is not a serious problem. After all, people will keep talking during a phone call, and the reserved channels (GSM uses two channels for each connection) facilitate reliable voice communication for the two sides; however, data communication is characterized by bursty transmissions rather than steady streams. For example, if a user visits a website, it is very likely that a few hyperlinked web pages will be downloaded from the web server sequentially but not continuously, as the user will take some time to read a page and decide which page to download next. Allocating a channel for a series of intermittent, bursty data transmissions would obviously be wasteful in terms of system utilization and costs.

On the other hand, the maximum data rate of a channel is significantly lower than what the wired Internet can provide, thereby affecting the acceptance of mobile data applications such as wireless application protocol (WAP), which is a standard set of protocols enabling wireless web applications on a cell phone. Simple aggregation of multiple channels for a data transmission will increase the data rate, but it also means that more system capacity is underutilized due to the characteristics of data transmission.

In contrast, GPRS takes a packet-oriented approach for data transmission and a dynamic, on-demand, bundled time-slot allocation approach for a higher data rate. Because it is unnecessary to reserve a channel, no connection has to be established before data transmission and the air interface for data appears to be always on as long as the mobile station remains active in a GSM network. This is one of the most important and marketable features of GRPS. A data-transfer operation from a mobile station to a specific destination could be spread to multiple time slots within a TDM frame, and several such data-transfer operations could possibly share the same eight time slot. Uplinks and downlinks are allocated separately. The GSM standard defines four channel coding schemes: CS-1 of 9.05 Kbps, CS-2 of 13.4 Kbps, CS-3 of 15.6 Kbps, and CS-4 of 21.4 Kbps. The lower the data rate, the higher the error detection and correction capability. If all eight logical channels are used with CS-1 for a GPRS data transmission, the aggregate data rate will reach 72.4 Kbps. GRPS is designed to coexist with traditional voice services, thus in reality the number of logical channels available for GPRS is always restricted by system load in addition to demanding data transmission. The characteristics of asynchronous data traffic further limits the number of logical channels for uplink data transmission. Furthermore, power consumption will increase rapidly as the data rate of the transmitter on the mobile terminal increases.

3.4.2.2 GPRS Architecture

Two new components have been added to the GSM network architecture for GPRS: Serving GPRS support node (SGSN) and gateway GPRS support node (GGSN) (see Figure 3.4). Like MSCs and BSCs, a GSM network generally has many SGSNs and GGSNs, which constitute a GPRS core network. All other existing GSM components are also

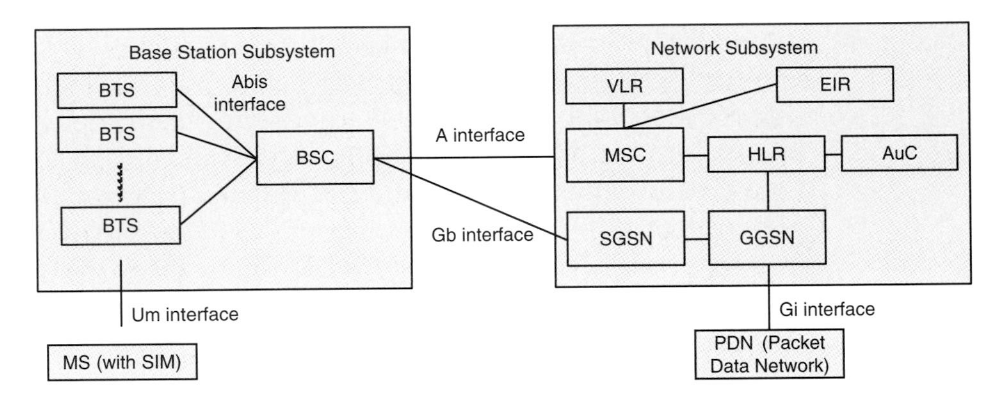

Figure 3.4 GPRS architecture.

involved. Additionally, to support GPRS, the mobile station has to incorporate the GPRS terminal functionality. Ideally, for a BTS, only a software upgrade is necessary to enable voice and data communication via the same radio interface, and it can be done remotely. This way the large amount of BTS hardware and the number of antennas do not have to be changed.

On the BSC, a packet control unit (PCU) device is added to deal with data packets from BTSs. It has two functions: (1) separate packet-switched traffic from circuit-switched traffic originating from mobile stations, and (2) multiplex circuit-switched traffic from the GSM core network and packet-switched traffic from the GPRS core network into an intermingled data frame to serving cells. Packet-switched traffic from a PCU on an MSC is sent to the corresponding SGSN, which keeps track of all serving mobile stations. GGSNs directly connected to packet data networks (PDNs) such as X.25 or the Internet. To the external networks, a GGSN appears to be an IP router that takes responsibility for converting and forwarding data packets to destination mobile stations. It achieves this by performing translation between external network node addresses and GRPS mobile addresses and forwarding data to the SGSN responsible for the mobile station in question according to a GPRS tunneling protocol (GTP). Because GRPS data transmission is always initiated from a mobile station, for those mobile stations currently engaged in GPRS data transmission, a GGSN is able to maintain a list of records that map these mobile stations to their serving SGSNs. In reality, IP dominates both the external PDNs and the GPRS core network. GPRS supports both IPv4 and IPv6. The GTP is an IP-based protocol with TCP (Transmission Control Protocol) and UDP (User Datagram Protocol) as transport. As a consequence, the address translation conducted on a GGSN is actually IP network address translation (NAT) that allows mobile stations to use private IP addresses within the GPRS core network. For all mobile stations known to a GGSN, upon leaving the GPRS core network their private IP addresses will be translated into globally unique, routable IP addresses. The GPRS core network may also implement DHCP (Dynamic Host Configuration Protocol) servers and DNS (Domain Name System) servers for dynamic IP assignment and domain name mapping, respectively. The GGSN also performs authentication and collects traffic volume of each subscriber for billing.

3.4.2.3 GPRS Services

General packet radio service supports both point-to-point (PTP) packet transfer services and point-to-multipoint (PTM) services. The PTP service provides both connectionless mode of PTP connectionless network service (PTP-CLNS) for IP-based data applications and connection-oriented mode of PTP connection-oriented network service (PTP-CONS) for X.25. The PTC-CONS supplies a virtual circuit regardless of the mobile station's location, realized by serving SGSNs and GGSNs. The PTM service essentially offers multicasting among mobile stations in a certain geographic area. To support point-to-multipoint features, a point-to-multipoint service center (PTM-SC) must be added as well.

3.4.2.4 GPRS Terminals

For mobile stations to use GPRS services, a GPRS terminal that handles packet data transmission via the radio interface is required. A GSM/GPRS cell phone has built-in GPRS terminals. Depending on how these two types of services are offered, three classes of terminals have been defined: class A, which can handle both data and voice simultaneously; class B, which can handle either one but not both at any given time; and class C, which can only attach to one aspect of the services. To date, due to the high cost of class A terminals, most GRPS cell phones are class B terminals on which GPRS data service is suspended during voice calls and short message service (SMS) and resume afterwards. Data-only terminals do not support voice, such as GPRS PCMCIA cards for laptop computers and PDAs.

To identify the capability of a GPRS terminal to use multiple logical channels, 12 multislot classes of GPRS terminals have been defined. Each multislot class specifies the maximum achievable data rates in both uplink and downlink of a data transmission. Written as d+u, the first number (d) indicates the number of downlink timeslots, and the second number (u) represents the number of uplink timeslots. A third number, the active slots, specifies the total number of time slots that can be used simultaneously for both uplink and downlink data transmission. For example, class 10 represents 4 downlink slots, 3 uplink slots, and 5 active slots. A multislot class and a terminal service class make up a specification of a GPRS terminal. Following the abovementioned example, the class 10/class B designation means that the terminal supports class 10 as the multislot class and can handle both data and voice services but not at the same time (class B).

3.4.2.5 Packet Data Protocol Context

For a mobile station to start a new GPRS session such as WAP browsing, it must obtain a packet data protocol (PDP) address used in the PDN. If the external PDN is the Internet, the mobile station is likely to contact a DHCP server to obtain a dynamic IP address. The key data structure used to map a PDP to the identification of the mobile station is referred to as a PDP context, which consists of the PDP type (*e.g.*, IPv4, or x.25); the PDP address assigned to the mobile station; the requested QoS profile, which specifies the desired service level in terms of delay, throughput, and reliability; and the address of a GGSN that serves as the access point to the PDN for the mobile station. A PDP context is session specific, and is stored in the mobile station, the SGSN, and the GGSN. The mapping between the PDP and IMSI of a mobile station allows the GGSN to locate the mobile station with the help of MSCs. For each data session, a different PDP context must be created to allow parallel access to multiple PDNs.

3.4.2.6 Enhanced Data Rates for Global Evolution

Enhanced data rates for global evolution (EDGE) is a further step toward 3G cellular systems by GSM. The objective of EDGE is a high data rate compared to GRPS. EDGE is also known as GSM384, as it can provide a data rate up to 384 Kbps if all eight time slots are used. Recall that GSM/GPRS uses GMSK modulation. EDGE is based on a new modulation scheme, called 8-PSK, that allows a much higher bit rate. In 8-PSK, each symbol transmitted through the air interface carries three bits instead of only one, as in GMSK, thereby greatly improving the data rate. EDGE defines nine coding schemes of different bit data rates. By monitoring the channel-to-interference ratio (C/I), EDGE can automatically switch coding schemes in favor of a higher data rate or reliable transmission. EDGE is also designed to make the convergence of GSM and IS-136 TDMA (D-AMPS) smoother. A major obstacle of this foreseeable convergence is channel bandwidth mismatch. TDMA channels are 30 KHz wide, whereas GSM channels are 200 KHz wide. Compact EDGE was introduced to solve this problem. It uses much fewer frequencies than classic EDGE but has a wider 200-KHz channel. Compact EDGE uses the same modulation scheme as EDGE classic does, with some key exceptions that allow it to be deployed in the spectrum of less than 1 MHz.

3.4.2.7 High-Speed, Circuit-Switched Data

High-speed, circuit-switched data (HSCSD) is an evolutionary technology for GSM systems moving towards 3 G UMTS. It can provide a data rate up to 43.2 Kbps using multiple time slots simultaneously. HSCSD allows various error correction methods to be used, whereas GSM has only one error correction method to deal with transmission errors in the worst case. HSCSD can also be an option in EDGE and UMTS systems. A major issue that wireless operators must consider while upgrading GSM/GPRS to EDGE or to 3G UMTS/WCDMA is to make the most out of the existing GSM infrastructures. GPRS and EDGE are considered intermediate solutions for early adopters before 3G.

3.5 Code-Division Multiple Access

The notion of CDMA has two distinct meanings: It can refer to the multiplexing scheme of code-division multiple access, or it can refer to second-generation cellular systems that use DSSS as the spread spectrum scheme and code-division multiplexing as the underlying multiplexing technology. Unlike GSM, which is also the name of the standard, the standard for CDMA cellular systems is IS-95.

3.5.1 Code-Division Multiple Access Concept

Section 3.2 introduced CDM. CDMA is actually DSSS utilizing CDM. Like GSM, CDMA uses a dedicated frequency band for multiple simultaneous signal transmission, but what underlies this frequency use scheme is spread spectrum, which essentially spreads a single signal from a transmitter over the entire shared frequency band in such a way that signals will not interfere with each other, thanks to a spreading code assigned to each signal. A single data bit of 1 from a mobile station is mapped to a chip sequence that identifies the mobile station. For a data bit of 0, the complement of the chip sequence is used. The chip sequence is normally 64 or 128 chips long and is pairwise orthogonal, meaning that the normalized inner product (*i.e.*, dot product) of any two distinct chip sequences (they are considered vectors of +1 and –1 in mathematical terms) is 0. After the mapping, multiple data bitstreams from different mobile stations are added linearly and transmitted. The intended receiver knows the chip sequence of the individual mobile station and uses it, along with the received aggregated bitstream, to compute data bits of that mobile station. The computation is quite straightforward: Simply compute the normalized inner product of the chip sequence of the desired mobile station and the received bitstream. In this way, the data bits sent by that mobile station will be recovered. Without knowing the correct chip sequence of a transmitter, the computation will yield some pseudorandom bits like noise. An implicit assumption of the decoding procedure is that the receiver and the transmitter are well synchronized in time, which allows the necessary computations for the correct portion of the transmitted bitstream. This is often done by utilizing a special synchronization bit sequence.

The chip sequences assigned to mobile stations can be generated by the Walsh code, an algorithm that produces mathematically orthogonal codes derived from the Walsh matrix. The Walsh-encoded chip sequences appear to be random noise to mobile terminals. Initially, the chip sequences are of equal length. To increase the number of usable chip sequences in the coding space, variable-length chip sequences have been devised and used in today's CDMA systems. Interested readers can refer to A. J. Viterbi's book "CDMA – Principles of Spread Spectrum Communication" for more details of CDMA codes.

3.5.2 IS-95

IS-95 is the underlying standard of CDMA systems. It is worth noting that CDMA is primarily designed and promoted by Qualcomm Inc., which holds key intellectual property rights related to CDMA technology. IS-95 is also commonly referred to as cdmaOne. Table 3.3 outlines key parameters of IS-95.

The forward link refers to the link from a base station to a mobile station, whereas the reverse link is the link from a mobile station to a base station. For both types of links, voice is encoded at a rate of 9600 bps after some error correction code is added. In a forward link, both data and voice are encoded by a forward error correction (FEC) scheme, resulting in a doubled bit rate of 19.2 Kbps. In a reverse link, because a different FEC scheme is used, the resulted data rate is 28.8 Kbps. For each forward link, 64 logical channels, each corresponding to a mobile station, are scrambled to prevent repetitive patterns. A reverse link is comprised of up to 32 logical access

Table 3.3 IS-95 Key Parameters.

IS-95 Parameter	Description
Multiple access method	**CDMA with FDM**
Frequency range of down links	869–894 MHz
Frequency range of up links	824–849 MHz
Number of channels of the frequency range	20
Channel spacing	1.25 MHz
Number of logical channels in forward link	64 (of which 55 can be traffic logical channels)
Number of logical channels in reverse link	94 (of which up to 62 can be traffic channels)
Number of users per channel	798
Voice traffic bit rate	9600 bps
Encoded traffic bit rate in forward link	19.2 Kbps
Encoded data bit rate in reverse link	28.8 Kbps
Chip sequence size	64 bits
Digital modulation scheme of forward link	Quadrature phase shift keying (QPSK)
Digital modulation scheme of reverse link	Orthogonal quadrature phase shift keying (OQPSK)
Channel bit rate	1.2288 Mbps

channels for paging and 62 logical traffic channels. For both types of links, the DSSS function spreads data of the logical channels over the available frequency range, resulting in an overall 1.228-Mbps data rate. Specifically, a 42-bit-long mask code is used on a reverse link to identify logical traffic channels that are dedicated to connecting mobile stations to a base station. The same mask code is also used to produce a bitstream that will be modulated onto the carrier using orthogonal QPSK or offset QPSK (OQPSK). OQPSK differs from QPSK in that in the implementation of OQPSK one of the two half-rate bit-streams of the original input signal is delayed for one bit period to reduce phase shift at a time. Because of duplex communication, the total number of reverse-link logical channels for traffic must be the same as the total number of forward-link channels.

3.5.3 Software Handoff

Global system for mobile (GSM), as well as other TDMA or FDMA systems, uses hard handoff, in which a mobile station will not switch to a new base station until connection to the current mobile station is released. Generally, only when the signal strength of the new base station is sufficiently stronger than that of the current one plus a threshold does the mobile station proceed to connect to the new base station. The reason for utilizing signal strength for hard handoff is frequency reuse among neighbor cells. As a mobile station moves toward a neighboring cell, the signal strength from the new one tends to increase gradually. Unfortunately, such spatial frequency reuse does not exist in CDMA at all. In CDMA, a different handoff approach is used, namely soft handoff. Because all cells essentially use the same frequency band, it is not possible to switch frequency for the handoff. In fact, a mobile station will connect to more than two base stations at a time and constantly monitor the signal strength of them. The handoff will take place only when one of them shows fairly stronger signal strength than others. Before that, all the base stations service the mobile station independently. Soft handoff is advantageous over hard handoff because the mobile station does not lose contact with the cellular network during handoff execution, increasing the possibility of successful handoff. The standard allows up to six base stations to be connected from a mobile station during handoff execution, although in real systems not that many can actually be connected.

3.5.4 Road to 4G

IS-95, or cdmaOne, has been designated by Qualcomm as the second-generation digital CDMA cellular system standard. The next generation of cdmaOne is cdma2000, and others are in various stages of development, such as cdma2000 1x RTT, cdma2000 1x EV, cdma2000 1x DV, cdma2000 1x DO, and cdma2000 3x RTT (explained shortly). As mentioned before, the second-generation GSM systems are evolving to a different type of CDMA system: UMTS/WCDMA. It is quite clear that the concept and underlying technologies of CDMA finally dominate the air interface of the future cellular world, after a long round of debates and remarkable business practices.

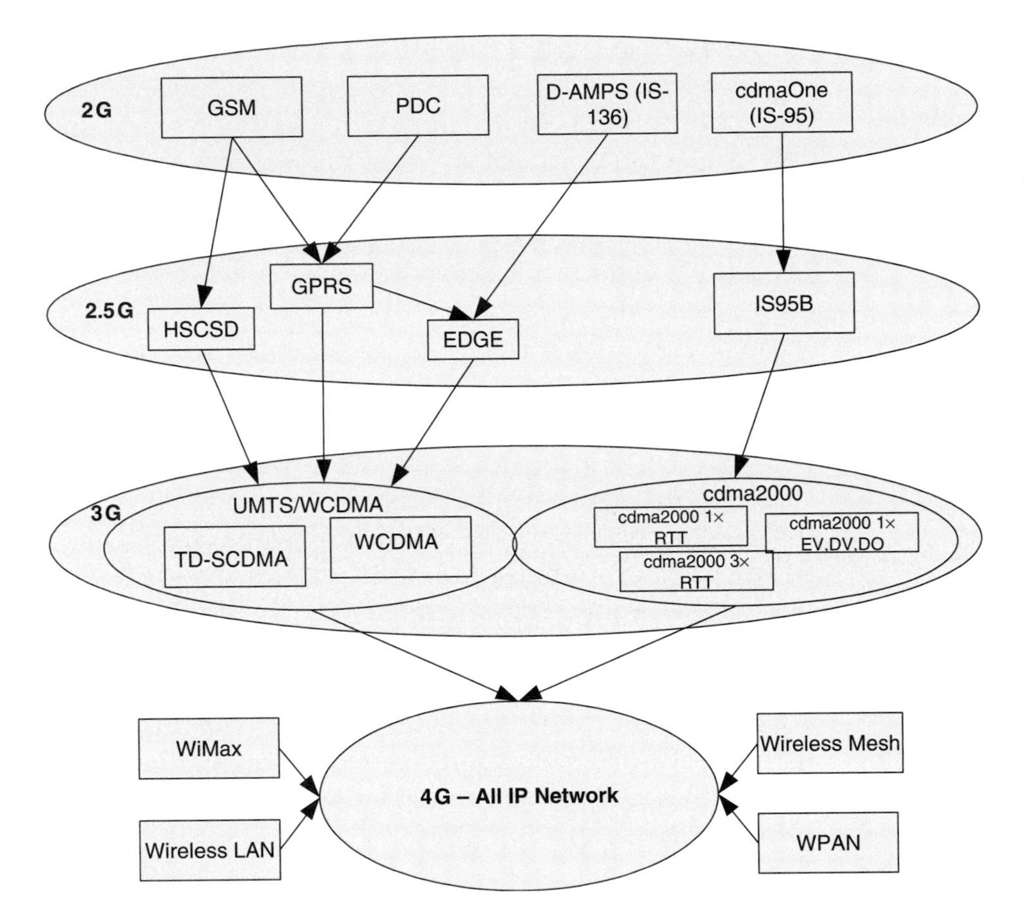

Figure 3.5 Road to 4G.

Figure 3.5 shows the evolution of 2G cellular systems toward 3G. On the road to 3G, TDMA systems such as GSM, PDC, and D-AMPS may take different paths involving GPRS, EDGE, or HSCSD as 2.5G solutions. Things are much clearer on the CDMA side: cdmaOne (IS-95A) will be replaced by IS-95B as a 2.5G system, then by cdma2000 systems. The standardization body supporting UMTS/WCDMA is 3GPP, whereas the counterpart for cdma2000 is 3GPP2. Figure 3.5 also shows wireless networks beyond 3G — the so-called All IP 4G networks.

3.6 GSM Versus CDMA

The two major 2G cellular systems are not compatible with each other. GSM and CDMA networks are both widely used worldwide. For example, in the United States, Verizon Wireless, Sprint PCS, and ALLTEL are CDMA operators, whereas Cingular Wireless, AT&T Wireless (merged with Cingular in 2004), and T-Mobile USA are GSM operators. Another wireless operator, Nextel, uses iDEN, a TDMA technology developed by Motorola. While the Europeans enjoy continent-wide, GSM-dominated wireless services, the world's largest mobile wireless market, China, with a total number of about 398 million subscribers as of 2005 (source: Computer Industry Almanac Inc.), is basically shared by two companies: China Telecom, which operates a GSM network, and China Unicom, which operates both GSM and CDMA networks. Table 3.4 provides a technical comparison of GSM and CDMA with an emphasis on the data services for next-generation mobile computing. Most differences between the two types of systems have been extensively discussed in the preceding sections, except speech encoding and power control. CDMA employs variable-rate codec for speech encoding, which is more efficient than GSM's fixed-rate codec. Power control of CDMA systems requires a closed-loop approach, thus it is faster than GSM's open-loop approach.

3.7 3G Cellular Systems

There is some debate over which cellular systems should be considered are so-called 3G systems, especially with regard to EDGE and some cdma2000 systems. Because GPRS and EDGE are GSM based, it is fairly intuitive to put them into the 2.5G category. In the CDMA camp, one cdma2000 system, called cdma2000 1xRTT, has been arguably considered a 3G system. Generally, all cdma2000 and UMTS/WCDMA systems may be considered 3G systems, as shown in Table 3.5.

Table 3.4 Comparison of GSM and CDMA.

Feature	Global System for Mobile (GSM)/General Packet Radio Service GPRS	Code-Division Multiple Access (CDMA)
Multiple access scheme	Time-division multiple access (TDMA)	CDMA
Duplexing	Frequency-division duplex (FDD)	FDD
Frequency bands	900 MHz, 1800 MHz, 1900 MHz	800 MHz, 1900 MHz
Channel bandwidth	200 KHz shared by 8 time-slotted users	1250 KHz shared by 64 users (codes)
Data rate	Initially 9600 bps, now 38.4 Kbps or 115 Kbps shared by 8 users	9.6–14.4 Kbps
Carrier RF spacing	1.25 MHz	200 KHz
Handoff	Hard handoff	Soft handoff
Speech encoding	Fixed rate codec	Variable rate codec
Power control	Open-loop and slow power control	Close-loop and faster power control
Identification	SIM card	Hardwired in the handset
3G	UMTS/WCDMA	cdma2000
Road to 3G	GPRS, EDGE, or HSCSD	cdma2000 1x (IS-95B)
Market	Europe, Asia, Australia, South America, North America, including some US MNOs such as Cingular/AT&T wireless, and T-Mobile	Asia (South Korea and China), Canada, United States; mobile network operators (MNOs) such as Verizon Wireless and Sprint PCS

The International Telecommunication Union (ITU) made a request for proposal (RFP) in 1997 for cellular technologies for the International Mobile Telecommunication (IMT)-2000 program. A proposal for a universal mobile telecommunications system (UMTS) was submitted by the European Telecommunication Standards Institute (ETSI) to ITU. Its radio interface is universal terrestrial radio access (UTRA). In addition to the proposals and systems outlined in Table 5, other 3 G radio access technologies are

Table 3.5 The 3G Landscape.

3 G Systems	Key Features	IMT Proposal	Radio Interface
Universal mobile telecommunications system (UMTS)/wideband code-division multiple access (WCDMA)	144 Kbps satellite and rural outdoor, 384 Kbps urban outdoor, 2048 Kbps indoor and low range outdoor	IMT 2000 CDMA direct spread by 3GPP	Direct-sequence spread spectrum (DSSS) CDMA; both frequency-division duplex (FDD) and time-division duplex (TDD) are used
cdma2000 1x RTT	Also known as cdma2000 1x MC (multiple carriers); data rate up to 144 kbps	IMT 2000 CDMA multicarrier by 3GPP2	Uses a 1.25-MHz band, coexistent with IS95
cdma2000 1x EV-DO	Downlink (forward-link) data rates up to 3.1 Mbps and uplink (reverse-link) rates of 154 Kbps		Backward compatible with cdmaOne
cdma2000 1x EV-DV	Downlink (forward-link) data rates up to 3.1 Mbps and uplink (reverse-link) rates of up to 451 Kbps		Backward compatible with cdamOne
cdma2000 3x RTT	2–4 Mbps		Uses three 1.25 MHz bands
TD-SCDMA (Time Division – Synchronous Code Division Multiple Access)	2 Mbps	IMT-2000 CDMA TDD by China	TDMA and CDMA combined

listed as follows:

- IMT-2000 TDMA single carrier, originally promoted by the Universal Wireless Communications Consortium (UWCC); EDGE is one of the IMT-2000 TDMA SC technologies
- IMT-2000 FDMA/TDMA, also known as digital enhanced cordless telecommunications (DECT), the enhanced version of the cordless phone standard.

3.7.1 UMTS/WCDMA Versus cdma2000

A UMTS system works in two modes; its frequency-division duplex (FDD) mode is the well-known wideband CDMA (WCDMA), whereas its time-division duplex (TDD) mode seems to remain unnoticed by the public. cdam2000 is the evolution of cdmaOne, the current CDMA system in the United States. In fact, strictly speaking, WCDMA only refers to the radio interface aspect of the entire UMTS system. The same radio interface technology is used by NTT DoCoMo and J-Phone (a subsidiary of Vodafone) as well. As an FDD system, WCDMA does not require time synchronization among base stations. It allows a bit rate up to 384 Kbps, compared with the maximum rate of 2 Mbps in TDD UMTS system. In particular, China has proposed a TDD UMTS system, called TD-SCDMA, and is vigorously promoting this technology among Chinese telecommunications device manufacturers and wireless operators.

cdma2000 is a general term representing technical specifications such as cdam2000 1x RTT, cdma200 1xEV-DO, cdma2000 1xEV-DV, and cdma2000 3xRTT. RTT stands for radio transmission technology, EV-DO for evolution-data optimized, and EV-DV for evolution-data and voice. 1xRTT can provide a peak rate of 153.6 Kbps, while 3xRTT may theoretically offer a peak rate of 3.09 Mbps. The first commercial system of cdma2000 1xEV-DO was launched in South Korea in January 2002. cdma2000 is backward compatible with existing IS95/cdmaOne systems, whereas WCDMA requires an overhaul of existing base stations. There is no synchronization in WCDMA systems, thus sophisticated protocol designs and handoff mechanisms are not required. On the other hand, cdma2000 requires base-station synchronization. In some sense, WCDMA can be seen as an opportunity for operators to challenge Qualcomm's CDMA technology monopoly. Readers interested in the evolution of mobile networks are encouraged to refer to Vriendt *et al.* [2].

3.7.2 UMTS/WCDMA

A UMTS system is comprised of three components and two interfaces (see Figure 3.6). The components are the user environment (UE),

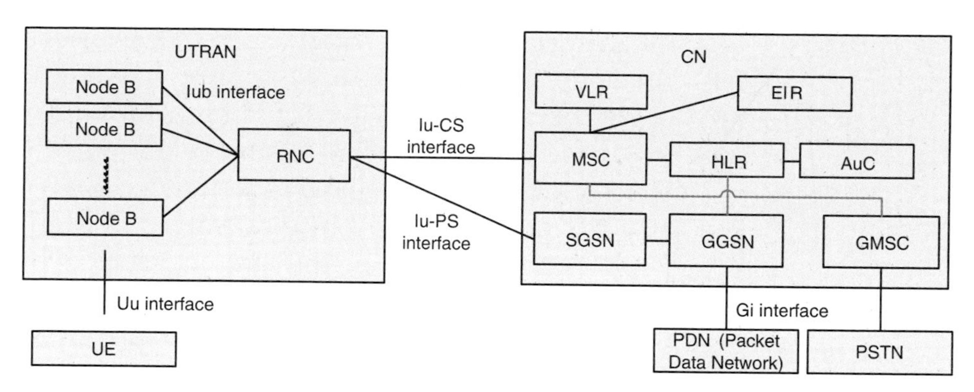

Figure 3.6 UMTS architecture.

the UMTS terrestrial radio access network (UTRAN), and the core network (CN). The interface between UE and UTRAN is referred to as Uu. The interface between UTRAN and a Node B is Iub. UMTS introduces Node Bs as base stations (BTSs in GSM) and radio network controllers (RNCs) as BSCs in GSM. Similar to GSM and GPRS, MSCs and SGSNs control RNCs through the Iu interface. In particular, an MSC connects to an RNC through an Iu–CS (circuit-switching) interface, whereas a SGSN connects to an RNC through an Iu-PS (packet-switching) interface. In UMTS, GMSCs and GGSNs connect to PSTN and PDNs. Other components such as HLR and VLR are the same as in GSM but with enhanced functionality for UMTS.

UMTS uses a pair of 5-MHz channels, one in the 1900-MHz range for uplink and one in the 2100-MHz range for downlink. In contrast, cdma2000 uses one or more arbitrary 1.25-MHz channels for each direction of transmission. UMTS is expected to deliver a user data rate of 1920 Kbps, although in reality 384 Kbps is probably what the system can really offer. A future version of UMTS/WCDMA, high-speed downlink packet access (HSPDA), will offer data speeds up to 8 to 10 Mbps and 20 Mbps for multiple-input, multiple-output (MIMO) antenna systems. The data modulation scheme is QPSK for uplink and BPSK for downlink. The chip rate is 3M chips per second.

As a spread spectrum radio interface, WCDMA uses soft handoff just as cdmaOne does for the same reason: It is quite difficult to control power beyond the hysteresis if hard handoff is employed because in CDMA systems forcing a mobile station to operate over some hysteresis level will cause large interference.

The first UMTS network went into operation in the United Kingdom in 2003. AT&T Wireless in the United States deployed UMTS in selected cities in late 2004. Japan's largest telecommunication service provider NTT DoCoMo launched the first WCDMA-based 3G network, dubbed FOMA (Freedom of Mobile Multimedia Access), in 2001.

3.7.3 cdma2000

cdma2000 is another standard under the ITU-2000 program (see Figure 3.7). It comes in two stages: 1X and 3X. Using the existing cdmaOne infrastructure, cdma2000 1X can supply a maximum user data rate of 207 Kbps and a typical data rate of 144 Kbps in general. It doubles the voice capacity of cdmaOne systems and offers six times the capacity of GSM or TDMA systems. cdma2000 3X further improves the user data rate to 2 Mbps.

In cdma2000, three major components exist in the overall network architecture: mobile station, radio access network, and core

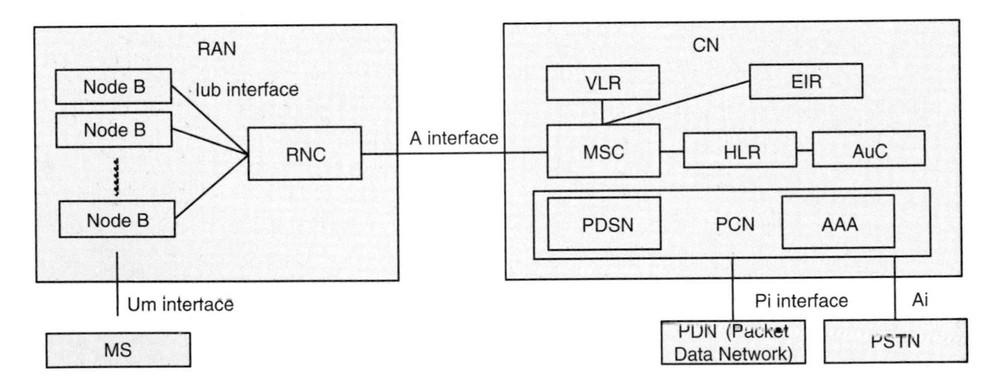

Figure 3.7 cdma2000 architecture.

network. The interface between a mobile station and radio access network is called Um, and the interface between radio access network and core network is called A. In addition, the core network can be further decomposed into two portions: One portion, the packet core network (PCN), connects to external IP networks via a Pi interface, whereas the other connects to PSTN via an Ai interface. Similar to a UMTS network, the core network of cdma2000 also has MSCs, HLRs, and VLRs. The principle difference between the core network of cdma2000 and those of other cellular systems is the PCN that provides IP network access to mobile stations. A component in PCN, the packet data service node (PDSN), performs roughly the same task as a SGSN in UMTS or GPRS; however, in cdma2000, two IP access methods are provided: simple IP access and mobile IP access. Simple IP access is the traditional way to obtain and retain an IP address within a geographically located subnet. When the mobile station moves to another subnet, it has to redo the DHCP procedure and obtain a new IP address. This is the case when a mobile worker uses a laptop computer to connect to an enterprise network across several buildings. Mobile IP access enables a mobile station to use the same IP address across different regions. In this case, a *home agent* of the mobile station will assume the responsibility of maintaining the same IP address for the mobile station. A *foreign agent* that is part of the PDSN is used to assign a temporary address to the mobile station that just moved in, and tunnels packets from the home agent to the mobile station. Note that GPRS has only a single IP access method: the simple IP access. It has to be emphasized that cdma2000 has better IP support. This is indeed a tremendous advantage of cdma2000 over UMTS, as in the long run the core cellular network will be interoperable with other wired or wireless networks with IP as the underlying network protocol.

Another major task of the PCN is authentication, authorization, and accounting (AAA). Three parties are involved: home AAA (HAAA), broker AAA (BAAA), and visited AAA (VAAA). HAAA stores a subscriber's profile information. Once requested by a VAAA, it will authenticate and authorize a subscriber and send the response to back to the VAAA. For accounting, VAAA is able to receive accounting information from HAAA and provides the subscriber's profile to the PDSN. BAAA is used as an intermediate server when VAAA and HAAA are not directly associated with each other.

3.7.4 4G Cellular Systems

As of 2004, UMTS/WCDMA and cdma2000 3G services have been rolled out in a number of countries and continue to gain some ground among business professionals. It is widely agreed that 3G will replace 2G and 2.5G systems in the next several years, providing a seemingly high throughput of several megabits per second for a mobile station. While this data rate seems sufficiently large for popular applications such as text messaging and web browsing on a cell phone, it cannot meet the relentless demand of emerging applications such as full-motion video broadcasting and videoconferencing. In response, researchers have moved on to 4G cellular systems, which provide even higher data rates of 20 Mbps to 100 Mbps. It has to be noted that wireless LANs can now provide data rates of up to 54 Mbps, much higher than current and future variants of 3G systems, whether UMTS or cdma2000. On the other hand, a significant effort has been made to coalesce voice and data communication in all 3G systems to leverage the legacy systems as much as possible. As voice over IP technology matures, an all-IP wireless network that supports voice and data over the same packet-switching infrastructure will become technically feasible to build and will be more cost effective than current 3G frameworks.

Emerging technologies for 4G wireless networks are summarized as follows; note that the application of these technologies is not limited to cellular systems:

- Smart antenna technologies exploit spatial separation of signals to allow an antenna to focus on desired signals as a way to reduce interference and improve system capacity.
- MIMO (multiple-in, multiple-out) utilizes antenna arrays at both the transmitter end and receiver end to boost the link data rate and system capacity. MIMO takes advantages of multi-path

propagation of signals by which more data can be sent in a single channel by splitting and recombining data onto multiple paths.

- Orthogonal frequency-division multiplexing (OFDM), multiple-carrier-code-division multiple access (MC-CDMA), modulation, and multiplexing technologies will improve the robustness of signal transmission and the data rate.
- Software radio or software-defined radio will make it possible to reconfigure channel modulation and multiplexing on the fly.

In a broader vision, 3G cellular systems are merely one type of wireless access in the world of mobile computing. Other wireless access technologies, such as wireless LANs and WiMax, have demonstrated great potential to become a primary means of network access, as shown in Figure 3.5. These technologies may complement each other and may certainly compete head-on with each other in a variety of industry segments, leading to coexistence and integration of these systems and spurring new services and applications. The following section talks about a galaxy of new services and applications in the wireless arena.

3.8 2G Mobile Wireless Services

Aside from traditional voice services, other cellular wireless services have proliferated in some countries. For example, Short Message Service (SMS) is a vastly successfully service in some Asian countries, and has become a major revenue source to MNOs. Customized mobile web surfing also gain some ground as WAP and iMode mature. In this section, we discuss these two types of 2G mobile wireless services.

3.8.1 WAP and iMode

Wireless application protocol (WAP) is an open-application layer protocol for mobile applications targeting cell phones and wireless terminals. It was developed by the WAP Forum, which has been consolidated into the Open Mobile Alliance (OMA). The current release is WAP 2.0. WAP is intended to be the World Wide Web for cell phones. It is independent of the underlying cellular networks in use. To a cell phone or personal digital assistant (PDA) user, WAP is perceived as a small browser application that can be used to browse some specific websites, quite similar to the web browsing experience on a desktop computer but with significant constraints due to the form factor of the mobile terminal. A WAP system employs a proxy-based architecture to overcome the inherent limitations of mobile devices with respect to low link bandwidth and high latency. Below is a list of features that separate WAP from other application protocols:

- Wireless markup language (WML), WML script, and supporting WAP application environment — Together, they are referred to as WAE. WML is an HTML-like markup language specifically devised for mobile terminals that have limited bandwidth, fairly small screen size, limited battery time, and constrained input methods. WSL is a scaled-down scripting language supported by the WAP application environment. In addition, WAP 2.0 supports XHTML language, which allows developers to write applications for both desktop computers and mobile terminals.

- WAP protocol stack — WAP Version 1.0 includes wireless session protocol (WSP), wireless transaction protocol (WTP), wireless transport layer security (WTLS), and wireless datagram protocol (WDP). Version 2.0 incorporates standard Internet protocols into its protocol stack, such as TCP, transport layer security (TLS), and HTTP (Hyper Text Transport Protocol). Both TCP and HTTP are optimized for wireless environments.

- WAP services, such as push and traditional request/response, user agent profile, wireless telephony application, external functionality interface, persistent storage interface, data synchronization, and multimedia messaging service.

WAP 1.0 has proved to be a technological hype; it has been intensively promoted by wireless operators and content providers but has received little, if any, positive feedback from users. Because of that, WAP has sometimes been referred to as "wait and pay." Interestingly, it is not only the protocol but also the applications utilizing WAP that, as a whole, push users away because of application performance, input methods, and the GUI interface, among other reasons. Moving toward standard IP protocols rather than specialized wireless protocols, WAP 2.0 addresses most of the problems of the protocol stack and the application environment, thereby giving the technology a brighter future.

iMode is a successful wireless application service provided by NTT DoCoMo. It is very similar to WAP in that it defines an architecture of web access on mobile terminals, primarily cell phones. Like WAP 2.0, iMode adopted standard Internet protocols as transport for applications, but iMode does not use any gateways. Instead, it utilizes overlay packet network on top of a cellular network for direct communication. The fundamental difference between WAP and iMode is that iMode requires mobile terminals to be designed to adapt to the services and applications of iMode, while WAP focuses on adapting itself to fit into general mobile terminals. Furthermore, NTT DoCoMo's effective WAP initiative has managed to attract many satisfied providers who can offer a wide array of services and applications to users [4].

3.8.2 Short Message Service

No one ever expected that short message service (SMS) would be such a tremendous success, one that exemplifies the perfect marriage of a business model with a wireless technology. European and Asian subscribers have been using SMS for years. More than a billion SMS messages are sent each month in some countries. Finally, as of 2004, SMS began to take off in North America. SMS allows two-way transmission of 160-character alphanumeric messages between mobile subscribers and external computing systems such as e-mail systems and paging systems. Because of its increasing popularity, SMS has been extensively combined with many new types of information services in addition to traditional usage. For example, both Google and

Yahoo offer Internet searching via SMS. SMS was initially designed to replace alphanumeric paging service with two-way guaranteed messaging and notification services.

Two new types of SMS components have been added to the cellular network: short message service center (SMSC) and signal transfer point (STP). An SMSC is a central controller of SMS services for the entire network. It interfaces with external message sources, such as voice-mail systems, e-mail systems, and the web. Messages sent from a mobile subscriber will also be stored and forwarded by the SMSC. An STP is a general network element connecting two separate portions of the network via SS7 signaling protocol. In the case of SMS, numerous STPs interface with the SMSC, each handling SMS transmission and delivery to and from a large number of mobile stations. No matter where the messages come from, the SMSC will guarantee delivery and inform the transmitter. For the SMSC to locate a mobile station for message delivery, it must utilize the cellular network, especially the HLR, VLR, and MSC of the mobile station.

Short message service has been enhanced with new capabilities to support enhanced message service (EMS) and multimedia message service (MMS). If you consider SMS to represent very early plain-text e-mails, you might think of EMS as the fancier HTML e-mails containing pictures, animations, embedded objects such as sound clips, and formatted text. MMS is the next-generation messaging service that supports rich media such as video and audio clips. The wide use of picture messages sent from a camera cell phone is merely one example of MMS in action. MMS consumes more bandwidth so it requires a high data rate for the underlying network and considerable computing capability of the mobile handset. The multimedia service center (MMSC) performs similar tasks as the SMSC for SMS. The following list outlines the necessary steps of an MMS procedure:

- The transmitter sends a message to the MMSC from a cell phone, PDA, or networked computer.
- The MMSC replies to the transmitter with a confirmation of "message sent." In fact, it is not sent to the receiver yet, as the message is stored at the MMSC.

- The MMSC locates the receiver with the help of a number of cellular network elements, such as MSCs, HLRs, and VLRs. If the mobile station of the receiver is on, the MMSC sends a notification of a new message to it, along with a URL to the new message. Otherwise, it waits and tries again later.
- The receiver can choose to download the message right away or save the URL to download it later.
- The MMSC will be notified by the receiver that the message has been downloaded and presumably read, then the MMSC notifies the transmitter that the message has been delivered.

MMS is the natural evolution of SMS, with EMS as an optional intermediate messaging service, but it is very unlikely that MMS will replace SMS completely as plain text messages are preferable in many cases. Additionally, MMS does not require 3G; it can be done in 2.5G systems such as GPRS and EDGE. Problems that may hinder the widespread use of MMS include digital rights management of content being exchanged among many mobile subscribers, development of a user-friendly interface design, and sufficiently large bandwidth for message delivery.

3.9 Wireless Technologies Landscape

In the wireless world, aside from cellular technologies, myriad wireless technologies have emerged and matured. At the eve of the new millennium, "wireless" typically referred to the use of cell phones. After only a few years, the dramatic growth of new wireless communication and computing technologies has fundamentally changed our perception of wireless technologies. This section discusses these technologies from an overall perspective. Once again, the emphasis of the discussion is on mobile data access in the greater domain of mobile computing, rather than on wireless communication. Figure 3.8 depicts the landscape of existing and emerging

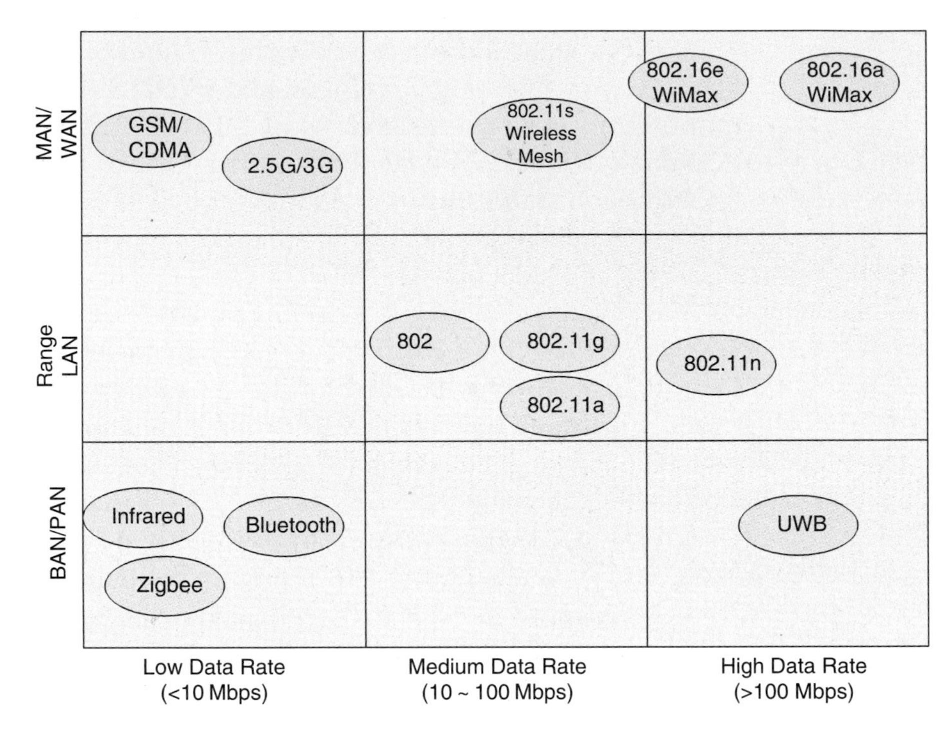

Figure 3.8 Wireless technology landscape.

wireless technologies with respect to two significant characteristics pertaining to mobile computing: data rate and signal transmission range. As the figure suggests, cellular systems are positioned in a grid of low data rate and high signal range. Wireless LANs (the 802.11 family) provide medium and high data rates in a local area range. Within the body area network (BAN)/personal area network (PAN) range, ultra-wideband (UWB) is expected to supply a quite high data rate, whereas Bluetooth, ZigBee, and infrared fall into the low data rate range. For point-to-point and multipoint wireless communications, 802.16a and 802.16e offer a high data range for communication over a metropolitan area network (MAN). A broad set of applications has been created to make use of the data rates in each range for various cases where wireless communication is preferred.

3.10 802.11 Wireless LANs

A wireless LAN is a local area network that utilizes radio frequency communication to permit data transmission among fixed, nomadic, or moving computers. Wireless LANs can be dividend into two operational modes: infrastructure mode and *ad hoc* mode, depending on how the network is formed. Most wireless LANs operate in infrastructure mode. In many cases, a wireless LAN is used to avoid the hassle of establishing a wired LAN (*e.g.*, cabling in a multiroom building or a large open space such as a warehouse or a manufacturing plant). Several computers are connected over the air to a central access point that in turn links to the wired network. At the same time, a laptop computer with a wireless LAN interface is able to access the backend wired network across different access points in an intermittent or real-time fashion. In all these scenarios, a wireless LAN infrastructure of networked access points is needed. These access points may connect directly to each other via wireless links or rely on the wired network for interconnection.

Ad hoc mode is more flexible than infrastructure mode in that it does not require any central or distributed infrastructure devices or computers to operate. Instead, computers in an *ad hoc* wireless LAN temporarily self-organize into a group to serve each other in a peer-to-peer manner. In some cases when it is not feasible to build a network infrastructure for technical or other reasons (*e.g.*, troops on the battlefield or sports spectators in a huge stadium), an *ad hoc* wireless LAN seems a good solution.

Today, the dominant radio frequency technology used to build a wireless LAN is a spread spectrum on the unlicensed 2.4-GHz frequency band, as defined in the IEEE 802.11 standards and ETSI HIPERLAN (High Performance Radio Local Access Network). Other radio frequency technologies such as infrared wireless LANs and narrowband microwave LANs have faded away following the explosive growth of spread spectrum wireless LANs. The following is a list of advantages of radio frequency wireless LANs over infrared; narrowband microwave LANs are not considered because they are primarily used for point-to-point wireless communication rather than group communication:

- High bandwidth—802.11 wireless LANs support a link bandwidth up to 11 Mbps for 802.11b and 55 Mbps for 802.11a and HIPER-LAN2, much higher than that of infrared, which is only up to several megabits per second.
- No line-of-sight (LOS) restriction — Infrared requires LOS for transmission, but radio does not as long as the frequency in use is not too high. This is the major reason why 802.11 wireless LANs are the number one choice for home networking.
- Easy to set up and use — The 802.11 protocols are designed to allow almost zero configuration of the network and the interfaces. Of course, the default setting is by no means secure but it does work.

3.10.1 Architecture and Protocols

According to the IEEE 802.11 standard, a basic service set (BSS) is a number of computers equipped with wireless LAN interfaces connecting to an access point (AP). Multiple access points can be connected to a wired or wireless distribution system (DS), whereas several BSSs interconnected via the distribution system comprise an extended service set (ESS) uniquely identified by an extended service set identifier (ESSID) or SSID. The APs may broadcast the ESSID such that anyone within its coverage is able to discover it and configure the wireless LAN interface to participate in the ESS. If the ESSID is not broadcast, users have to know it from other sources in order to access the network.

In many ways, wireless LANs are designed to be the wireless equivalent of LANs such as Ethernet; consequently, similar to other 802 LAN standards, these IEEE 802.11 standards define two bottom layers of protocols: PHY (physical layer) and MAC (medium access layer), retaining the upper layers of the TCP/IP stack. These two layers essentially hide the underlying low-level details of data transmission and medium access, supplying the same interface to the logical link control (LLC) sublayer of data link as that of a wired LAN. Hence, applications will not detect any difference when running in a wired LAN or a wireless LAN. The various 802.11 wireless LAN standards differ only in physical layer (*i.e.*, frequency band being used), signal multiplexing schemes, modulation schemes, and data rates.

According to the MAC layer of 802.11, to ensure reliable frame transmission between two stations an ACK (acknowledgement) frame will be sent from a destination station to the source station when the destination receives a data frame from the source. If no ACK is received at the source, it may simply retransmit the data frame. The MAC layer of 802.11 also employs the carrier sense multiple access/collision avoidance (CSMA/CA) medium access mechanism to provide reliable frame transmission service, formally known as the distribution coordination function (DCF), which was defined as a sublayer of MAC. On top of it is a partial sublayer, the point coordination function (PCF), as shown in Figure 3.9. Two traffic services are supported by 802.11: asynchronous data service and time-bounded service. The first is commonly the best effort service, while the latter can guarantee a maximum delay but relies on a centralized polling master to offer contention-free service. DCF supports

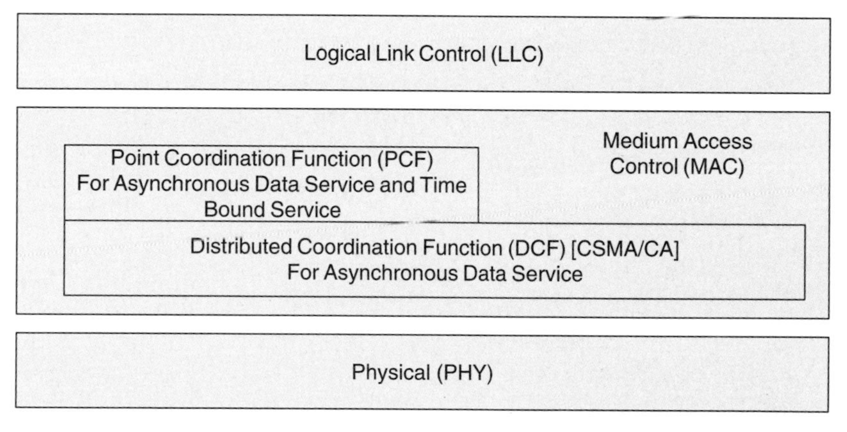

Figure 3.9 802.11 protocol architecture.

asynchronous data services only in either infrastructure mode or *ad hoc* mode, whereas PCF is used for time-bounded service only in infrastructure mode. In particular, in addition to a general CSMA/CA mechanism, DCF has an optional MAC mechanism that addresses the hidden terminal problem and exposed terminal problem (discussed shortly). On the other hand, the contention-free service offered by PCF is realized by a centralized polling master called the point coordinator. 802.11 defines a time interval called superframe that consists of two separated stages, first for the poll and second for regular asynchronous contention-based access. The resulting effect resembles time-division multiple access, where each station receives an evenly distributed share of the bandwidth.

Radio communication is via a shared medium. To use the shared medium, a station must first sense the communication channel (the carrier) and make sure it is not occupied, a procedure called CSMA. If the channel is idle, it can begin to transmit; otherwise, it will wait for a random amount of time with a contention window and sense the channel again. In addition, the radio communication of wireless LANs is a half-duplex operation in that it cannot transmit and receive at the same time. This is because when the station is transmitting the strength of its own transmission will mask all other signals nearby in the air. Recall that, for Ethernet, collision detection (CD) is used (by monitoring abnormal current of the wire) when a station is sending data. In wireless LANs, such an approach is not possible because a station has no way to detect a collision while sending data; therefore, collision avoidance is used instead of CD and is implemented by a three-way handshake protocol described as follows.

Due to spatial limitations of signal strength, a station may not draw a correct conclusion on channel usage, leading to signal interference or channel underutilization. Usually, collisions occur when a station is receiving two signals at the same time. This "hidden terminal" problem is depicted in Figure 3.10a. In the figure, station B initiates a transmission to A, and station C also begins transmitting to A at the same time because C, after sensing the channel, is unaware of B's transmission. Hence, two simultaneous transmissions cause interference at A. In this case, C is a hidden terminal to B.

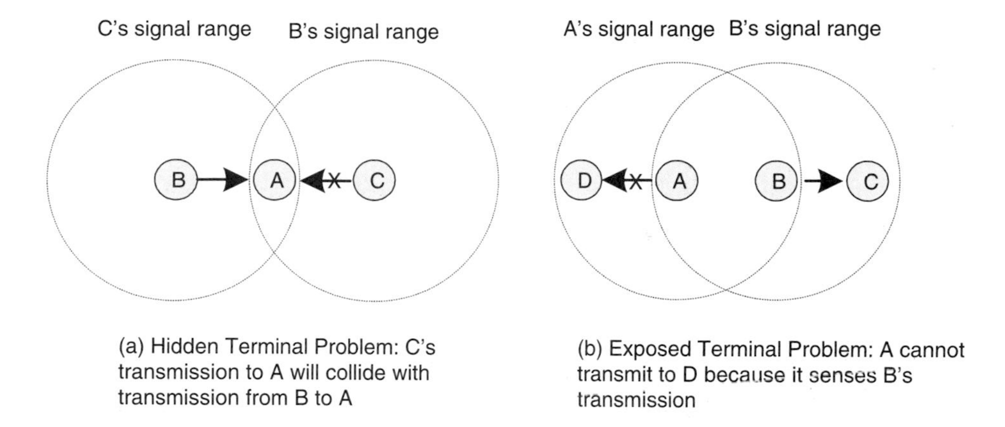

Figure 3.10 (a) Hidden terminal problem and (b) exposed terminal problem.

Another problem is the "exposed terminal," where the transmission of station B to station C effectively prevents B's neighbor (A) from transmitting because A senses the channel and finds B is using it; however, A, the exposed terminal, should be able to transmit to D without any problem, as A cannot reach C and the transmission from A to D should not affect the ongoing transmission from B to C. The exposed terminal problem causes channels to be underutilized and thereby reduces the data throughput of the system. Figure 3.10b shows the exposed terminal problem.

To address these problems, some sort of coordination of channel use among stations is needed. The MAC layer of 802.11 solved the problems by introducing a handshake protocol along with small RTS and CTS frames. A request-to-send (RTS) frame is first sent from the source to the destination. Then, if the destination is not receiving, it replies with a clear-to-send (CTS) frame. Upon receiving the CTS frame, data frame transmission can begin. With this enhancement, both the hidden terminal and exposed terminal problems are avoided. For the hidden terminal problem, C will not receive CTS until the transmission from B to A is over. Thus, it will refrain from transmission to A during that period. For the exposed terminal problem, between B and C, B first sends an RTS to C. A also receives this RTS. Then C sends a CTS message to B indicating that B's transmission to C can begin. This CTS message will not reach A, meaning that the RTS it received belongs to a transmission beyond its coverage (B to C, in this case). Thus, A can safely proceed to send frames to D. Note that RTS and its CTS both contain the time duration of the data transmission plus ACK transmission. Stations that must refrain from this imminent data transmission will have to wait until it is over and then contend for the channel (CSMA). Collisions may only occur when multiple stations access the channel by sending RTSs or data frames.

3.10.2 Frame Format

The 802.11 frame format is depicted in Figure 3.11. The figure on the top shows the frame structure of nine fields in bytes. The figure on the bottom shows the first field, the frame control field, in bits. Depending on the control field, some of the other fields may contain specific information:

- Frame control — Indicates the frame type and provides related control information.
- Duration/connection ID — Duration indicates the time in microseconds that the channel is occupied which can be used for network allocation vector computation during the RTS/CTS exchange. The field may also be used as a station ID for power-save poll messages.

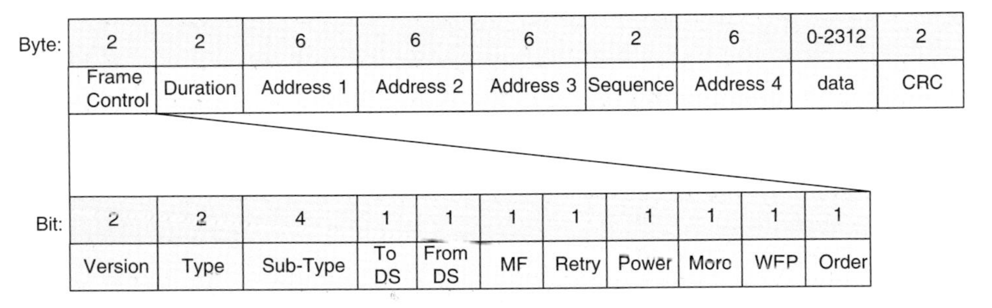

Figure 3.11 802.11 frame format.

- Address 1–4: Up to four 48-bit addresses can be specified in these four fields. The usage of these four fields depends on the To DS and From DS bits in the frame control for both *ad hoc* mode and infrastructure mode. These addresses may specify a frame source station, a frame destination station, a transmitting station or access point, and a receiving station or access point.
- Sequence control — Contains a 4-bit fragment number for fragmentation and reassembly and a 12-bit sequence number to detect duplicates.
- Frame data — Up to 2312 bytes of data can be transmitted with one frame.
- CRC — Contains a 32-bit cyclic redundancy check of the frame.

The following fields are defined in the frame control:

- Version — Indicate 802.11 version number (0).
- Type — These 2 bits indicate frame type: management, control, or data. Management frames are used for communication between stations and access points such as association, dissociation, probing, and beaconing. Control frames are used to provide reliable frame delivery service.
- Subtype — Depending on the type of field, subtype further identifies the function of a frame.
- To DS — This field is set to 1 if the frame is destined to a DS from a station or an access point to another access point.
- From DS — This field is set to 1 if the frame is leaving from a DS for a station or an access point. This field is set when a frame is sent from an access point.
- MF — Indicates that more fragments follow the current one.
- Retry — Indicates that this frame is a retransmission of a previous one that does not has an ACK.

- Power control — This field is set to 1 if the transmission station is in sleep mode; it is set to 0 if it is in active mode.
- More data — Indicates that the transmitter has more data for the intended receiver. If the transmitter is an access point, the field informs a station in sleep mode that more data are buffered at the access point.
- WEP — This field is set to 1 to indicate that the frame body is encrypted using the WEP algorithm (described later).
- Order — This field is set to 1 to inform the receiver that a number of fames are in strict order.

3.10.3 Beacon Frame

802.11 beacon frames are used for a number of purposes including time synchronization, power saving, and access point discovery. Like other types of frames, a beacon frame contains source MAC address and destination MAC address. The source MAC address is the physical address of the access point that sends the beacon, whereas the destination MAC address consists of all 1s for broadcasting. A number of fields make up the beacon frame body, such as beacon interval, timestamp, SSID, supported data rates, wireless LAN capability, and traffic indication map (TIM). Beacon interval and TIM are used for power management. A station may sleep to save power and only wake up periodically to check if the associated access point has some buffered frames destined for it. The beacon interval indicates the next time a beacon from the access point will be sent and whether there are buffered frames at the access point. The TIM will further inform listening stations which ones have buffered unicast frames at the access point. Stations will go back to sleep if no buffered unicast frames are intended for them. Others will stay awake to receive the buffered unicast frames. The timestamp in a beacon frame allows a station to synchronize its local clock with the access point, as time synchronization is needed for DCF and physical layer signal transmission. SSID identifies the underlying ESS. A station can automatically associate to an access point by reading beacon frames from it. Some vendors have an option to disable SSID in beacon frames to enhance security. The supported rates field in a beacon frame tells stations which data rates are supported by the access point. The wireless capability field indicates the requirements of stations that wish to participate in the wireless LAN.

3.10.4 Roaming in a Wireless LAN

Roaming in wireless LANs is conduced in a nomadic way: Connection to the current access point will be lost before a new one is discovered and associated with it. If a mobile station is moving within a set of access points sharing the same ESSID, what is needed is a layer-2 roaming procedure that does not involve IP change. The procedure is also called roaming in a roaming domain or a broadcast domain and is briefly outlined as follows.

- The station senses a roaming operation is necessary based on monitored signal strength, frame acknowledgment, and missed beacons, if any. The client then starts the roaming procedure by looking for adjacent access points.
- The station scans for access points. It can perform active scanning by sending out probe messages on each channel or passive scanning in which it silently listens to beacon messages of an access point. Either approach can be taken before or after the decision to roam. Beacon messages and probe response messages have necessary information used to associate to the new access point.
- The station tries to associate itself with a new access point by sending an association request message. The access point will reply with an association reply message, and the station can then participate in the new BSS.

The 802.11 standard does not specify the algorithm to determine when roaming is needed and the algorithm to scan access points.

Products from different vendors are likely to be incompatible in their support of roaming among different access points. To address this issue, the IEEE 802.11f (Inter Access Point Protocol, or IAPP) working group has been formed and a standard for roaming among APs in the same network segment has been devised. The IAPP protocol defines a set of operations for access points to handle roaming stations. It can be implemented using UDP/IP or SNAP (Sub Network Access Protocol). Two basic operations are *announce* and *handover*. The 802.11f does not support cross-subnet roaming, meaning that vendors will still have to design proprietary protocols for wireless handoff and routing.

Roaming across different roaming domains requires layer 3 operations in addition to layer 2 operations because a mobile station will obtain a new IP address while moving to a new network segment. The association process and authentication based on 802.11i (introduced in Chapter 6) may introduce a delay of hundreds of milliseconds. For ongoing data connections, this does not result in noticeable downtime, but for voice traffic it is noticeable. As voice over wireless LAN is gaining some traction, fast roaming that eliminates noticeable disconnections primarily for voice traffic is needed. To speed up authentication, 802.11i has been enhanced with *key caching*" and *preauthentication*. Key caching caches a user's credential on the authentication server so users do not need to reenter that information when they return to the system. Preauthentication allows a station to be authenticated to an AP before moving to it. On the other hand, IEEE 802.11r working group was formed to solve the fast roaming problem.

Another imminent problem in this realm is roaming among WiFi hotspots operated by wireless internet service providers (WISPs). The Wi-Fi Alliance has published *WISPr Best Current Practices for Wireless Internet Service Provider (WISP) Roaming*. WISPr recommends that WISPs adopt a web-browser-based universal access method (UAM), which essentially specifies some log-on/log-off web pages for user authentication using a web browser. RADIUS (Remote Authentication Dial in User Service) is the preferred AAA protocol for Wi-Fi roaming. AAA data exchange among WISPs allows users to be authenticated and billed for services by their home entities (the users' WISPs). For details of WISPr, please consult the WISPr document at http://www.wi-fialliance.org/opensection/wispr.asp.

3.10.5 IEEE 802.11 Family

Within the IEEE 802.11 family, a few wireless LAN technologies represent the evolution and refinement of wireless LANs. Table 3.6 provides a comparison of these technologies. Note that, to the general public, Wi-Fi is probably the term that links to 802.11 wireless LANs. The Wi-Fi Alliance is a nonprofit industry association formed in 1999 to certify the interoperability of wireless LAN products based on IEEE 802.11 specifications. It has over 200 member companies. The goal of the Wi-Fi Alliance is to enhance the user experience through product interoperability and, understandably, promote the wireless technology and products for business interest.

The initial 802.11:1997 standard contained three incompatible options — infrared, FHSS, and DSSS — to support data rates of 1 to 2 Mbps. For FHSS, 79 channels are allocated in the 2.4-GHz ISM band in the United States and Europe. For DSSS, an 11-chip Barker sequence is used. A Barker sequence is a special binary sequence of −1

Table 3.6 802.11 Wireless LAN Technologies.

	802.11b	802.11a	802.11g	802.11n
Maximum data rate (Mbps)	Up to 11	Up to 54	Up to 54	108–320 Mbps
Frequency band	2.4 GHz	5 GHz	2.4 GHz	2.4 GHz
Modulation	CCK (Complementary Code Keying)	OFDM	CCK and/or OFDM	OFDM
Number of nonoverlapped channels	3	8-12	3	Not yet specified
Range (meter)	100	50	100	100
Power consumption	Low	High	Low	Low

and +1 possessing mathematical characteristics that can be utilized to improve a coding scheme's robustness and error-correction capability. Only a few Barker sequence are known. The one that is 11 in length is used in the initial 802.11 DSSS, which only supports data rates of 1 and 2 Mbps. The updated 802.11b:1999 standard discontinued further specification of infrared and FHSS, focusing instead on only enhancements to DSSS WLANs. 802.11b added new 5.5- and 11-Mbps data rates, based on CCK modulation, a new chip sequence using 8-chip complementary code keying. Wi-Fi-certified products implement DSSS as defined by 802.11b:1999, supporting both 1 and 2 Mbps with the Barker code and 5.5 to 11 Mbps with CCK. 802.11b defines a total number of 14 channels separated by a 5-MHz gap, from 2414 to 2484 MHz, but only 11 are usable due to FCC regulations in the United States. Furthermore, for DSSS to operate, the bandwidth of these channels should be 22 MHz apart in the frequency domain. As a result, only channel 1 (2412 MHz), channel 6 (2437 MHz), and channel 11 (2462 MHz) can be used at the same time. In Europe, these channels are channel 1 (2412 MHz), channel 7 (2442 MHz), and channel 13 (2472 MHz).

In 802.11a, OFDM is used instead of DSSS. 802.11a operates on the UNNI 5-GHz band with a total number of 12 nonoverlapping channels. Channel spacing is 20 MHz. Recall that OFDM leverages multiple carriers (52 in the case of 802.11a) of different frequencies to transmit the same bitstream. Each channel of 802.11a leverages 52 subcarriers that are evenly separated by a distance of 312.5 KHz, plus some virtual subcarriers that are not used. The data rates of 802.11a are 6, 9, 12, 18, 24, 36, 49, and 54 Mbps, each of which is realized by a combination of a specific PSK or QAM digital modulation scheme and OFDM symbol setting.

802.11g wireless LANs operate at the 2.4-GHz band but can offer much higher data rates up to 54 Mbps. To be backward compatible, 802.11g incorporates 802.11b's CCK to achieve bit transfer rates of 5.5 and 11 Mbps in the 2.4 Ghz band. To obtain higher data rates at the 2.4-GHz band, it adopts 802.11a's OFDM scheme. Use of the 2.4-GHz ISM band permits 802.11g to have almost the same signal coverage as 802.11b.

802.11n is the latest wireless LAN standard and promises to offer data rates up to 108 to 320 Mbps at the 2.4-GHz ISM band. As of this writing, no official release has been made by the IEEE 802.11n working group. Two proposals are being considered, and it is unclear which one will finally win. One group, TG nSynch, advocates using a 40-MHz bandwidth for each channel. The competing World Wide Spectrum Efficiency (WWiSE) group wants to retain the 20-MHz bandwidth (as in 802.11b, a, and g) and utilize 2X2 MIMO (two transmitters and two receivers in each device) and OFDM. Recall that MIMO is in essence a spatial-division multiplexing technology that leverages multipath propagation to generate quasi-independent paths in space in order to boost the capacity of the system.

In addition to new wireless LANs, some other 802.11 working groups are focusing on specific issues of general wireless LANs. For example, 802.11c and 802.11d work on wireless switching that enables extension of wireless LANs, while 802.11e emphasizes providing QoS support at the MAC layer for audio and video services. Probably the most notable one is 802.11i, the new security mechanism to replace WEP and intermediate WPA (see below).

High-performance radio LAN (HIPERLAN) is a wireless standard developed by ETSI. HIPERLAN version 1 offers up to 10 Mbps of data rate within a range of 50 meters, targeting the wireless home networking market. HIPERLAN version 2 was actually codeveloped with 802.11a. As a result, HIPERLAN/2 uses the 5-GHz UNNI band and provides data rates up to 54 Mbps. An interesting component of the HIPERLAN/2 is the so-called *convergence layer* defined in its protocol stack. The convergence layer unifies the data link layer (*data link control* layer in HIPERLAN terminology) functionality of various wireless access technologies and provides a unified interface and services to the network layer. This enables a HIPERLAN/2 node to interconnect with heterogeneous networks such as UMTS and the Internet. The standard specifies a cell-based convergence layer for ATM networks and a packet-based convergence layer for general packet-switching networks.

3.10.6 Security in Wireless LANs

Security was undoubtedly the biggest problem of wireless LANs. It has been shown that air serves as an excellent field for network-based hacks and attacks targeting wireless LANs. The topmost security issue of wireless LAN is no security at all. Wireless LANs without any security configuration account for most home wireless networks and many enterprise wireless LANs. In most cases, these open wireless LANs allow any computers with a wireless LAN interface to join the network with little or no trouble. Some security professionals have demonstrated this problem by driving around an area in a city and detecting available wireless LANs along the road, an activity called "war driving" [5]. Another major problem of wireless LANs is the security mechanism of 802.11 protocols: wired equivalent privacy (WEP), the first security mechanism implemented in 802.11b, has proved to have a serious flaw in the key scheduling algorithm that may result in unauthenticated access. New security standards have been devised and incorporated in 802.11. Chapter 6 (Mobile Security and Privacy) presents a detailed discussion of these topics.

3.11 Bluetooth

3.11.1 Architecture and Protocols

Bluetooth is probably the most widely used Wireless Personal Area Network (WPAN) technology now. A WPAN is a small-scale wireless network that connects a few computing or communication devices in the range of several meters. A WPAN may be comprised of a wide range of fixed or mobile devices that have been equipped with radio interfaces, such as computers, cell phones, PDAs, mp3 players, portal game devices, digital cameras, digital camcorders, and so on. The vision of mobile computing encompasses a more rapid and broad proliferation of WPAN technologies in our daily lives that renders convenient and high-performance data access among any intelligent electronic devices. This section and the following section introduce two wireless PAN technologies: the widely used Bluetooth and the emerging ultra-wideband (UWB).

3.11.2 Bluetooth Overview

Harald I. Bluetooth (Danish Harald Blåtand), King of Denmark between 940 and 985 A.D., conquered Norway in the year 960 AD. His "bluetooth" was a result of eating too many blueberries. More than 1000 years later, in 1994, his nickname was used to name a wireless technology that connects cell phones or other devices without using cables. The company that took the initiative to invent the short-range, low-power, and low-cost radio technology is Ericsson. In February 1998, an industry consortium, called SIG (Special Interest Group) of Bluetooth was formed by five companies across three different sectors of the industry: Ericsson, Nokia, IBM, Toshiba, and Intel. Ericsson and Nokia were major cell phone manufacturers, IBM and Toshiba were major laptop computer manufacturers, and Intel's strength was signal processing (in addition to computer processors). In July 1999, the Bluetooth SIG released a 1500-page specification (Bluetooth 1.0). In 2001, the first Bluetooth-enabled products, primarily cell phones and PDAs, were announced. More than 1500 companies adopted Bluetooth for their products. To date, Bluetooth has become the *de facto* short-range wireless technology for mobile devices. The IEEE 802.15 working group for personal area networks has adopted Bluetooth as one of the IEEE 802.15 standards: IEEE Std 802.15.1-2002. Other 802.15 standards are 802.15.2 for the coexistence of WPAN and wireless LAN, 802.15.3 and 802.3a for UWB, and 802.15.4 for ZigBee. The features of Bluetooth are summarized as follows:

- Short range — 10 to 100 m
- Low cost — less than $5
- Low power — 10 to 100 mW
- Low data rate — 1 to 2 Mbps

The interoperable applications of Bluetooth fall into the following categories:

- Cable replacement — Computers are notorious for having cluttered cables for various peripherals such as a keyboard, a mouse, speakers, and a headset. More and more people use an earpiece connected to a cell phone while making a call. It would be far more convenient to have wireless connections between the peripherals and the devices. Bluetooth can be used for this purpose.
- *Ad hoc* data networking — As more mobile devices are used by the general public, *ad hoc* networking capability is often desired to facilitate occasional data transfer and interaction. Bluetooth is designed to allow effortless network setup of a number of compatible devices in a short range.

It is worth noting that Bluetooth and wireless LANs are not exactly targeting the same application scenarios of wireless connectivity, even if both of them (i.e., 802.11b wireless LANs and Bluetooth) operate at the same 2.4-GHz band. Bluetooth by and large is used for power-limited mobile devices for data transfer within a person's reach, which is why it is considered a WPAN technology. Wireless LANs, on the other hand, provide much higher bandwidths over a longer distance but consume more power.

3.11.3 Bluetooth Architecture

The basic unit of a Bluetooth network is a piconet consisting of up to eight devices. One of the devices is the designated master, which actually controls all the other slave devices in terms of radio communication, data transfer, and security mechanisms. Bluetooth uses frequency-hopping spreading spectrum for its radio interface. Each slave will be notified with the same frequency hopping sequence by the master; thus, all devices in a piconet hop simultaneously on the same sequence. Communication between any two of the eight devices has to go through the master in a time-division manner, sharing the 1-Mbps data rate. Multiple piconets can connect together via some overlapping devices, forming a scatternet.

How do two Bluetooth devices communicate with each other? To answer this question, one needs to first look at the Bluetooth protocol stack, which makes up the core portion of the Bluetooth specification. Figure 3.12 shows a simplified protocol stack of Bluetooth. Keep in mind that the official Bluetooth specification has incorporated many new protocols that can be divided into two categories: core specification and profile specification. The core specification is mainly concerned with the physical layer and data link layer, whereas the profile specification covers the applications and functions required for Bluetooth to support data or voice wireless applications. In addition, the Bluetooth protocol stack architecture has been designed to facilitate the operation of many existing commonly used protocols on top of the core protocols. The basic functions of each component in the Bluetooth protocol stack are summarized in Table 3.7. The "Adopted" category refers to existing protocols.

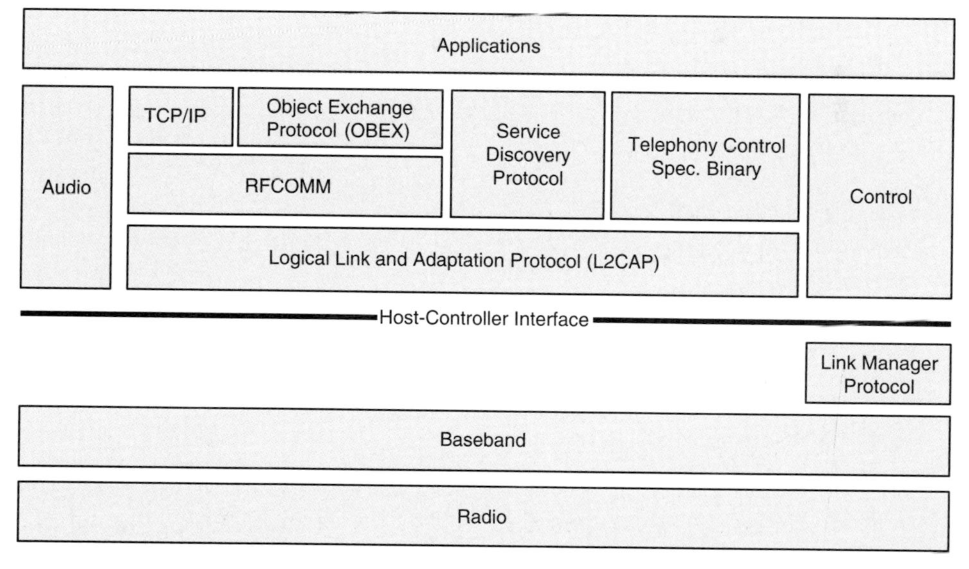

Figure 3.12 Bluetooth protocol stack.

Table 3.7 Bluetooth Protocol Stack Components.

Protocol Stack Component	Category	Description
Radio	Core	Radio signal modulation and transmission
Baseband	Core	Frequency-hopping and time-division multiplexing between a master and its slaves
Link manager protocol (LMP)	Core	Establishment of logical channels, authentication, and power management
Host controller interface	N/A	Software interface to baseband and link manager, as well as Bluetooth hardware status and registers
L2CAP (logical link control and adaptation protocol)	Core	Adaptation of upper layer services and application to the Bluetooth; key component of Bluetooth stack from application developer's point of view
Control	Adopted	Control over link manager protocol
RFCOMM (Radio Frequency Communications)	Cable replacement protocol	Emulation of serial line interface following the widely used EIA-232 (formerly known as RS-232) standard
Service discovery protocol (SDP)	Core	Discovery of services provided on Bluetooth devices
TCS-BIN (telephony control specification, binary)	Cable replacement protocol	Establishment of speech and data calls between a master and a slave
OBEX (object exchange protocol)	Adopted	Exchange of structured data objects such as calendar, vCard (electronic business card), and calendar entries
Audio	Adopted	Supporting audio communication directly on top of baseband

Key components are discussed to explain how Bluetooth supplies desired data services as follows.

3.11.4 Radio and Baseband

The radio layer of Bluetooth utilizes the 2.4-GHz ISM band, the globally free available frequency band, for spread spectrum communication. Seventy-nine MHz of bandwidth is used for frequency hopping, with 1-MHz carrier spacing. The modulation scheme is GFSK at a rate of 1 bit per Hz, providing a data rate of 1 Mbps. The frequency-hopping rate is 1600 hops per second with a dwell time of 625 μsec. As a WPAN, the radio interface of Bluetooth imposes strict emitted power control. Three classes of transmitters are defined based on power and signal transmission range.

- Class 1 outputs a maximum of 100 mW and a minimum of 1 mW for the greatest distance (around 100 m without obstacles). Power control is mandatory.
- Class 2 outputs power between 0.25 mW and 2.4 mW for a range of about 10 m without obstacles. Power control is optional.
- Class 3 outputs around 1 mW with range of a few meters or less.

The power control algorithm can be implemented in the link control protocol and controlled by the *control* component in the protocol stack.

The baseband layer controls transmission of frames in association with frequency hopping. The master in a piconet takes the channel to transmit in even-numbered hops, and slaves transmit in odd-numbered hops, reflecting a time-division duplex for all devices in a piconet. A single frame can be transmitted in the duration of 1, 3, or 5 hops. Depending on the nature of the logical link between a slave and the master, two types of links are offered. One is the asynchronous connectionless (ACL) link for best-effort packet data transmission. The other is synchronous connection oriented

(SCO) for time-critical data such as voice. Frames sent on ACL links may have to be transmitted if lost, whereas frames sent over SCO links will never be retransmitted, necessitating upper layers for error correction.

The baseband layer has defined some types of frames that correspond to various purposes of the baseband frames. Different types of frames can carry different sizes of payload data and error correction schemes. In particular, the access code field in a baseband frame indicates the purpose of the frame in a special state. For example, a frame with the inquiry access code (IAC) will be sent when a device elects to scan for other devices within the radio range in a series of 32 frequency hops. Bluetooth devices can be configured to periodically hop according to the inquiry scan hopping sequence to scan inquires. When an inquiry is detected, the device, now the slave, will reply with its address and timing information to the master, then the master and the slave begin the paging process to determine a common hopping sequence to establish a connection. Eventually, both the master and slave will hop on the same sequence of channels for the duration of the connection.

3.11.5 L2CAP and Frame Format

The L2CAP layer loosely matches the data-link layer of the OSI model. Apart from framing and multiplexing of packet streams, it also supplies QoS for the ACL links. Two alternatives services are provided:

- Connectionless service — Datagram-like service without establishing a connection
- Connection-mode service — A connection is required before a data exchange between the master and a slave

Three types of channels are provided by the L2CAP layer. For each type, a channel identification (CID) is assigned to identify the channel in use:

- Connectionless — Unidirectional channel used primarily for broadcasting by the master. A slave can only have one connectionless channel to the master. Its CID is 2. Connectionless channels are used to implement connectionless service.
- Connection-oriented — Full-duplex channel for connection-mode service. Between a slave and the master there can be multiple connection-oriented channels, each of which is identified by a unique CID larger than 63.
- Signaling — This is not for data exchange but for signaling between L2CAP entities.

Protocol Data Units (PDUs) handled by the L2CAP layer are of a similar format across the three types of channels. In addition to the payload data (in the case of a signaling command PDU, the payload is the command representation), a field of PDU length and a field of CID are encapsulated. CIDs of connection-oriented channels are used to conduct multiplexing and demultiplexing of upper layer data sources. For connectionless channels, a PDU that is carried by the channel has a protocol/service multiplexing (PSM) field to indicate its upper layer source. On the transmitter side, the L2CAP-layer PDUs may be fragmented into small segments if the underlying logical channel cannot send packets of that length.

3.11.6 RFCOMM

RFCOMM (Radio Frequency Communications) is a cable replacement protocol that can be used to connect two Bluetooth devices using a virtual serial line interface. It emulates the 9-pin circuit of an RS-232 interface. Multiple emulated serial connections (up to 60) can be multiplexed into the same Bluetooth connection, while the actual number of connections supported is implementation specific. A complete communication path involves two applications running on two devices with a communication segment between them. The applications utilizing RFCOMM treat the connection as a regular serial line connection via one of its serial ports.

In Bluetooth, a profile is a set of interrelated protocols and pertinent parameters that are chosen for a specific user case. The profile that accounts for virtual serial line communication is the serial port profile, which includes RFCOMM, SDP, LMP, and L2CAP in addition to baseband and radio. The serial port profile essentially defines a point-to-point wireless link between two Bluetooth devices that can be used by the general network layer. The two Bluetooth devices are called *endpoints*, each identified by a unique address. SDP is the protocol used to obtain the address of the other endpoint.

3.11.7 SDP

A service is a shared function that provides some data and performs an operation on behalf of a consumer. Service discovery is a key issue in an *ad hoc* network environment such as Bluetooth piconet or point-to-point direct communication. The service discovery protocol (SDP) in Bluetooth defines a simple request-and-response mechanism that uses service records and service classes for service discovery and browsing. A Bluetooth device that is configured to offer a service should implement an SDP server. The service is described in a service record with a number of service attributes. A service record is identified by a 32-bit handle. A service attribute consists of an attribute ID and a value. SDP defines the following attributes: ServiceRecordHandle, ServiceClassIDList, ServiceRecordState, ServiceID, ProtocolDescriptionList, BrowseGroupList, LanguageBaseAttributeIDList, ServiceInfoTimeToLive, ServiceAvaliability, BluetoothProfileDescriptorList, DocumentationURL, ClientExecutibleURL, IconURL, ServiceName, ServiceDescription, and ProviderName. Semantics of the attributes are further structured into service classes. As a result, a service record must have a ServiceClassIDList attribute that contains a list of service classes representing the general and exact descriptions of capabilities of the underlying service. Each class ID is a universally unique identifier (UUID) that is guaranteed to be unique in space and time.

Before service discovery between two Bluetooth devices occurs, they must be powered-on and initialized such that a Bluetooth link between them can be established, which may require the discovery of the address of the other device by initiating an inquiry process and the paging of the other device, as introduced in the previous section. Then a client can search for desirable services using a list of service attributes or browse the services offered by an SDP server by issuing a specific UUID of the BrowseGroupList attribute that represents the root browse service group of the SDP server. All services that may be browsed at the top level are members of the root browse group.

3.11.8 Bluetooth Evolution

Since its inception, the Bluetooth SIG has made significant effort to improve and promote the technology. In response to feedbacks on Bluetooth specification 1.1, the Bluetooth SIG has released verions 1.2 and 2.0 of the specifications. Enhancements of Bluetooth include high data rates, interference resistance, and security. As shown in the protocol stack, Bluetooth supports both voice and data, and audio communication can be built directly on top of baseband. Bluetooth audio communication provides two types of encoding schemes: PCM and continuously variable slope delta (CVSD). The voice channels support 64 Kbps. A piconet can have up to three simultaneous full duplex voice channels. For asymmetric data transmission, the data rate can be as high as 721 Kbps one way and 57.6 Kbps the other way. For symmetric data transmission, the maximum data rate is 432.6 Kbps. Bluetooth 1.2 and 2.0 are expected to support a maximum data rate of 2.1 Mbps. The 2.4-GHz ISM band is used by many wireless enabled devices; thus, the potential interference between Bluetooth devices and others such as wireless LANs has to be addressed. Bluetooth 1.2 incorporates adaptive frequency hopping (AFH), which allows the selection of idle frequencies for frequency hopping, thereby improving resistance to interference.

Bluetooth security has been criticized to some extent due to the user's lack of total control over wireless connections and data transmission. Bluetooth provides link-level authentication and encryption using unit address, a secret authentication key, a secret privacy key, and a random number. A number of concerns have been raised over Bluetooth security mechanisms as a result of a few proof-of-concept attacks on communication and user data, such as Bluesnarfing and Bluejacking. This topic is discussed in more detail in Chapter 6.

3.12 Ultra-Wideband

Ultra-wideband (UWB) is a disruptive short-range radio frequency wireless technology that could provide a potential solution to many problems in the WPAN communication and computing domain, such as low data rate and insufficient frequency. Despite the standardization controversy with regard to UWB, commercial UWB products were demonstrated at the Consumer Electronics Show in early 2005. Prototypes of UWB-enabled cell phones, HDTVs, DVD players, and music players are expected to hit the market very soon. One example of such an effort is the wireless USB technology, a short-range wireless connectivity technology resembling the wired USB standard. UWB was initially developed in the 1960s for high-resolution radar communication. The primary inventor of UWB was Gerald Ross, who held several patents for this technology. UWB was originally referred to as "baseband," "carrier-free," or "impulse." In 1978, Bennett and Ross published a seminal paper on UWB titled *Time-Domain Electromagnetics and Its Applications*. The year 1986 saw the birth of the first UWB system, and the FCC approved the marketing and operation of UWB in 2002.

The FCC's First Report and Order [6] defines a UWB device as any device emitting signals over a bandwidth that is 20% greater than the center frequency or a bandwidth of at least 500 MHz at all times of transmission within a frequency band between 3.1 and 10.6 GHz, as shown in Figure 3.13. UWB devices operate by emitting

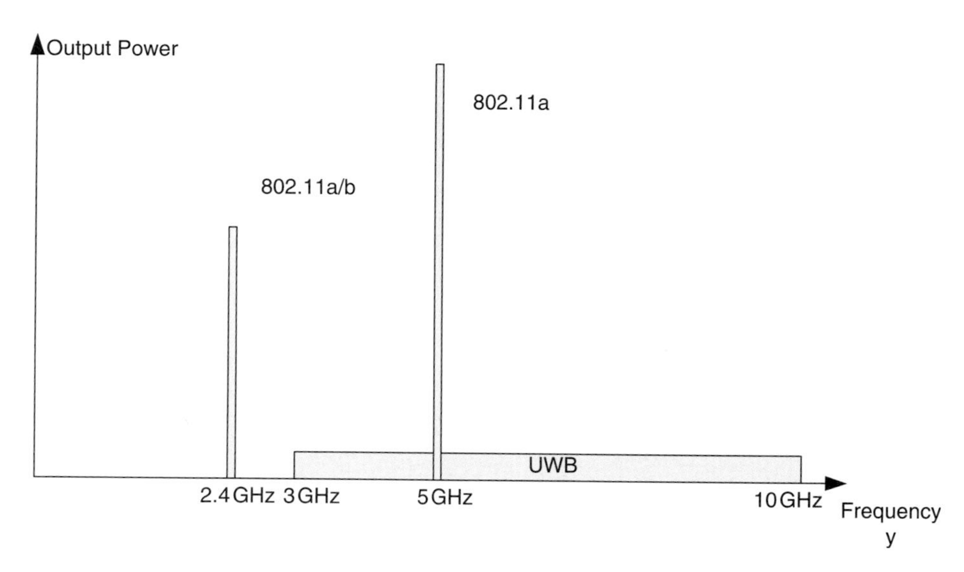

Figure 3.13 Ultra-wideband.

a large number of very short pulses (often of a duration of only nanoseconds or less) of signals over a wide bandwidth within a range of 10 m, resulting in an unprecedented data rate on the level of several hundred megabits per second. UWB does not require any dedicated frequency allocation. Instead, it is designed to operate in frequency spectrum occupied by existing radio technologies. The channel capacity of UWB is linearly proportional to the bandwidth occupied for signal transmission. The advantages of UWB include:

- High data capacity — Due to the use of wide bandwidth, UWB offers very high data capacity up to several gigabits per second.
- Use of a license-exempt frequency band — As a short-range wireless technology, UWB does not require any licensed frequencies to operate.
- Low power — The output power of UWB is at the level of less than 1 mW, compared with tens to a few hundred milliwatts of wireless LAN access points; typically 3 mW is allowed for a cell phone.

- Resilient to multi-path fading and distortions — Because the signal is transmitted over a wide bandwidth with sufficient redundancy, fading and distortion are significantly reduced.
- Security — UWB is inherently secure. Like other spread spectrum technologies, the signal appears to be random noise to outsiders.

3.12.1 UWB Standards

The first standard for UWB was IEEE 802.15.3, which was released in 2002. This standard does not offer many advantages for wireless services and applications, as the data rate can be easily obtained in a wireless LAN (See Table 3.8). A new workgroup, IEEE 802.15.3a, was formed to address high data rate UWB standardization. Two major proposals from different groups of companies are under review: multiband OFDM (MB-UWB) and DS-CDMA. Table 3.8 provides a comparison of the 802.15.3 UWB system and the forthcoming 802.15.3a UWB system.

3.12.2 UWB Applications

In addition to military use of the UWB technology, many companies are working to bring UWB to industrial operations and to people's daily lives. The application scenarios of UWB for the consumer

Table 3.8 Ultra-Wideband System Comparison.

	802.15.3	802.15.3a Proposals
Frequency allocation	3.1–10.6 GHz	3.1–10.6 GHz
Channel bandwidth	15 MHz	500 MHz–7.5 GHz
Number of radio channels	5	1–15
Spreading	—	Multiband OFDM or DS-CDMA
Digital modulation	QAM and DQPSK	BPSK and QPSK
Range	10 m	2–10 m
Data rate	11 Mbps (QPSK) to 55 Mbps (64 QAM) with a minimum of 22 Mbps	110 Mbps at 10 m, 200 Mbps at 4 m, up to 1300 Mbps at 2 m
Emitted power	200 uW	100 uW or –41.3 dBm/MHz

market can be summarized as follows:

- High-speed data transfer between mobile devices in a WPAN — Given a data rate of 100 to 500 Mbps at a distance of 1 to 10 m, computers, PDAs, cell phones, and consumer electronic devices are able to exchange data much faster than via other wireless technologies. The first wave of UWB products will probably target wireless home networks, where interconnection between a wide range of computing, communication, and consumer electronics devices has always been a troublesome problem.
- Cable replacement — It may be possible to link an LCD screen or a television to a computer or any other UWB-enabled electronic devices without using a video cable. The data rate may be further improved to 1 to 2.5 Gbps but only within a short distance of several meters. This would allow video streaming over a number of wireless devices, ranging from handheld mobile devices to computers and HDTVs.
- Wireless measurement in a short distance — An example of this application scenario would be measuring the oil level in a storage tank.
- Location and movement detection — Used in vehicular radar systems, UWB devices can detect the locations of fixed or moving objects near a vehicle. Such information can be used for various applications such as collision avoidance in a parking lot.
  Inventory tracking and supply chain management — Products in a warehouse or a store can have embedded UWB RF tags containing a small amount of data, permitting any UWB readers to access such information.

3.13 Radio Frequency Identification

Radio frequency identification (RFID) is a wireless radiofrequency technology that allows objects, persons, and spaces to be remotely identified using low-cost electromagnetic tags. In its simplest form, an RFID tag attached to an object can store data that can be used to identify the existence of the object or maintain other information regarding the object. The RFID technology has been in place for more than 40 years, primarily being used in a very narrow range of industrial and military applications and remaining unnoticed by the mass market. In the last several years, RFID technology has matured in many ways such as a longer signal range, faster data transfer rate, and shorter tag reading intervals. The reduced cost of RF tags has fostered mass deployment and use of this technology. The retail chain company Wal-Mart was arguably the strongest driving force behind the application of RFID technology. The company requires its top 100 suppliers to have RF tags attached to pallets and cases by 2005 following the Electronic Product Code (EPS).

3.13.1 RFID System

The form factor of RF tags varies largely, but the components that make up a tag are usually the same: a transponder, an antenna, and a tiny integrated circuit (IC). For example, some tags are in the form of planar labels with an aluminum spiral coil on a polymer substrate and delivered in reels, and a flip-chip at its center. Figure 3.14 shows 4 such tags. The antenna of a tag (the dark lines circling the tag in Figure 3.16) is printed on the tag. Depending on how they are powered, RF tags can be divided into two categories: passive tags and active tags. A passive tag does not have a battery; it relies on the interrogation signal from the reader to transmit data back to the reader using a transponder and a small antenna. An active tag has a battery and an onboard transceiver. Compared with passive tags, active tags provide a long signal range but at a high cost.

A typical RFID system is composed of readers and tags, as shown in Figure 13.15. A reader is electronic equipment composed of a transceiver and an antenna. Some reader can even be integrated into a mobile device such as a PDA or a cell phone. A reader is able to interrogate many tags at a time within its transmission range. Once interrogated, a tag will respond with a unique digital ID. The reader

Figure 3.14 High-frequency passive RFID tags. (Photograph courtesy of Texas Instruments; © Copyright 1995–2005 Texas Instruments Incorporated.)

then uses the ID to retrieve corresponding information from a back-end database. Examples of potential IDs that can use RFID include bar codes, license plates, student and employee IDs, conference attendee IDs, passports, ISBNs of publications, and software license keys. Some tags allow the reader to "write" or modifying the stored information.

The frequency bands used by RFID communication are 100 to 500 KHz, 10 to 16 MHz, 850 to 950 MHz, and 2.4 to 5.8 GHz. Modern RFID systems choose to operate on high-frequency bands in order to achieve a higher data rate and longer signal range. The downside of choosing a high-frequency band is the line-of-sight requirement. Collision will occur if two tags respond to the same interrogation from a reader; thus, some readers can only read one tag at a time. Others implement anticollision mechanisms to schedule the tags' responses, making it possible to read multiple tags simultaneously. The standard of RFID radio interfaces for different frequency bands is ISO/IEC 18000 created by ISO/IEC/JTC1/SC31.

3.13.2 RFID Applications

Radio frequency identification technology has already been used by many industry sectors ranging from security systems in a building

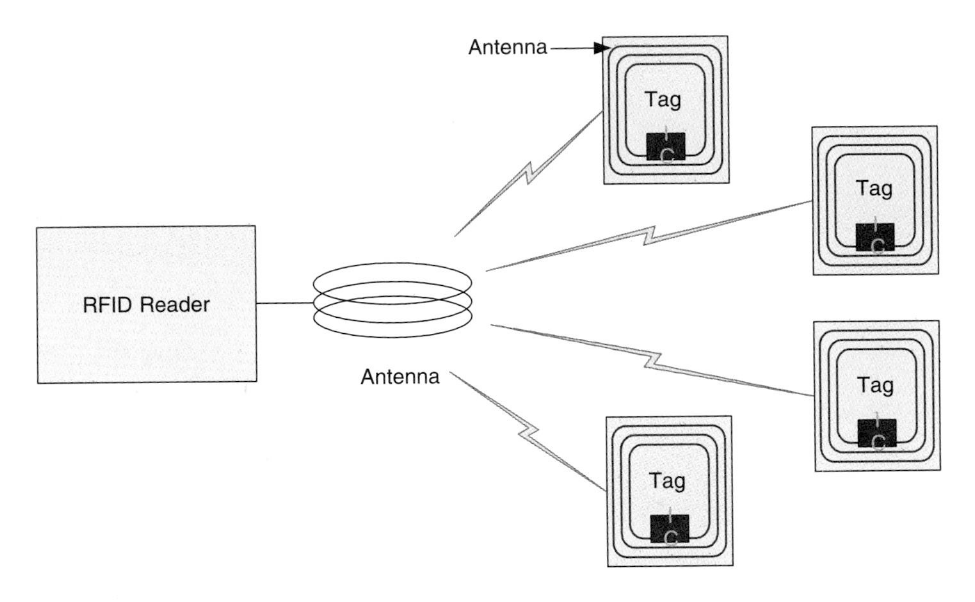

Figure 3.15 An RFID system.

to automatic toll systems on highways. The cost of each passive tag is expected to eventually drop to less than 5¢, thereby making it cost effective to deploy tags for a vast number of objects for tracking and monitoring that otherwise would be highly error prone and labor intensive. Active RFIDs can be used in the following applications:

- Warehouse inventory tracking for logistics and supply chain management (tags can be embedded into pallets, cases, and containers)
- Equipment maintenance in a hospital and part tracking in a factory
- Vehicle tracking and toll payment, such as EZPass on some U.S. highways
- Product and book positioning in a store, warehouse, manufacturing plant, etc.

Passive tags are powered by signals from a reader via magnetic induction. Passive tags can be used in the following applications:

- Automated entry in security systems that currently utilize pass-code entry pads and magnetic card swipe machines (tags would be embedded into personal IDs that are scanned wirelessly at the point of entry)
- Product tagging in a store, replacing bar codes and UPCs (*e.g.*, smart grocery stores and smart department stores)
- Wild animal, livestock, and pet tracking on farms and by pet-control organizations
- Vehicle antitheft systems
- Luggage tagging in airports
- Pallet and case tracking for retailers
- Passport control in customs, airports, and government offices (U.S. passports will soon have embedded RFID tags)

Needless to say many potential applications overlap the two categories. Depending on specific application scenarios and business objectives, users may choose either active tags or passive tags to fulfill their needs in a cost-effective manner. In addition, a smart phone or a smart handheld device with built-in or attached RFID readers can replace specific reader devices in many circumstances. Conversely, a smart phone may contain an RF tag that is mapped to the phone user.

Radio frequency tags may be combined with wireless sensors to extend the capability of the radio frequency system. Embedded sensors in tagged objects allow real-time monitoring of some environmental factors within the proximity of the objects. In addition, in the long run, RF tags may use tiny microprocessors and store more valuable data in local memory. They can even self-organize into a network with collaborative ambient intelligence and in-network information processing without readers.

As is true for many pervasive computing technologies, RFID raises some privacy concerns with regard to tagging and our daily lives. Imagine a world where everything is tagged, and readers are everywhere. From the point of view of a manufacturer or retailer, real-time supply-chain visibility boosts productivity and aids future product development by providing extensive information regarding consumers (*e.g.*, who they are, where they are, and when and how they use the products). But consumers, for the most part, may not want to reveal this information to those manufactures and retailers. Suppose a consumer purchases an electronic shaver that has an embedded RF tag uniquely identifying the shaver. When the consumer travels with the shaver, he may be tracked by any RF readers that can reach the shaver. These privacy issues are very common in the mobile computing domain, as wireless technologies tend to pervade our lives to an unimaginable level of imbedding and integration. Chapter 6 discusses mobile privacy issues in further detail.

3.14 Wireless Metropolitan Area Networks

Wireless MANs refer to a set of wireless data networks that provide wireless data access in a metropolitan area. The principle advantage of building wireless MANs for data access as opposed to establishing a wired network infrastructure is the cost of copper-wire or fiber optic cable, installation, and maintenance. In rural areas and developing countries where telephone lines and cable televisions are not in place, a wireless data access solution is more cost effective than a wired network solution. Depending on how wireless technologies are used in the infrastructure, wireless MANs can be categorized into the following types:

- Wireless "last mile" (fixed broadband wireless access)
- Wireless data access for mobile terminals
- Wireless backbones or wireless mesh

The first type is still based on a wired network infrastructure; that is, base stations connect directly to a backend wired network.

Point-to-multipoint wireless communication replaces wired network communication between a base station and the end-user's computer, the so-called "last mile." Telephone-line-based last-mile access allows dial-up data access and ADSL (with necessary modems), whereas cable-television-based last-mile access permits higher bandwidths and an always-on connection. Dedicated T1 is commonly used by businesses. For the general public, these Internet service providers coined the terms "broadband Internet" or "high-speed Internet access" in order to differentiate high-speed data access services such as ADSL and cable television from traditional dial-up service. In fact, one of the driving forces behind the wireless last-mile technology is that the broadband Internet access of ADSL and cable has grown rapidly in recent years.

The second type of wireless MANs target mobile data access. In a sense, 2.5 and 3G cellular networks could be considered wireless MANs or wireless WANs as they have provided wide-area mobile data access for cell phone users. On the other hand, it would be natural to speculate on extending wireless LANs to cover a larger area and to allow roaming across areas covered by these base stations. Still, this type of wireless MAN relies on a wired network infrastructure to function, as the base stations connect directly to a wired network. Many proprietary wireless MANs have been in operation for years. They mainly target a very narrow business market such mobile professionals, rather than the general public.

The third type of MAN is a pure wireless network, in which backbones as well as the means of access are both wireless. Base stations are not connected to a backend wired network; instead, they coordinate with adjacent base stations, forming a mesh for data forwarding over a wide area. This is a significant development with regard to providing data access services to underdeveloped areas where no fixed networks exist.

3.14.1 Wireless Broadband: IEEE 802.16

The most noticeable technological development in wireless MANs and wireless WANs are embodied by the IEEE 802.16, 802.20, and ETSI HIPERMAN standards. Based on the open IEEE 802.16 and

HIPERMAN, a commercialized technology called WiMax has been devised. The WiMax Forum, an industry consortium of over 100 companies, has been formed to promote the technology and provide certified, interoperable WiMax products. IEEE 802.16 specifies the PHY and MAC layers. It will support higher network layers and transport layer protocols such as ATM, Ethernet, and IP. Characteristics of IEEE 802.16 are listed in Table 3.9.

It is noteworthy that the frequency band of 10 to 66 GHz specified by the initial 802.16 standard requires line-of-sight transmission. Some other frequency bands are also specified in later versions of the standard in order to provide indoor wireless access. The MAC layer portion of 802.16 addresses QoS by introducing a bandwidth request and grant scheme. Terminals can be polled or actively signal the required bandwidth, which is based on traffic QoS parameters. 802.16 employs a public-key infrastructure in conjunction with a digital certificate for authentication.

Extensions of IEEE 802.16 include:

- 802.16a, which specifies a data rate up to 280 kbps per base station over the 2- to 11-GHz frequency band reaching a maximum of 50 km and mesh deployment
- 802.16b, which addresses QoS issues surrounding real-time multimedia traffic
- 802.16c, which defines system profiles that operate at 10 to 66 GHz for interoperability

Table 3.9 IEEE 802.16 Summary.

Feature	Description
Frequency band	10–66 GHz, 2–11 GHz
Range	UP to 40 km (about 30 miles)
Multiplexing/modulation	OFDM, adaptive modulation
Channel data rate	75 Mbps for both uplink and downlink
Antenna	Directional antenna, point-to-multipoint
Multiple access	Demand-assignment multiple access–time-division multiple access (DAMA-TDMA)

- 802.16d, which represents system profile for 802.16a devices
- 802.16e, which standardizes handoff across base stations for mobile data access.

The ETSI HIPERMAN standard is similar to 802.16a. It has been developed in very close cooperation with IEEE 802.16, such that the HIPERMAN standard and IEEE 802.16a standard can work together seamlessly. As a result, many of the characteristics of 802.16 are available in HIPERMAN, such as QoS support, adaptive modulation, and strong security. HIPERMAN supports both point-to-multipoint and mesh network configurations. The differences between HIPERMAN and 802.16 are primarily on the PHY layer. In order to create a single interoperable standard for commercialization, as well as product testing and certification, several leaders in the wireless industry formed the WiMax Forum. Another IEEE working group, called IEEE 802.20 Mobile Broadband Wireless Access (MBWA), uses the 500-MHz to 3.5-GHz frequency band for mobile data access, an application also targeted by 802.16e; however, 802.20 does not have as strong industry support as 802.16 does.

3.14.2 WiMax

The WiMax Forum harmonizes IEEE 802.16 and ETSI HIPERMAN into a WiMax standard. The core components of a WiMax system include the subscriber station (SS), also known as the customer premise environment (CPE), and the base station (BS). A BS and one or more CPEs can form a cell with a point-to-multipoint (P2MP) structure, in which the BS acts as central control over participating CPEs. The WiMax standard specifies the use of licensed and unlicensed bands within the 2- to 11-GHz range, allowing non-line-of-sight (NLOS) transmission, which is highly desired for wireless service deployment, as NLOS does not require high antennas in order to reach remote receivers, which reduces site interference and the deployment cost of CPE. NLOS raises multi-path transmission issues such as signal distortion and interference. WiMax employs a set of technologies to address these issues [7]:

- *OFDM* — As discussed earlier in this chapter, OFDM uses multiple orthogonal narrowband carriers to transmit symbols in parallel, effectively reducing intersymbol interference (ISI) and frequency-selective fading.
- *Subchannelization* — The subchannelization of WiMax uses fewer OFDM carriers in the upstream link of a terminal, but each carrier operates at the same level of the base station. Subchannelization extends the reach of upstream signals from a terminal and reduces its power consumption.
- *Directional antennas* — Directional antennas are advantageous in fixed wireless systems because they are more powerful in picking up signals than are omnidirectional antennas; hence, a fixed CPE typically uses a directional antenna, while a fixed BS may use directional or omnidirectional antennas.
- *Transmit and receive diversity* — WiMax may optionally employ a transmit and receive diversity algorithm to make use of multi-path and reflection using MIMO radio systems.
- *Adaptive modulation* — Adaptive modulation allows the transmitter to adjust modulation schemes based on the SNR of the radio links. For example, if the SNR is 20 dB, 64 QAM will be used to achieve high capacity. If the SNR is 16 dB, 16 QAM will be used, and so on. Other NLOS schemes of WiMax, such as directional antenna and error correction, are also used.
- *Error correction techniques* — WiMax specifies the use of several error correction codes and algorithms to recover frames lost due to frequency-selective fading or burst errors. These codes and algorithms are Reed Solomon Forward Error Correction (FEC), convolutional encoding, interleaving algorithms, and Automatic Repeat Request (ARQ) for frame retransmission.
- *Power control* — In a WiMax system, a BS is able to control power consumption of CPEs by sending power-control codes to them.

The power-control algorithms improve overall performance and minimize power consumption.

- *Security* — Authentication between a BS and an SS is based on the use of X.509 digital certificates with RSA public key authentication. Traffic is encrypted using Counter Mode with Cipher Block Chaining Message Authentication Code Protocol (CCMP) which uses AES (Advanced Encryption Standard) for transmission security and data integrity authentication. WiMax also supports 3DES (Triple Data Encryption Standard).

Initially, the WiMax Forum has focused on fixed wireless access for home and business users using outdoor antennas (CPEs), and indoor fixed access is under development. A base station may serve about 500 subscribers. WiMax vendors have begun to test fixed wireless broadband access in metropolitan areas such as Seattle. Due to its relatively high cost, the major targets of this technology are business users who want an alternative to T1, rather than residential home users. A second goal of the forum is to address portable wireless access without mobility support, and another is to achieve mobile access with seamless mobility support (802.16e). Recall that a WiFi hotspot offers wireless LAN access within a limited coverage of an access point; the WiMax Forum plans to build *MetroZones* that allow portable broadband wireless access. A MetroZone is comprised of base stations connected to each other via line-of-sight wireless links, and 802.16 interfaces for laptop computers or PDAs that connect to the "best" base station for portable data access. This aspect of WiMax seems more compelling in terms of potential data rate compared with 3G cellular systems.

Like the WiFi forum, the WiMax forum aims at providing certification of WiMax products in order to guarantee interoperability. In March 2005, Alvarion, Airspan, and Redline began to conduct the industry's first WiMAX interoperability test. WiMax chips for fixed CPEs and base stations developed by Intel will be released in the second half of 2005, and WiMax chips for mobile devices will be released in 2007. As the time of this writing, some WiMax systems were expected to go into trial operation in late 2005.

3.15 Satellite

Global wireless communication is comprised of two elements: terrestrial communication and satellite communication. Cellular networks are primarily terrestrial based, consisting of a vast number of base stations across heavily populated areas. In some circumstances, such as research laboratories established in the Antarctic, satellite communication is the only means of communication. Some other applications of satellite communication include military satellite espionage, global television broadcast, satellite radio, meteorological satellite imaging, and global positioning system (GPS). In addition, satellite complements cellular networks in reaching far rural areas and have been integrated into worldwide GSM and CDMA systems.

3.15.1 Satellite Communication

Despite the advantage of providing global coverage, satellite communication is known to have significant drawbacks. For one thing, satellite links introduce greater propagation latency than fiber optic links due to the much longer distance a signal must travel back and forth between a terminal and a satellite. A delay of even half a second when using a geostationary satellite phone is noticeable. Bandwidth is another downside of satellite communication compared to terrestrial wired or wireless communications. Although a single satellite may cover a large geographical area (known as the "footprint"), the cost of the entire system remains extremely high, making its acceptability by the general public economically impossible.

3.15.2 Satellite Systems

Satellites orbit the Earth at different heights in various periods. The higher the satellite, the longer the period of the satellite will be.

Table 3.10 Satellite Systems.

	Geostationary (GEO)	Medium Earth Orbit (MEO)	Low Earth Orbit (LEO)
Orbit height (km)	36,000	5000–20,000	1000–2000
Orbit period (hours)	24	6	1.5–2
Number of satellites required to cover the Earth	3	12	>66
Frequency band	L, S, C, Ku, Ka[a]	L[a]	L[a]

See text for description of frequency bands.

The orbits can be circles or eclipses. Earlier satellites were composed of transponders that received signals on one frequency and transmitted them on another. Digital technologies were introduced later to allow improved quality of the signals and more reliable communication. Signals transmitted from a satellite to the Earth attenuate proportional to the square of the distance. A variety of atmospheric conditions also influence satellite signal transmission, such as rain absorption and meteors in the space.

Communication satellites can be divided into four categories based on the orbit of the satellite in space: geostationary (GEO) satellite, medium Earth orbit (MEO) satellite, and low Earth orbit (LEO) satellite, as shown in Table 3.10.

Geostationary satellites remain relatively stationary at a height of about 36,000 km. Three of them are required to cover the entire surface of the Earth. The frequency bands allocated for GEO satellite communication by the ITU are L band (1.5-GHz downlink, 1.6-GHz uplink, 15-MHz bandwidth), S band (1.9-GHz downlink, 2.2-GHz uplink, 70-MHz bandwidth), C band (4.0-GHz downlink, 6.0-GHz uplink, 500-MHz bandwidth), Ku band (11-GHz downlink, 14-GHz uplink, 500-MHz bandwidth), and Ka band (20-GHz downlink, 30-GHz uplink, 3500-MHz bandwidth). GEO satellite systems are primarily used for television broadcasting, such as Direct TV and Dish Networks, and mobile communications. The newest member of this family is satellite digital radio, which provides CD-quality music over more than 1000 channels.

Medium Earth orbit satellites orbit the Earth at heights of around 10,000 to 20,000 km. GPS systems use MEO satellites to provide precise location identification with a range of several meters. 24 GPS satellites operated by U.S. Department of Defense orbit the Earth twice a day at a height of about 19,320 km. The civilian use of GPS operates at 1575.42 MHz, part of the L band. A GPS receiver must communicate with at least three GPS satellites in order to compute a specific two-dimensional location via triangulation. With four or more signals from GPS satellites, the receiver is able to calculate a three-dimensional location.

Low Earth orbit satellites are much closer to the surface of the Earth than MEO and GEO satellites. Their period can be as short as 1 or 2 hours. Because of the considerably shorter distance between LEO satellites and receivers, propagation latency is reduced down to about 10 msec; however, to offer global coverage, many more satellites are needed. For example, the Iridium system was originally designed to have 77 satellites in space (element 77 is iridium). The Teledesic project planed to launch 840 LEO satellites. These numbers had to be scaled back in order to keep costs under control. Aimed at reducing the cost of satellites, another system, Globalstar, has 48 satellites and a large number of ground base stations. (It must be noted that Iridium went bankrupt in 1999 as a result of a small user base and high operational cost.) The data rate offered by LEO satellite systems varies from kilobits per second to megabits per second, depending on the target applications.

3.16 Wireless Sensor Networks

Data communication continues to expand in both scope and complexity, from internal communication among the hardware components of an individual computer to intercomputer network communication via wired or wireless BANs, PANs, LANs, MANs, and WANs. At the same time, computers are becoming more closely related to the physical world and human beings, gathering, monitoring, processing, and analyzing data to allow instrumentation and automation and to facilitate decision making. Wireless sensor networks (WSNs) represent networks that are embedded into our physical environments. A sensor is a tiny electronic device that can respond to a physical stimulus and convert it into numeric data. A wireless sensor network is composed of many low-power, low-cost, autonomous sensor nodes interconnected with wireless communication of sensory data. A myriad of measurements can be done by sensors, including environmental properties such as temperatures, humidity, and air pressure; presence, vibration, and motion detection of objects; chemical properties; radiation levels; GPS; light; and acoustic and seismic activities. Data gathering is conducted intermittently at a specified frequency. A sensor node in a WSN possesses sufficient computing power to process sensory data gathered locally or transmitted from other sensor nodes via wireless links. Furthermore, sensor nodes in a WSN self-organize into a network topology, thereby improving robustness and reducing maintenance costs.

3.16.1 WSN Applications

The wide range of sensors and collective instrumental functionality of WSNs, coupled with the underlying wireless networks, make it possible to provide unprecedented levels of data access and associated intelligence, bringing about a new dimension of application for different industry sectors. WSN applications can be divided into three categories [8]: monitoring space, including objects as part of the space; monitoring operation states of objects; and monitoring interactions between objects and space. The first category represents the most common and basic use of WSNs (dealing with physical environments), whereas the second is mainly concerned with a specific entity rather than its surroundings. The third category encompasses more sophisticated monitoring and control over communications and interactions between objects and between an object and its surroundings. Some pilot projects have explored WSNs for a number of different application scenarios. Many potential applications are being developed to leverage WSNs. Some examples are introduced as follows.

3.16.1.1 Environmental Sensing

Using a large number of sensor nodes deployed in a target geographic location, it is possible to derive useful patterns and trends based on datasets collected over time. Examples of environmental sensing are light sensing, microclimate monitoring, traffic monitoring, pollution level monitoring, indoor climate control, and habitat monitoring. Very often users are only concerned with independent characteristics of an entity, such as the number of vehicles passing by during a time period or the propagation speed of some contaminant in a river.

3.16.1.2 Object Sensing

Aside from environmental sensing, sensors can be attached to objects and collect data regarding motion, pressure, or any mechanical, electronic, or biological characteristics of the host. Object sensing is predominantly used in industrial control and maintenance. Examples include structural monitoring of buildings, bridges, vehicles, and airplanes; sensing machinery wear in a factory; industrial asset tracking in warehouses and stores; surveillance in parking lot and streets; crop monitoring; and military-related object sensing in battlefields. In particular, RFID, a scaled-down wireless sensing technology, utilizes small tags of very limited local computing power and storage to identify and inventory objects. Section 13 has presented a detailed introduction to RFID.

3.16.1.3 Sensing with Intelligence

More challenging application scenarios require embedded intelligence that goes beyond raw data sensing, thus requiring the simultaneous sensing of multiple related quantities and in-network processing so as to detect internal interactions between objects. Examples of this category are monitoring wildlife habitats, telemedicine sensing, context-aware pervasive computing using sensors, and disaster management. For instance, researchers at the

University of California, Berkeley, and Intel have developed a successful experimental WSN to monitor petrels on an uninhabited island off the coast of Maine [9,10]. The birds being observed are Leach's store petrels, a type of tiny reclusive seabirds that burrows in sandy soil and emerges only at night. To ornithologists, monitoring and understanding the comings and goings of these birds in a wild area are not simple tasks, as they would have to dig into the birds' burrows for more information. The WSN deployed on the island consists of 190 wireless sensor nodes called *motes*, some of which are located in burrows and others on the ground, and a solar-powered central computer station that collects sensory data from a gateway mote and reports back to a remote site in real-time via satellite links. Sensors on the motes monitor temperature, humidity, barometric pressure, and ambient light. The temperature reading within a burrow can be used to infer if a petrel is present or not. Other data also contribute to our understanding of the behavior of these petrels and their responses to changes in their surroundings.

3.16.2 Wireless Sensor Node

A sensor node is made up of four basic components: sensing unit, processing unit, transceiver unit, and power unit. The sensing unit usually consists of two components: a sensor and an analog-to-digital converters (ADC). The processing unit acts as a tiny computer: a microprocessor and some RAM. The processing unit runs an embedded operating system and executes WSN applications that control the operations of the sensor and communication between sensor nodes. The transceiver unit is a low-power radio operating on an unlicensed frequency band. The power unit is a battery for regular sensor nodes. Note that in most cases a WSN will have a special sensor node that acts as the gateway for other sensor nodes with respect to ultimate data delivery. The gateway node interfaces to computers via RS232 or Ethernet links. As a result, the gateway node is different from other regular nodes, in both size and processing functionality, thus requiring more power supply.

Following is a list of sensor node characteristics that affect the design of WSN system architectures and applications:

- *Size* — Sensor nodes are very small, due to advancements in semiconductor technologies.
- *Low power* — Sensor nodes are expected to operate for a long time before the battery drains out. In many cases, it is prohibitive to replace batteries.
- *Autonomous, unattended operations* — Once deployed, sensor nodes should self-organize to work as programmed. Remote reprogramming is sometimes possible.
- *Inexpensive* — Their low cost makes it possible to deploy a large number of sensor notes at a moderate cost.
- *Adaptive to environments and themselves* — Sensor nodes are able to adapt to environmental and status changes.

3.16.3 Self-Organized Networks

The physical layer of a WSN is nothing new: radiofrequency transmission at unlicensed bands. Line-of-sight is not required. The data link layer monitors the channels and transmits frames only when the channel is idle. The network layer and transport layer require more discussion. Like *ad hoc* networks, the routing paths between each two nodes cannot be determined and configured prior to deployment because there is no predefined fixed infrastructure in WSNs. Sensors nodes have to discover multi-hop routes to relevant nodes themselves. This is often done via routing data dissemination, in which packets that contain the transmitter and the distance to the root are flooded in the network. A sensor node, upon receiving such packets, will be able to find a "parent" who is closer to the root; hence, a distribution tree can be generated. Data collection from sensor nodes can be routed back the root following the distribution tree.

Task or query dissemination throughout a sensor network is data-centric in association with data aggregation, a routing scheme known as *directed diffusion* [11]. Sensor nodes are not addressed uniformly using numeric identifications; instead, the addressing and naming schemes are correlated with the application. They are identified by "attribute–value" pairs in their data. A task in the form of some attribute inquiry, is sent out from some nodes in the hopes of obtaining relevant data from other nodes, and then all participating nodes form a routing gradient toward the originators. In the case where a WSN is used as a platform of the sensory database, the applications and underlying routing schemes must support declarative queries, thereby making the detail of in-network query processing and optimization transparent to the user. Power consumption is another crucial factor when it comes to in-network aggregation support of query processing. Sophisticated power-aware query processing and packet routing schemes have been devised to reduce the overall power consumption of a WSN.

Sensory data delivery can be performed in several ways. Sensor nodes can actively report readings periodically to its parent or only report when an event occurs. The delivery procedure can also be initiated by a user issuing a command that is diffused across the network. Depending on the design objectives, a WSN may apply different data delivery models to different sensor nodes. For example, some high-level roots in the distribution tree may employ a request-and-response mechanism for queries, whereas some low-level sensor nodes may simply report data continuously.

Compared with mobile *ad hoc* network, network communication over WSNs imposes additional constraints other than node mobility and power consumption. Sensors node are more prone to failure, and their computational capability and memory capacity are greatly limited. When designing a protocol stack of a WSN, these constraints have to been taken into account. Specifically, because complete raw data forwarding is not necessary in many circumstances, data aggregation may be conducted at various levels of the distribution tree to reduce the amount of data being transferred upward to the gateway while still providing sufficient information to other nodes. Furthermore, data aggregation can be combined with applications of the WSN to further improve the efficiency of data collection and dissemination schemes. This reflects one of the most important characteristics of WSNs: cross-layer design. The well-known sensor operating system is TinyOS [12], which is an open-source, event-based embedded operating system developed at the University of California, Berkeley. TinyOS provides a set of components for networking, memory management, and power management, as well as data acquisition and query processing tools. The programming language supported by TinyOS is nesC, a C-like language for embedded network system development. Issues of wireless sensor networks are further discussed in Chapter 5.

3.16.4 ZigBee

One of the emerging applications of WSN is wireless monitoring and control. ZigBee is a such an application that uses low-power and low-data-rate networked sensors. It was developed by the ZigBee Alliance, an industry association of semiconductor companies and network equipment companies such as Ember, Honeywell, Mitsubishi Electric, Motorola, Samsung, and Philips. It has to be noted that the term *ZigBee* refers to the silent communication between honeybees where the bee dances in a zig-zag pattern to tell others the location, distance, and direction of some newly found food. Wireless sensor network communication somewhat resembles the ZigBee principle in that they must be simple and effective.

The idea is to take advantage of wireless sensors to monitor environments, objects, and human beings and control devices, appliances, and facilities. Wireless sensors make it possible to remotely and conveniently monitor or be notified of operational states or crucial state change of an object, such as a dying battery in a smoke detector and rapidly increasing temperature in a truck carrying frozen goods. Wireless sensor networks in ZigBee are not designed to carry large data transfer due to the limited capability of wireless communication; however, these sensors are able to form a fully functional networks, self-organize for efficient data routing and in-network processing, and self-heal in case of node failure. The initial target markets of ZigBee products are home control, building control, industrial automation, personal healthcare, consumer electronics, PC and peripherals control, etc. Key specs of ZigBee include the following:

- Frequency bands: 868 MHz, 915 MHz, and 2.4 GHz
- Data transfer rate up to 250 Kbps
- Signal transmission range of 10 to 100 m, depending on the sensors being used
- AES encryption of data
- Various ZigBee applications can work with each other
- Low power usage

Unlike UWB or Bluetooth, ZigBee specifications do not define radio interface and data link layer protocols; ZigBee simply uses the IEEE 802.15.4 physical radio standard, as shown in Figure 3.16. The ZigBee network application support layer and application profile are the major components that make up the ZigBee specification. Because ZigBee is a proprietary protocol rather than an open standard like those ratified by IEEE, its fate hinges on how it refines itself to become the *de facto* industry standard. To this end, standardization battles seem inevitable.

3.17 Standardization in the Wireless World

The advent of next-generation mobile computing calls for open standards and platforms to enable interoperability. As has been discussed in this chapter, a full spectrum of wireless technologies is set to be integrated to allow roaming in on unprecedented level. Proprietary technologies do not fit into this new era of convergence, as it would be difficult for them to gain ground to a great extent due to

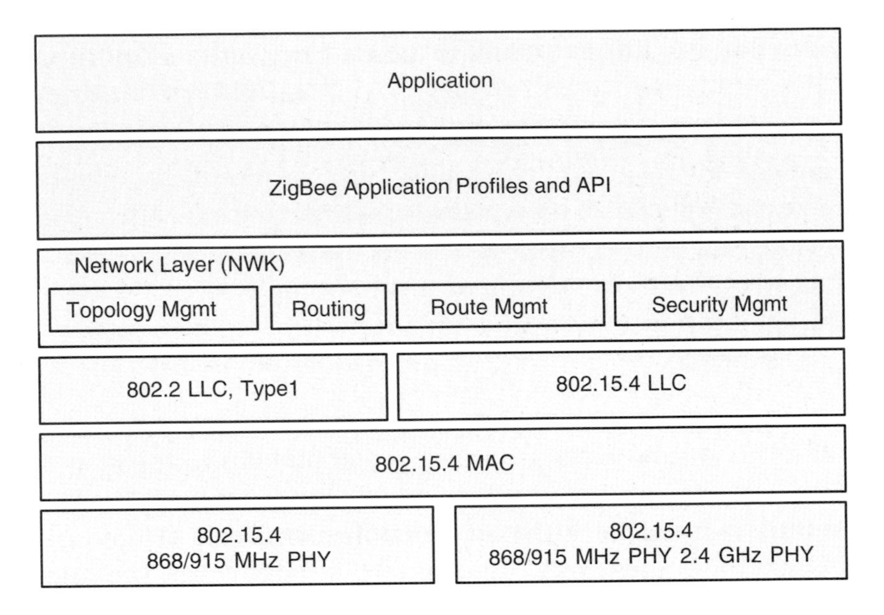

Figure 3.16 ZigBee protocol stack.

the limited number of vendors and compatible products. On the contrary, open, well-crafted standards for the technology will enable and encourage any interested business parties to engage in developing and manufacturing products or offering services that are guaranteed to be interchangeable or compatible. Open standards essentially provide a solid foundation of framework of a technology as well as design constraints, thereby boosting the spread and acceptance of the technology.

A standard is a specification or definition that has been approved by a recognized standards organization such as ITU, IEEE, and ETSI, or is generally accepted as a *de facto* model by the industry. In the context of computing, standards exist for computer hardware, communication protocols, programming languages, operating systems, and some applications. Network communications have a wide range of standards, such as IEEE 802.3 Ethernet standard for LANs, IEEE 802.11 and ETSI HIPERLAN for wireless LANs, GSM, and IS-95 and IS136.

In addition to communication standard bodies such as ITU, IEEE, and ETSI, some other standards bodies have been founded for specialized technological fields. The American National Standards Institute (ANSI) is primarily responsible for software and programming language standardization; it has created ANSI C and C++. HTML and XML have been adopted by the International Organization for Standardization (ISO) and the World Wide Web Consortium (W3C). The Internet Engineering Task Force (IETF) has released a number of requests for comments (RFCs) that serve as the basis of many network protocols. Many computer peripheral standards such as the Personal Computer Memory Card International Association (PCMCIA), Universal Serial Bus (USB), and compact flash have been created by industrial forums or associations.

3.17.1 Cellular Standard Groups

The two standards bodies behind competing cellular technologies are the Third Generation Partnership Project (3GPP) and Third Generation Partnership Project 2 (3GPP2). 3GPP is an international organization supporting the development of UMTS/WCDMA systems. 3GPP partners include ETSI of Europe, ATIS of the United States, ARIB and TTC of Japan, TTA of Korea, and CCSA of China. 3GPP has released two versions of UMTS standards, namely Release 99 and Release 2000. 3GPP2 is the parallel partnership project for cdma2000 technology. It consists of TIA of the United States, ARIB and TTC of Japan, TTA of Korea, and CCSA of China. ITU is a United Nations organization responsible for maintaining and extending worldwide coordination of different governments and private sectors and managing THE radiofrequency spectrum. 3GPP and 3GPP2 are formed under ITU.

3.17.2 IEEE Standards

The Institute of Electrical and Electronic Engineers (IEEE) has been the key standards organization in promoting networking technologies for many years. For wireless technologies, IEEE has established

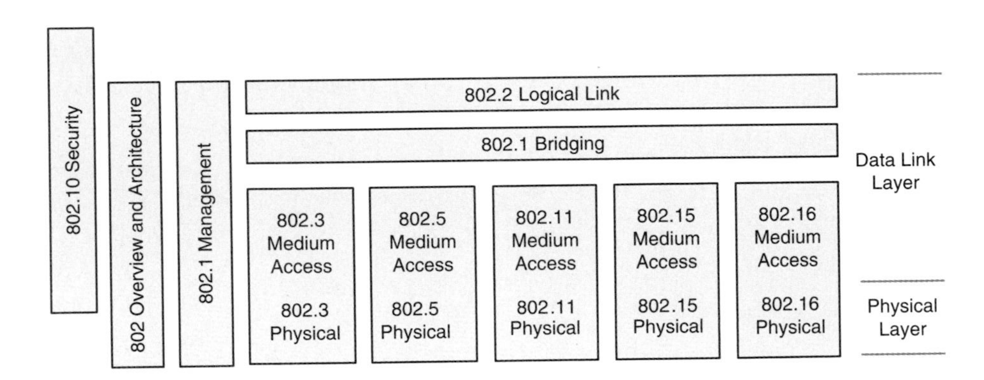

Figure 3.17 IEEE 802 Standards.

several working groups mainly under the 802 standard committee. Figure 3.17 shows an overview of 802 standards. Below is a list of such working groups and their tasks:

- IEEE 802.1 — LAN/MAN architecture with emphasis on internetworking and link security (inactive)
- IEEE 802.2 — Logical link control, part of the data link layer protocol of a LAN
- IEEE 802.3 — Ethernet, the dominating LAN technology
- IEEE 802.4 — Token bus, a LAN technology utilizing token rings over coaxial cables
- IEEE 802.5 — Token ring, another token ring LAN technology (inactive)
- IEEE 802.6 — Metropolitan area networks, a specification of MANs using Distributed Queue Dual Bus (DQDB) (inactive)
- IEEE 802.7 — Broadband TAG (Technical Advisory Group), a broadband LAN
- IEEE 802.8 — Fiberoptic TAG, a fiber optic LAN standard (inactive)
- IEEE 802.9 — Isochronous LAN, an Isochronous Ethernet (IsoEnet) (inactive)

- IEEE 802.10 — Security, specifying key management, access control, and data integrity for LANs and WAN (inactive)
- IEEE 802.11 — Wireless LAN, a set of protocols for wireless LANs operating on unlicensed 2.4-GHz and 5-GHz bands
- IEEE 802.12 — Demand priority, 100BaseVG-AnyLAN (inactive)
- IEEE 802.13 — Not used (for some reason)
- IEEE 802.14 — Cable data, a MAC layer specification for multimedia traffic over hybrid fiber and coaxial networks
- IEEE 802.15 — Wireless PAN, a set of protocols for short-range wireless networks, including Bluetooth (802.15.1)
- IEEE 802.16 — Broadband wireless access, PHY and MAC layer protocols for point-to-multipoint broadband wireless access; WiMax is based on 802.16
- IEEE 802.17 — Resilient packet ring (RPR), a protocol to improve resilience for packet data traffic over fiber rings
- IEEE 802.20 — Mobile Broadband Wireless Access, PHY and MAC layer protocols for mobile data access

3.17.3 Standards War

Emerging innovational technologies usually imply huge business opportunities. Different groups of industry alliances always attempt to influence the standardization of these technologies in favor of their own business interests. This sometimes leads to serious conflicts within a standardization body which inevitably put the technology in stalemate and affect the promotion of the underlying technology with respect to providing a unified, interoperable solution framework for interested parties. For instance, the IEEE standardization of UWB (802.15.3) has been deadlocked due to proposals from two rivalry groups: the MBOA Alliance (Intel and TI lead) and UWB Forum (Motorola leads). Each side claims its proposal is superior to the other. Seeing no immediate ratification of a standard, both groups are moving forward to advance their approaches in commercial developments, effectively creating a segmented UWB market. The evolution of cellular networks is another example of a standards war. The lack of a global standard of cellular networks has resulted in two dominating 2G GSM and CDMA systems and two ongoing 3G deployments: UMTS/WCDMA and cdma2000, backed up by two organizations, 3GPP and 3GPP2, respectively. If a united standard agreement cannot be reached by the different groups, it is very likely the market will make the final decision. The standards body will supposedly pick the approach that is the most popular in the marketplace. Interestingly and understandably, it is not always the technically superior approach or system that eventually wins the majority of the market. We have seen this happen with Betamax versus VHS, two competing videotape standards back in the 1990s. It would be interesting to see what will happen to those emerging wireless technology standards.

3.18 Summary

Mobile wireless is probably the most active area in the domain of computing and communications. For one thing, the boundary between data networking and voice communication continues to blur, as voice over IP begins to gain some ground in enterprise networks to replace PBX. The dominance of mobile voice communications in the wireless world, including cellular network services, walkie-talkies, and cordless phones, may begin to change as a result of evident growth of mobile data services and applications, enabled by a multitude of wireless technologies operating within the ranges of BAN, PAN, LAN, and WAN. In addition, mobile communications and computing are taking place in many forms, including interperson, intrasystem, intersystem, and person to system. This implies ubiquitous mobile access to a converged mobile wireless infrastructure from everywhere, all the time, and from all mobile devices.

This chapter talked about the technical details of the state of the art of many wireless technologies, including cellular, wireless LANs, wireless MANs, Bluetooth, UWB, satellite, wireless sensor networks, and RFID. The emphasis was on explaining how the technology works, how it compares to other competing or complementary technologies, and what are the possible applications are that can take advantage of the technology. To readers not familiar with these technologies, thorough basic understanding of these technologies is required before proceeding to the remaining chapters. The next chapter will focus on hardware and software platforms for smart phones. It will cover topics such as mobile processor, memory and storage, input method, display, existing software solutions, and application development environment.

Further Reading

Federal Communication Commission (FCC) Spectrum page, http://www.fcc.gov/oet/spectrum/; National Telecommunications and Information Administration (NITA) Spectrum page, http://www.ntia.doc.gov/osmhome/osmhome.html.

WAP 2.0, http://www.openmobilealliance.org/tech/affiliates/wap/wapindex.html.

3 GPP group, http://www.3gpp.org; MMS specification for stage 1 and stage 2, http://www.3gpp.org/ftp/Specs/html-info/status-report.htm

IEEE 802.11 work group, http://grouper.ieee.org/groups/802/11/; 802.11 wireless LAN standard download site, http://standards.ieee.org/getieee802/802.11.html.

A good survey on IEEE 802.11 standards is William Stallings' IEEE 802.11: wireless LANs from a to n, *IT Prof. Mag.*, 6(5): 2004, page 32–37.

Official Bluetooth membership website, http://www.bluetooth.org/; specifications download, https://www.bluetooth.org/spec/.

IEEE 802.15 working group for WPAN, http://standards.ieee.org/getieee802/802.15.html.

MBOA (Multi Band OFDM Alliance), UWB industry alliance, http://www.multibandofdm.org/.

UWB forum, another camp of UWB vendors, propos ed DS-CDMA, http://www.uwbforum.org/.

EPC Global (RFID code standard body), http://www.epcglobalinc.org/; RFID specifications version 1.0/1.1 can be downloaded from http://www.epcglobalinc.org/standards_technology/specifications.html.

WiMax Forum, http://www.wimaxforum.org; WiMax white papers can be downloaded from http://www.wimaxforum.org/technology/White_Papers/.

Intel's WiMax site, http://www.intel.com/netcomms/technologies/wimax/.

IEEE 802.16 working group, http://grouper.ieee.org/groups/802/16/; a list of 802.16 standards can be accessed at http://grouper.ieee.org/groups/802/16/published.html.

For a survey on wireless sensor networks and applications, see D. Estrin, D. Culler, and K. Pister, Connecting the physical world with pervasive networks, *IEEE Pervasive Comput.*, 1(1): 2002, page 59–69.

ZigBee Alliance, http://www.zigbee.org/; ZigBee whitepapers can be downloaded from http://www.zigbee.org/en/resources/#WhitePapers.

IEEE 802 LAN/MAN standards list, http://standards.ieee.org/catalog/olis/lanman.html.

References

[1] W. Stallings, *Wireless Communications and Networks*, Prentice Hall, Englewood Cliffs, NJ, 2002.

[2] J. D. Vriendt, P. Laine, C. Lerouge, and X. Xu, Mobile network evolution: a revolution on the move, *IEEE Commun.*, 40: 104–111, 2002.

[3] WAP Forum, *WAP 2.0 Technical White Paper,* Open Mobile Alliance, La Jolla, CA, 2002 (http://www.wapforum.org/what/WAPWhite_Paper1.pdf).

[4] S. J. Barnes and S. L. Huff, Rising sun: iMode and the wireless internet, *Commun. ACM*, 46(11):78–84, 2003.

[5] H. Berghel, Wireless infidelity. I. War driving, *Commun. ACM,* 47(9):21–26, 2004.

[6] FCC, Revision of Part 15 of the Commission's Rules Regarding Ultra-Wideband Transmission Systems, First Report and Order, Federal Communications Commission, Washington, D.C., 2002.

[7] WiMax Forum, WiMax's Technology for LOS and NLOS Environment, White Paper, The WiMax Forum, Hillsboro, OR, 2004.

[8] D. Culler, D. Estrin, and M. Srivastrava, Overview of Sensor Networks, *IEEE Computer*, 37(8):41–49, 2004.

[9] A. Mainwaring, J. Polastre, R. Szewczyk, D. Culler, and J. Anderson, Wireless sensor networks for habitat monitoring, in *Proceedings of the ACM International Workshop on Wireless Sensor Networks and Applications (WSNA'02),* Atlanta, Georgia, September 28, 2002.

[10] J. Kumagai, Life of birds: wireless sensor network for bird study, *IEEE Spectrum*, 41(4):42–49, 2004.

[11] C. Intanagonwiwat, R. Govindan, and D. Estrin, Directed diffusion: a scalable and robust communication paradigm for sensor networks, in *Proceedings of the Sixth Annual International Conference on Mobile Computing and Networking (MobiCom'00),* Boston, MA, 2000.

[12] TinyOS Forum, *TinyOS*, 2004 (http://www.tinyos.net).

4 Mobile Terminal Platforms

The previous chapter was dedicated to supporting wireless technologies for implementing next-generation mobile services and applications. Those technologies could be used in a network infrastructure or on mobile terminals. Here, the notion of *mobile terminal* refers to any networked mobile device in the context of general communication systems. As explained in Chapter 1, mobile terminals operate on inherently resource-constrained devices, thus an array of design challenges confronts smart phones for next-generation mobile computing. In order to better understand and address these issues, we must first possess a clear understanding of the constraints and how they influence the role and functionality of mobile terminals such as smart phones; consequently, this chapter will focus on the hardware and software platforms for mobile terminals. Hardware components that make up mobile terminals such as mobile processors, storage and memory, display, battery, and network interface are discussed in detail. In addition, a high-level overview of competing software platforms is presented, including Palm, Symbian, Microsoft Windows Mobile, Linux, J2ME, and BREW. Readers who are familiar with these topics may skip this chapter and go directly to the next chapter.

4.1 Mobile Hardware

As a small computer, a mobile device consists of integrated or interconnected hardware components and software. These hardware components include a microprocessor (referred to as *processor* hereafter), read-only memory (ROM), random access memory (RAM), expansion storage, network interfaces, an antenna, a battery, and a display. Some mobile devices also have a hard disk. Among these hardware components, we are particularly interested in the processor, memory and storage, network interfaces, and display. This section discusses these hardware components to present an overall picture of mobile terminals. Batteries used in mobile devices were introduced in Chapter 2.

4.1.1 Mobile Processors

A mobile device is controlled by a small, embedded computer system. The notion of an *embedded system* refers to a specialized computer system that performs a fixed set of functions to control almost any electronic appliance or equipment ranging from industrial systems to DVD players and microwave ovens. Mobile devices such as cell phones, PDAs, laptop computers, specialized handheld scanners, pagers, walkie-talkies, and wireless-enabled portable gaming devices constitute a large portion of embedded systems.

The number of microprocessors in embedded systems is far greater than processors for desktop computers. Almost all electronic devices now have one or more microprocessors as part of the embedded system, including consumer electronics devices (*e.g.*, mp3 players, digital cameras, and televisions), as well as computer peripherals and network devices such as keyboards, universal serial bus (USB) drives, printers, network switches, routers, wireless access points, and cable modems. The automobile industry consumes a vast number of 8-bit processors; each modern car has about more than a dozen embedded processors controlling the engine, transmission, lighting circuitry, stereo systems, etc. Many embedded processors are used in machine tools and other equipment in manufacturing plants. The microprocessors (CPUs) used in mobile/portable devices are generally referred to as *mobile processors*, a type of embedded processor. The number of mobile processors is huge — just consider the vast number of cell phones, PDAs, laptop computers, and mp3 players alone.

Mobile processors possess several distinct characteristics as follows [1]:

- *Limited programmability* — A mobile processor can only supply limited computing capability (compared to desktop processors) due to power consumption constraints.
- *High I/O to computation ratio* — Network and other I/O communications are more frequent on mobile terminals than on desktop computers.
- *Stream data processing* — Multimedia processing is indispensable on many mobile terminals; hence, mobile processors are required to incorporate those multimedia capabilities.

Just as a few types of processor families dominate the desktop computer processor market, a few mobile processor families have emerged to become the major players in this segment. Following is a summary of these processors.

4.1.1.1 ARM

The advanced reduced instruction set computer (RISC) machine (ARM) is a microprocessor architecture initially developed by a British company called Acron. The ARM Company (a spin-off of Acron), licenses ARM technologies to semiconductor manufacturers and electronics device manufacturers. The ARM core is the leading 32-bit embedded processor architecture for a variety of mobile devices. In particular, the StrongARM SA-100 processor, a processor based on a 32-bit ARM core developed by Digital Equipment Company (DEC), was used in Apple's Newton. Intel's XScale microarchitecture is largely, if not completely, based on StrongARM, as DEC sold its entire chip division to Intel in 1997. It consists of an ARM-compliant execution core with instruction and data memory management units; data caches and buffers; power management, performance monitoring, debug, and JTAGunits; coprocessor interfaces; multiply-accumulate unit coprocessors; and core memory bus. (JTAG commonly refers to IEEE Std 1149.1-1990, IEEE Standard Test Access Port and

Figure 4.1 Intel XScale PXA 270 processor (© 2005 Intel Corporation, Courtesy of Intel Corporation.)

Boundary-Scan Architecture, which was originally proposed by some electronics and semiconductor manufactures as a solution to the problem of board test.) The XScale architecture features Intel's Dynamic Voltage Management technology, which allows operating voltage and frequency scaling on the fly. Figure 4.1 shows an Intel XScale processor. Palm PDAs that previously used Motorola DragonBall processors have begun to switch to StrongARM processors.

4.1.1.2 MIPS

The MIPS family was original developed by the MIPS Company, founded by John Hennessy. Its core is the MIPS RISC architecture. MIPS embedded processors and application-specific integrated circuits (ASICs) are claimed to have the smallest silicon footprint and lowest power consumption. They are widely used in cable modems and videogame devices such as Nintendo Gameboy/DS, and Sony Play Station Portable. The MIPS family offers both 32-bit and 64-bit processors. RISC based PowerPC processors are the outcome of a joint alliance of Apple, IBM, and Motorola. As is true for many other processor lines, PowerPC processors mainly target desktop and laptop computers rather than small mobile devices.

4.1.1.3 x86 and Centrino

Intel's x86 family started with the 8086 microprocessor back in 1978, followed by 8088, 80286, 80386, 80486, Pentium, Pentium Pro™, Celeron™, Pentium II™, Pentium III™, and Pentium IV™. Because of the dominance of x86 processors in the desktop computer market, some other vendors such as AMD and Cyrix have been developing x86-compatible processors for years. These desktop system processors are not suitable for mobile devices, as they consume quite a bit of power, and the microarchitecture is not optimized for mobile computing. To that end, Intel provides mobile processors such as Pentium M as part of the Centrino platform and individual Mobile Pentium II, III, and IV, primarily for laptop computers. Most of these processors are complex instruction set computer (CISC) based, but some of the latest ones have incorporated a significant portion of RISC techniques. Intel's mobile hardware platform, Centrino, consists of Pentium M processors, mobile chipsets, and on-board wireless adaptors for wireless LAN and the upcoming WiMax. For example, at the time of this writing, the latest Centrino platform, code-named Sonoma, features Pentium M processors up to 2.13 GHz, 2-MB on-chip memory, 533-MHz font-side bus, Intel 915 Express Chipset, 802.11 a/b/g adaptors, dual-channel 400/533-MHz memory, graphic accelerators, and high-definition audio. Future versions of Centrino will include WiMax support. AMD's x86 compatible mobile processors, mobile Athlon 64 and Turion 64, differ significantly from Pentium M in their support for 64-bit computing. Despite its remarkable success in the desktop processor market, mobile x86 processors fall far behind ARM and Motorola 68K for mobile computing and are being replaced by Intel's XScale line of processors.

4.1.1.4 Transmeta Crusoe and Efficeon

Cruose™ and Efficeon™ are two mobile processor families developed by Transmeta. These processors utilize a LongRun power management technology that enables processors to change voltage and frequency dynamically to delivery high performance while considerably reducing energy consumption [2]. The LongRun power management technology continuously samples a processor's various sleep states and adjusts the CPU cycle frequency and operational voltage in response to the needs of the operating system and application software. In addition, LongRun power management does not require system BIOS, an operating system, device drivers, or existing applications to use software interfaces in order to take advantage of them. Crusoe and Efficeon are both x86-compatible processors that employ a so-called "code morphing" technique to translate x86 instructions into native instructions of the underlying very long instruction word (VLIW) engine. In this sense, Crusoe can be regarded as a virtual CPU engine. Transmeta's mobile processors are primarily used in systems in which the need for power saving holds a high priority, such as laptop computers, tablet PCs, ultrapersonal computers (handheld computers that run full versions of desktop operating systems), and embedded devices, as well as servers in data centers.

4.1.1.5 Other Mobile Processors

Other microprocessors for mobile devices include Sun Sparc, Motorola 68000, and SuperH, each of which has managed to gain some ground in the mobile device market. For example, SuperH processors from Hitachi are very common on Sega game devices. Some of the Palm PDA lines have used Motorola 68000-based DragonBall processors. Some companies such as Nazomi and ARM have licensed Java technology for mobile devices, notably Java 2 Micro Edition (J2ME) from Sun Microsystems, and have developed hardware/software combined Java chips (namely, Java accelerators) that offload execution of native and Java applications from baseband processors in a mobile device. J2ME is discussed in more detail in Section 4.7.

4.1.2 Mobile Processor Performance

Moore's law, which claims that the computing power of semiconductor chips will double every 18 months without additional cost, surely holds for desktop computer microprocessors, but it is a different situation for mobile processors. Usually, computing power is measured as the number of million instructions per second (MIPS, not to be confused with MIPS processors), but mobile devices have the constraint of limited power, which most likely is provided by a battery. Thus, when measuring the performance of a mobile processor, power consumption has to be taken into account as well. As a result, a new metric, MIPS per watt (MIPS/W), has been devised. Commonly, the MIPS/W values of mobile processors fall within the ranges of 1000 to 2000, and power usage is in the tens of milliwatts to several hundred milliwatts.

Mobile processors such as Intel's XScale offer a clock speed of several hundred MHz to meet the ever-growing need for high computing power of high-end mobile devices such as smart phones. In addition, Intel researchers have demonstrated 1-GHz mobile processors. On the other hand, mobile processors have embraced system-on-chip (SoC) technology, which allows a processor to incorporate a set of distinct functionalities in the same package. For example, Intel's XScale provides digital signal processing and wireless communication in addition to the microprocessor core.

4.1.3 Memory and Storage

Memory represents another dimension of constraints for mobile devices, as it requires a small program footprint for both mobile operating systems and mobile applications. Three types of memory are used in this domain: RAM, ROM, and flash memory. Flash memory is a special form of nonvolatile electrically erasable programmable read-only memory (EEPROM). The prime advantage of flash memory over other regular ROM is that it allows data to be read or erased on a block level, as opposed to the byte level of other ROM types.

Earlier mobile devices usually provide 4 to 16 MB of static RAM (SRAM) for user data storage and 8 to 32 MB EEPROM or flash memory for system code. Today's mobile terminals typically offer much larger memory capacity: 64 to 128 MB SRAM for application code, 128 to 256 MB flash memory for system code, and 128 to 256 MB

Figure 4.2 MultiMedia and Compact Flash memory cards. (Courtesy of Kingston Technology Co.)

flash memory for user data. Some mobile devices use battery-powered nonvolatile RAM (NVRAM) to store data files. Flash memory is also used as external removable storage cards, such as SmartMedia cards, Compact Flash (CF) memory cards, memory sticks, MultiMedia Memory Cards (MMCs), and Secure Digital (SD). Compact Flash memory cards can supply as much as 4 GB of capacity. Figure 4.2 shows a CF memory card of 256 MB and an MMC of 128 MB.

On desktop computers, physical memory is considered first-level storage and hard drives (hard disk) are secondary storage. Memory is much faster for data access but generally is more expensive, whereas a hard disk typically provides much larger storage space. Due to limited physical memory, operating systems employ a virtual memory mechanism to achieve efficient memory management and high overall performance. If desktop systems were able to have tens of gigabytes of physical memory, hard disks would be eliminated completely. In fact, some high-end, mission-critical server systems use huge amounts of physical memory to ensure high performance; however, this is not a cost-effective approach to building general-purpose desktop computer systems. For mobile computer systems, because of the size of the system code and because the application codes are comparatively smaller than those of desktop computer systems, a pure first-level memory design can be implemented at moderate cost to achieve good performance. As a result, today's PDAs and cell phones only use memory for code and data storage.

Mobile devices very often support one or more I/O extension interfaces, thus allowing the use of large flash cards with gigabyte capacity. There are several reasons to use flash memory instead of a hard disk:

- Flash memory allows faster access.
- Flash memory is far smaller and lighter than hard disks.
- Flash memory is quiet.
- Flash memory does not have mechanical parts, which are supposedly not suitable for mobile devices.

On the other hand, hard disks offer much greater capacity than flash memory can provide. The cost of a hard disk is also much lower than a flash memory of the same capacity, if available. As more and more multimedia applications begin to appear on mobile devices, the need for very large storage space will dominate. Eventually a small, high-performance, low-power, and low-cost mobile hard disk will probably be preferred to flash memory.

So far, hard disks are not considered a standard component of a mobile device. Their power consumption is prohibitively high, and data access speed does not satisfy the needs of mobile applications. Ordinary 3.5-inch or 2.5-inch hard disks used in desktop and laptop computers cannot fit into the small packages of mobile devices. Recent breakthrough in hard disk research promises a brighter future for mobile storage. Some Samsung cell phones are reported to have a 1.5-GB hard disk. Portable mp3 players such as Apple's iPod provide a large hard disk of more than 10 GB in a small form factor.

4.1.4 Extension Interfaces

A variety of mobile device extension interfaces have been developed by different industry associations so as to enhance the storage capacity and functionality of a mobile device. These interfaces can be divided into two categories:

- *Memory and storage extensions* — These extensions are only used for code and data storage. Examples are MMC, SD, SmartMedia, MicroDrive, and xD Picture Card.
- *General I/O extensions* — These allow both memory and network cards to be used through the same interface. Examples are SDIO, Compact Flash, and Personal Computer Memory Card International Association (PCMCIA).

Table 4.1 highlights some of the existing extension interfaces supported by cell phones, PDAs, smart phones, digital cameras, and digital video recorders. It should be noted that some vendors provide CF adaptors that permit memory cards to be used on mobile devices with CF interface slots.

4.1.4.1 Input Device

The input device of a mobile device is crucial to the adoption of the device. Below is a list of input devices used on cell phones, PDAs, and other handheld computing devices:

- A *cell phone keypad* is typically a 12-button keypad consisting of keys 0 to 9, each also representing some letters and characters, plus the * key and # key. In addition, a cell phone keypad usually has four function keys (call, hang up, menu, cancel), as well as up and down keys.
- The *QWERTY keyboard* is a tiny version of standard English computer keyboards or typewriters. "QWERTY" refers to the six letters on the top left of a standard keyboard.
- The *alphabetic keyboard* is another type of small keyboard, with keys arranged alphabetically.
- The *stylus-based virtual keyboard* and *handwriting recognition* are mostly popular on PDAs. A user can either use a stylus to write

Table 4.1 Mobile Device Extension Interfaces

Category	Name	Size	Spec
General I/O extensions	Compact Flash (CF)	Connector: 43 mm wide; case: 36 mm deep; thickness: 3.3 mm for CF 1, 5 mm for CF II	Identical to PCMCIA–ATA (Advanced Technology Attachment) interface; CF memory is based on NOR flash memory technology (NOR refers to the logic gate type); it supports up to 4 GB, as of 2004, operating at 3.3 V and 5 V.
	Personal Computer Memory Card International Association (PCMCIA)	85.6 mm long × 54.0 mm wide; thickness: 3.3 mm for type I, 5.0 mm for type II, and 10.5 mm for type III.	68 pins; used for memory and network interfaces; operate at 3.3 V and 5 V.
	Secure Digital Input/Output (SDIO)	Same as SD	SD expanded with network capability including wireless LANs, Bluetooth, Ethernet, scanner, global positioning system (GPS), etc.; also supports SD memory cards.
Memory and storage extensions	MultiMedia Memory Card (MMC)	24 mm × 32 mm × 1.5 mm (roughly the size of a postage stamp)	Based on NAND flash memory technology (NAND refers to the logic gate type); supports up to 2 GB, as of 2004.
	Secure Digital (SD)	24 mm × 32 mm × 2.1 mm; similar to MMC	Based on MMC; provides encryption to allow secure distribution of copyrighted content; supports up to 2 GB, as of 2004.
	SmartMedia	25.0 × 37.0 × 0.76 mm	Based on NAND flash memory technology; supports up to 128 MB.
	Memory stick	Standard: 50.0 mm × 21.5 mm × 2.8 mm; duo: 31.0 mm × 20.0 mm × 1.6 mm	Based on NAND flash memory technology; supports up to 256 MB.
	xD Picture Card (eXtreme Digital)	20 mm × 25 mm × 1.78 mm (about 2.8 g)	Supports 1 GB, as of 2004; only available from Toshiba.

on the device screen or click keys on a virtual keyboard displayed on the screen. This input method eliminates letter keys and number keys.

Cell phones predominantly use the 12-button keypad as their input device. The keypad does not require much space, but entering a single English letter may require two or three presses. Micro keyboards provide a similar experience as on desktop computers, but they also tend to make the device wider and the keys smaller, making it difficult to use. The stylus-based input method is well suited for frequent text input on PDAs but usually cannot be operated with one hand. Smart phones primarily use the cell phone keypad, as phone functionality dominates the usage of such devices. Some PDA-like smart phones use micro keypads or stylus-based input, such as Microsoft Pocket PC™ Phones and the Palm Treo™ series.

4.1.4.2 Display

The display of mobile devices has evolved in a number of directions at a fast pace for many years. The very preliminary low-resolution chromosome displays on early cell phones and PDAs have been replaced by high-resolution color displays. Display screen dimensions vary from device to device, predominantly determined by the purpose of the device. For example, PDAs used to have a much bigger display screen than that of cell phones, because the personal information management (PIM) user interfaces tended to require more display space than the early fairly rudimentary cell phone applications. Mobile consumer electronics devices such as mp3 players usually provide very simple user interfaces on a small screen. The convergence of these mobile devices will surely have a great impact on mobile displays. The aggregation of multiple functions on a single mobile device requires a display that can satisfy the diverse requirements of those functions while retaining a desirable form factor and optimized power consumption. As an example, smart phones often have a sufficiently large display screen for wireless Internet applications as well as for PIM applications. Design choices have to be made with regard to a number of factors:

- *Size* — Size certainly matters. Cell phones usually have 2.2-inch (5.588-cm) diagonal LCD screens with backlights. The diagonal screen sizes for PDAs range from 2 to 10 inches.
- *Resolution* — Low resolutions (176 × 220 or 128 × 160) are still very common among mobile devices; however, QVGA (320 × 240), a quarter of standard VGA (640 × 480), is being supported by more newer smart phones. Some even support VGA on a 2.2-inch screen. The points-per-inch (PPI) value is around 300.
- *Color depth* — It is very common to see consumer mobile devices with a color depth of 12 bit (4096 colors), 16 bit (65536 colors), or 18 bit (262144 colors).
- *Backlight* — For a better display effect, the display screens of cell phones or PDAs usually have a backlight.
- *Power consumption* — Thin-film transistor (TFT) displays tend to consume more power than earlier passive matrix displays. The typical power consumption of displays on mobile devices is around several hundred milliwatts.

Because the display screen of a mobile device accounts for a significant amount of the power consumption of the mobile system, it must be considered in the overall power management scheme. In addition to employing advanced low-power display technologies, mobile software, including the operating systems and applications, must be power aware and adaptively control the display.

A fundamentally different class of display, *flexible display*, is looming on the horizon. Like a piece of electronic paper, a flexible display can be rolled up and even folded. When used on a mobile device, a flexible display can provide a much larger screen size. Research on flexible displays is underway at several companies, such as Philips, Lucent Bell Labs, and Universal Display. These devices utilize advanced material fabricating technologies on plastics, organic transistors, and organic LEDs. For example, researchers at Philips have demonstrated a thin (100 μm), flexible, active-matrix plastic display. These products will target such niche markets as the military and scientific research. Tired of staring at the phone display, smart phone users may try eyewear, a glass-like device that provides a big-screen viewing effect.

4.2 Software Platforms

A mobile software platform is defined as the combination of an operating system for a collection of compatible mobile devices with a set of related software development libraries, application programming interfaces (APIs), and programming tools. A mobile software platform essentially provides a complete solution to application development on mobile devices. Mobile software platforms are either proprietary for special devices or open to all independent software providers. Mobile devices used by consumers such as cell phones and PDAs are generally powered by well-known operating systems in conjunction with a rich set of software development tools and resources. Industrial mobile devices usually have proprietary operating systems designated for a limited range of devices. As mobile hardware technologies and wireless technologies continue to advance, mobile operating systems are required to take advantages of those advancements and provide strong support for application developers. This section looks at six prevalent mobile software platforms:

- Symbian — Symbian OS (Figure 4.3a)
- Palm — Palm OS (Figure 4.3b)
- Microsoft Windows Mobile for Smartphone and Pocket PC with Net Compact Framework (Figure 4.3c)
- Embedded Linux — Linux (Figure 4.3d)
- J2ME — KVM and configurations
- BREW

Figure 4.3 Smart phones running (from left to right) Symbian (Nokia 6620), Palm OS (Treo 650), I-Mate SP3, and NEC N900iL. (Courtesy of Symbian, Ltd.; Palm Source, Inc., src http://www.palmsource.com/products/products.cgi?Cat=Smartphones; Microsoft Corp.; Ziff Davis Publishing.)

Technically speaking, J2ME is not a full-blown software platform but a runtime environment that relies on an underlying operating system to execute Java applications. Each of these mobile platforms is bundled with a software development kit. RIM Blackberry is a proprietary mobile software platform only available on devices from RIM, so we do not cover it here. Indeed, the mobile software platform sector is a highly dynamic market undergoing fundamental reorganization, especially after Microsoft decided to engage in the battle with their Windows Mobile offerings. Understandably, traditional cell phone and PDA software platform providers such as Symbian and Palm have already sensed the pressure from those newcomers. It is too still early to predict how this diverse sector will evolve toward next-generation mobile computing, but one thing is for sure: All of these mobile software platforms will invariably target smart phones.

4.3 Symbian

Symbian is a private, independent company that develops and supplies the open standard mobile operating system Symbian OS. It is owned by some large cell phone manufacturers including Nokia, Ericsson, Sony Ericsson, Siemens, and Samsung. The primary targets of Symbian OS are cell phones and smart phones. Initially, Symbian was founded by Nokia, Ericsson, Motorola, Matsushita, and Psion as a joint venture in an effort to extend a proprietary cell phone operating system into an open standard. As of May 2005, 48 Symbian-powered cell phone models were on the market with sales of 32 million units in total.

Symbian OS was originally based on the EPOC operating system and was mainly used for PDAs developed by Psion. Interestingly, Symbian OS began with version 6.0, following EPOC version 5.0. The latest version of Symbian OS is 9.0. Symbian OS defines a few platforms of user interface (UI) reference models to accommodate disparate mobile devices. For example, the Quartz model targets PDA-like devices, the Crystal model is for common cell phones, the UIQ model is for customizable pen-based touch-screen mobile devices, and the Series 60 model is for numerical keyboard high-end cell phones and smart phones.

Symbian OS is designed to support a wide range of voice and data services in 2G, 2.5G, and 3G cellular systems, as well as multimedia and data synchronization. Key features of Symbian OS are mobile telephony supporting wideband code-division multiple access (WCDMA), global system for mobile (GSM)/general packet radio service (GPRS), and cdma2000 1x RTT; messaging services supporting short message service (SMS), enhanced message service (EMS), and multimedia message service (MMS); Internet e-mail servers; multimedia recording, playback, and streaming; communication protocols supporting Bluetooth, USB, and general transmission control protocol (TCP)/Internet protocol (IP) suites; and security in terms of full encryption (*e.g.*, 3DES, RC5, AES, RSA, SHA1, HMAC) and digital certificates, as well as secured protocols including transport layer security (TLS)/secure sockets layer (SSL), wireless

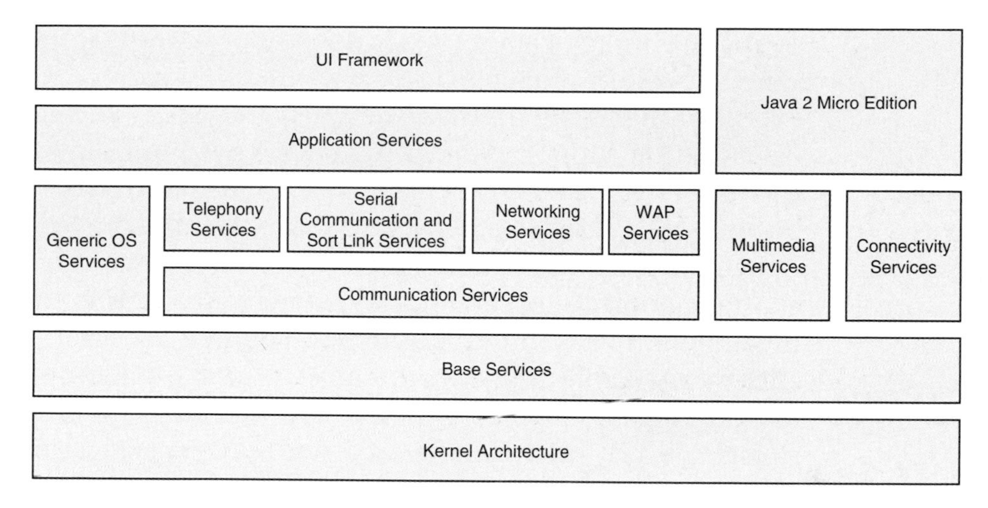

Figure 4.4 Symbian OS architecture.

transport layer security (WTLS), and IPSec. Symbian OS supports ARM processors and x86 emulation. Figure 4.4 depicts the architecture of Symbian OS version 8.1 [3].

Symbian OS is a real-time, multithreaded, preemptive kernel (versions prior to 8.0 do not provide real-time kernels) that performs memory management, process and thread scheduling, interprocess communication, process-relative and thread-relative resource management, hardware abstraction, and error handling. The kernel runs natively in the ARM core.

The basic services provide a programming framework for Symbian OS components, such as kernel and user API library, device drivers, file systems, and standard C++ library. On top of the basic services is a set of communications services, multimedia services, PC connectivity services, and generic OS services. Communication services act as the core to mobile telephony and data network access applications. Personal area network (PAN) connectivity such as Bluetooth, IrDA, and USB is also enabled by these services. Multimedia services deal with audio and video recording, playback, and streaming. Connectivity services are software components that implement PC synchronization. Generic OS services offer typical OS-related components such as memory management and file system access. Application services allow user programs to be executed in separated processes. An exception handling mechanism is also provided, as well as internationalization support. The UI framework is comprised of an array of UI components and an event-handling mechanism that permit easy porting of UI programs between different Symbian OS devices.

Symbian OS uses EPOC C++, a pure object-oriented language, as the supporting programming language for both system services implementations and application programming interfaces. It also allows Java applications for mobile devices (J2ME applications) to run on top of a small Java runtime environment. Symbian OS implements a CLDC/MIDP 2.0 profile of J2ME specifications. Further details on Java 2 Micro Edition are provided in Section 4.7.

4.4 Palm OS

Palm, Inc., is the company that created the PDA market. As explained in Chapter 2, the PalmPilot is believed to be the first successful PDA product (Palm has recently started to use Microsoft Windows Mobile for Smartphone on some of its smart phone products.). Palm OS is the underlying operating system for Palm PDAs. Palm also licenses Palm OS to other PDA manufacturers such as Handspring (merged with Palm in 2003), Sony, and IBM. Officially, the company that develops Palm OS is Palm Source, a spin-off from PalmOne, the new name of Palm, Inc., after it merged with Handspring. (Interestingly, PalmOne changed its name back to Palm, Inc., in 2005.) PalmOne is the PDA maker of a few product lines including Palm V, Tungsten, Zire, and Treo.

Palm OS up to version 4 follows a simplified design philosophy: making use of limited computing power to allow efficient operations. Applications on Palm devices are mostly consumer oriented and have an excellent user-friendly interface. The most notorious drawback of the old versions of Palm OS is a lack of multitasking support; the system can only run a single application at a time, like early DOS systems. This drawback did not seem to matter when only a few applications were available on the PDAs, and execution of those applications was generally fast without significant delay. As users began to demand more functionality, such as mobile telephony and PAN/LAN access, applications on Palm PDAs became more versatile and sophisticated, requiring a more advanced operating system to facilitate software development. In response to this trend, Palm designed Palm OS version 5 and 6, which have incorporated a wide range of improvements.

A major design goal of Palm OS 5 is to move from 68000 (or 68K) processors to several ARM processors such that Palm OS 5 can be executed on a broader hardware base. At the same time, Palm OS 5 is backward compatible with legacy Palm applications, thanks to the built-in Palm Application Compatibility Environment (PACE). In addition, Palm OS 5 supports multitasking, cryptographic provider management (CPM) with default crypto provider, mobile telephony, wireless PAN capabilities, and better graphical user interface (GUI) support. Palm OS 6 (Cobalt) is a milestone in Palm OS history. It is a complete rewrite of Palm OS and is the first Palm OS to support multithreading. It also provides more wireless capability, multimedia application support, and a variety of extension slots support. Some Palm OS applications are expected to be ported to Linux in the future, according to Palm Source, Inc.

The Palm OS 6 reference architecture is illustrated in Figure 4.5. Palm OS allows third-party hardware to be used as part of the system provided the hardware abstraction layer and system services support the third party hardware. The kernel of Palm OS is based on AMX, licensed from Kodak, which offers preemptive multitasking and protective memory management. System services are a set of modular components that provide communication, input method (Graffitti 2), GUI event handling, and multimedia processing. Palm OS 6 has built-in Bluetooth and Wi-Fi support. Core OS libraries supply low-level file system management functionality and TCP/IP. Third party software libraries can also be plugged into the system; for example, J2ME profile implementations can be added as third-party libraries. The PACE layer emulates non-ARM processors, thus allowing legacy applications to execute. A Palm device also provides a set

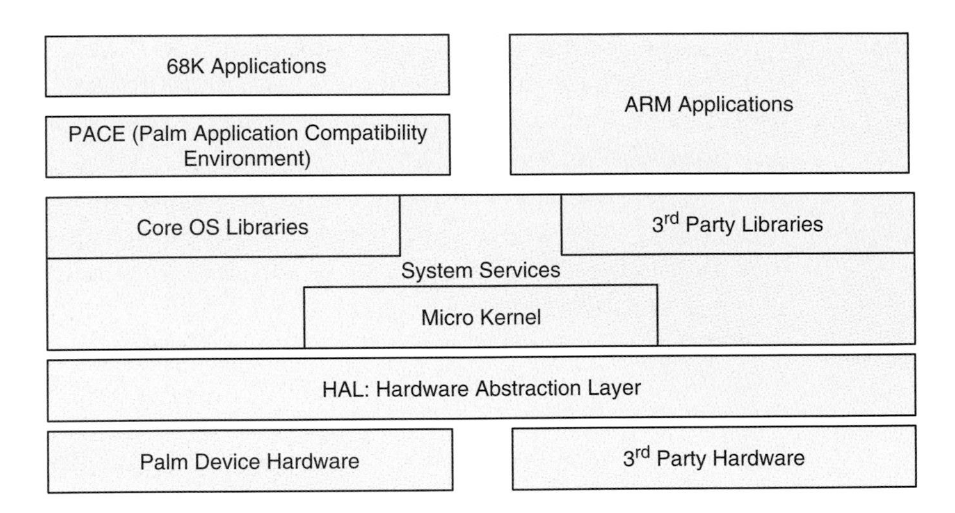

Figure 4.5 Palm OS 6 architecture.

of device applications including PIM applications, e-mail, messaging, mobile telephony, Internet browser, media player, hot sync, remote access, and file system access. Palm OS employs a database model to store files, rather than a block-based file model. Everything is a database: a PRC (Palm Resource Code)for a program (.exe) and a PDB (Palm DataBase) for program database. Synchronization between a Palm device and a computer is done using Palm HotSync Conduit.

We are particularly interested with the network communication capabilities of Palm OS 6 Cobalt. Multiple communication tasks can be performed at the same time on a Palm OS Cobalt device. Its I/O subsystem has been greatly enhanced to support several extension standards, such as SD, MMC, and SDIO. Palm OS Cobalt provides an integrated Bluetooth protocol stack and profiles including generic access profile, serial port profile, dialup networking profile, LAN access profile, service discovery application profile, generic object exchange profile, object push profile, headset profile, and hands-free profile. It has built-in Wi-Fi support along with wired equivalent privacy (WEP) and Wi-Fi protected access (WPA). The drawback of the Palm OS Cobalt is that the APIs, namely Palm OS Protein APIs, do not run on older versions of Palm OS; hence, devices powered by older Palm OS releases must be upgraded to Palm OS Cobalt to take advantage of the enhanced features.

Application development for Palm OS is based on the Palm OS 68K/Protein SDKs and developer suite in a programming environment. Developers can choose programming languages from C, C++, Visual Basic, and Java, although C is most widely used for Palm OS software development. A developer suite allows a developer to create both ARM-native Palm OS Protein powered applications for Palm OS Cobalt (Palm OS 6) and 68-K applications. Palm developers usually use Metrowerks CodeWarrior (www.codewarrior.com), PRC Tools (www.palmos.com/dev/tools/gcc), or Eclipse (http://www.eclipse.org/) as the cross-platform programming environment.

4.5 Microsoft Windows Mobile

Microsoft is a relatively newcomer in the mobile software platform market but has managed to achieve substantial growth with the Microsoft Windows Mobile platform. Windows Mobile refers to a complete solution to mobile software that consists of a Windows CE-based operating system, a programming framework, and supporting software development tools. Hoping to clone its success in the desktop system arena, Microsoft has evidently taken advantage of its experience with desktop Windows so as to lower the barrier of software development for mobile systems. Windows Mobile is clearly on the rise. For one thing, Windows Mobile for Pocket PC is gradually overtaking Palm OS as the number one platform for PDAs. As a major win, Microsoft Windows Mobile will replace Palm OS to power the Palm Treo™ smart phones.

As a general mobile software platform, Windows Mobile is tailored for PDAs and cell phones or smart phones, respectively. The latest device added to the Windows Mobile family is a portal media center, which can display recorded television programs, home videos, music, and photographs. For each type of device, a platform has been defined, such as Pocket PC 2003 (*i.e.*, Windows Mobile 2003 for Pocket PC), Smartphone 2003 (*i.e.*, Windows Mobile 2003 for

Smartphone), and Windows Mobile 5.0 for Smartphone. A Windows Mobile platform is actually a combination of the underlying operating system (an optimized variant of Windows CE), APIs, software development tools, and a set of standard applications. Notice the lack of a space between "Smart" and "phone" in the word "Smartphone." This usage more or less exclusively represents the Microsoft Windows Mobile platform for smart phones.

4.5.1 Windows Mobile

Figure 4.6 illustrates the architecture of Windows Mobile. On top of the original equipment manufacturer (OEM) hardware is the operating system, a version of Windows CE that was designed to support a variety of low-capability mobile devices such as PDAs, cell phones, smart phones, bar code readers, handheld computers, and embedded devices in automobiles, among others. The .Net Compact Framework

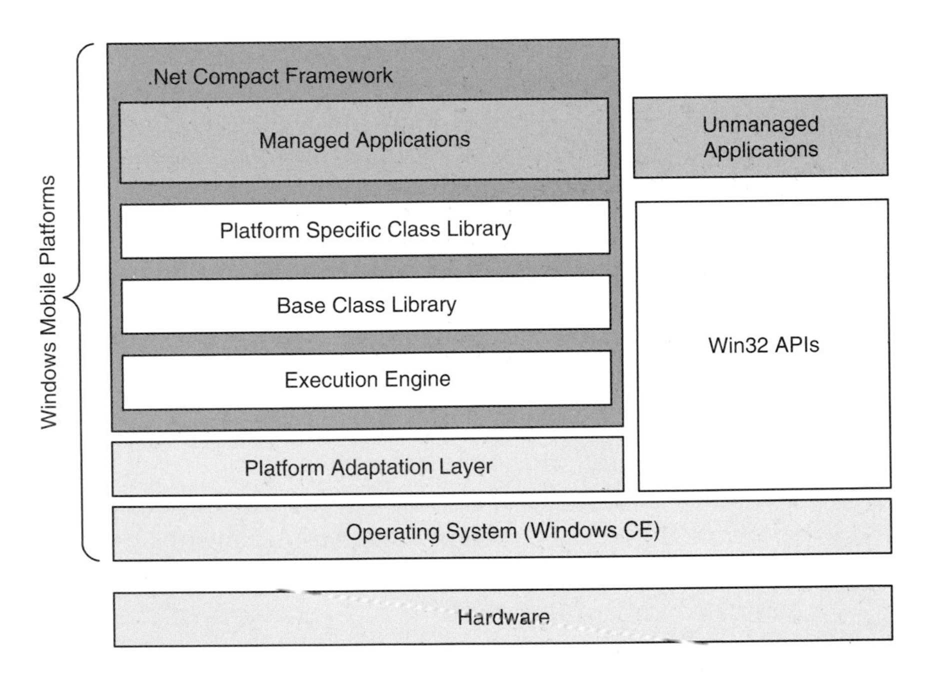

Figure 4.6 Windows Mobile.

provides a facility and classes that allow one to develop managed applications. The traditional Wind32 application design paradigm utilizing Win32 APIs or Microsoft Foundation Classes (MFC) is still supported.

It is noteworthy that prior to the strategic brand "Windows Mobile" Smartphone and Pocket PC were initially two different software platforms both based on Windows CE. As the convergence of communication and computing becomes more imminent, the distinction between these two types of mobile devices has begun to disappear. In fact, some people envisage a single "Windows Mobile" software platform for next-generation smart phones in the next several of years.

Windows Mobile 5.0 for Smartphone is based on the Windows CE.Net 5.0 operating system. The .Net Compact Framework is loaded into the ROM of the underlying smart phone device. So, what is .Net, exactly? .Net is Microsoft's general software infrastructure of Internet-based computing. Its core is eXtensible Markup Language (XML) Web services. XML Web services are widely accepted as the solution to enabling seamless, robust, and secure collaboration among heterogeneous Internet services and applications. What makes it possible to achieve this goal is the open-standard XML that is able to integrate data and its structure into a self-explanatory format such that they can be organized, edited, programmed, and exchanged between any applications, services, websites, and smart devices. The mechanisms that define how applications and service interoperate with each other using XML are simple object access protocol (SOAP) and universal data description interface (UDDI). With such a vision in mind, Microsoft has provided a wide variety of building blocks to enable .Net based Internet application development, as shown in Figure 4.7.

The heart of the .Net infrastructure is .Net Framework on most Windows-based operating systems, including Windows desktop systems, Windows servers, tablet PCs, Pocket PCs, and Smartphones. .Net Framework consists of two components: a common language runtime (CLR) and the .Net Framework class libraries. CLR is a layer between .Net application code and the underlying operating system.

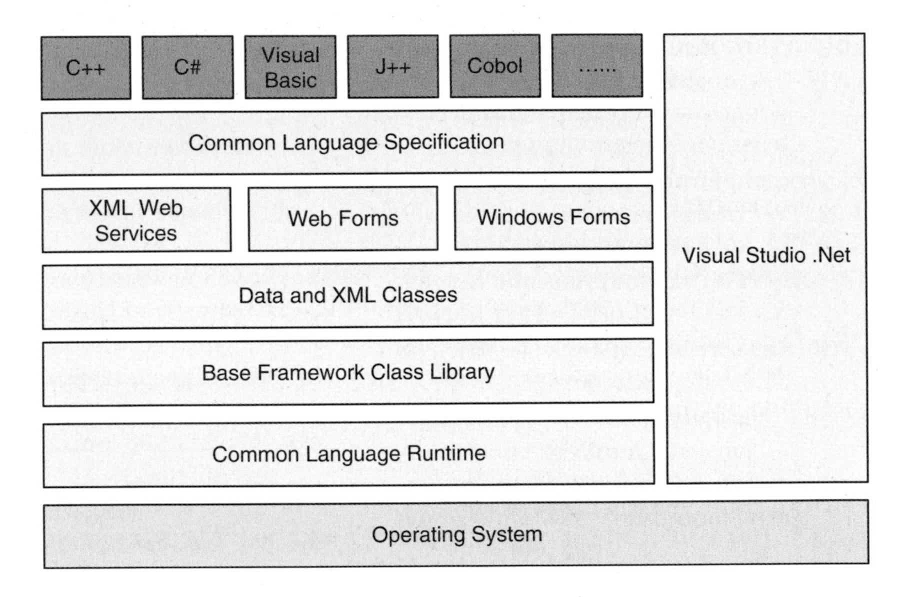

Figure 4.7 Microsoft .Net framework for desktop and server systems.

Developers can choose from a number of .Net-compatible programming languages, including C, C++, C#, and Microsoft Visual Basic, as well Fortran and Perl, for .Net application development. Programs developed using different languages will be compiled into platform-independent intermediate language (IL) code. When the code is executed on a .Net system, CLR performs a just-in-time compilation at run time to generate native code of the application based on the IL code. In a sense, CLR works like a virtual machine to the application in handling memory management including garbage collection, type checking, exception handling, and security enforcement. It supports common type system (CTS) and common language specification (CLS), two components of the Microsoft Common Language Infrastructure (CLI). Windows systems have .Net Framework as part of the operating systems. Some efforts have been made to port .Net Framework to non-Windows systems, such as the Mono Project [4]. The other component of the .Net Framework, .Net class libraries, is a package of classes and APIs available for any .Net-compatible programming languages. The libraries can be divided into three categories: ASP.Net for web application and web services development, ADO.Net for database access, and Windows Forms for building GUI applications for smart client (*i.e.*, desktop) applications.

4.5.2 .Net Compact Framework for Mobile Devices

For mobile devices such as PDAs, cell phones, and smart phones, a stripped-down heavily optimized version of .Net, the .Net Compact Framework, is introduced. Similar to its desktop cousin, the .Net Compact Framework also has two components: a CRL specifically designed for mobile devices and the .Net Compact Framework class libraries. The CLR is extremely (small less than 2 MB) and efficient compared to the desktop CLR .Net Framework due to restrictions of target mobile devices such as CPU power, memory, storage, networking capability, and power consumption. The .Net Compact Framework class libraries are composed of a subset of optimized desktop .Net Framework classes, plus some new classes specially designed for mobile device applications and services.

Mobile software development based on the .Net Compact Framework is facilitated by Microsoft's flagship programming tool, Visual Studio .Net, which provides support for smart device projects including the Pocket PC, Smartphone, and Windows CE projects. Developers can choose either C# or Visual Basic as the programming language. (According to Microsoft, managed C++ will be supported by .Net Compact Framework in the near future.) .Net Compact Framework is still undergoing intense development. Features and functionalities are being added to it on a regular basis along with newer versions. This suggest that, if .Net Compact Framework does not support a specific function, then developers must resort to traditional native Windows CE APIs, thus bypassing the .Net Compact Framework runtime environment. As part of the Smartphone platform offering, Microsoft also provides a Smartphone software development kit (SDK) for independent software vendors (ISVs). Developers can use the Smartphone SDK in conjunction with Visual Studio .Net for .Net Compact Framework-based managed code application development. Additionally, the Smartphone SDK also allows C or C++

unmanaged application programming using Win32 (WinCE) APIs, bypassing .Net Compact Framework runtime. Windows Mobile for Smartphone applications can leverage the following features provided by the underlying Windows CE system and the .Net Compact Framework:

- Platform invoke (P/Invoke) — This allows a managed program to call a platform-dependent API provided by the Windows CE operating system.
- ADO.NET — This allows convenient data access in connected or disconnected mode.
- Web services — These make it possible for a piece of software to communicate and operate with other entities on the network via a secure, unified interface.
- Rich Forms Designer — This GUI designer coupled with Visual Studio .Net is an indispensable tool for highly efficient user interface design.
- XML support — This leverages XML as the general structured data representation for both local and remote data interchange.

4.6 Embedded Linux

Linux is a free, open-source, UNIX-like operating system. It actually refers to a combination of two portions: Linux kernels and Linux applications. Linux kernels are developed and maintained by a group of people led by Linus Torvalds, creator of the Linux operating system. The latest Linux kernel can be obtained from http://kernel.org. A vast number of free, open-source applications have been developed and are in use on various Linux systems. Source code of Linux kernels and applications are mostly available under Gnu Public License (GPL). The GPL states that anyone making use of the GPL-protected source code for public release must also ensure free and open access to the code under the same terms of the GPL. A number of Linux distributions (*i.e.*, bundles of a Linux kernel and selected applications) have been provided by some companies such as RedHat and SuSe Linux. Thanks to the support of such companies as IBM, Oracle, and Novell, Linux is generally viewed as a serious server OS alternative to Windows and proprietary UNIX such as SunOS. It also has potential for the desktop OS market.

Linux is making giant strides in the embedded system market as well, powering up a broad range of network equipment, consumer electronics, industrial facilities, and mobile devices. In particular, several Linux distributions for PDAs are free for download on the Internet. Aside from those offerings from Linux enthusiasts, commercial Linux systems for mobile devices such as Monta Vista Linux are available in the market. Some PDAs and cell phone manufactures, especially those in East Asia, have begun to release Linux-based cell phones and smart phones. For example, NTT DoCoMo will use Linux on its 3G cell phones, and Chinese telecom equipment manufacturers have selected Linux as the embedded operating system for 3G TD-SCDMA cell phones. Numerous cell phone manufacturers such as Motorola and NEC are developing Linux smart phones.

Embedded Linux systems could be either hard real time or soft real time. *Hard real time* means the system must respond in a deterministic way every time a relevant event occurs. Examples of hard real time systems are autonomous control systems that require a real-time operating system (RTOS). In *soft real time systems*, quick responsiveness is desired but not guaranteed. Mobile devices used by consumers generally fall into the soft real time category. The latest Linux kernels support both hard real time and soft real time applications, due to the preemptive kernel design and the so-called O(1) process scheduler that utilizes a special task queue to achieve constant time scheduling. Below is a list of characteristics of embedded Linux systems:

- The monolithic kernel supports multitasking and multithreading and can be tailored for different applications scenarios.
- The open-source community provides support for the latest network technologies, including Bluetooth, wireless LAN, and wireless sensor networks.

- Applications under GPL licenses are modifiable and extensible.
- License fee is low or nonexistent.

Because Linux can be customized to fit into different mobile devices with various hardware configurations, it is of great importance to standardize hardware and software interface and interoperability between conforming devices. To this end, some companies have formed the Consumer Electronics Linux Forum (CELF) as an effort to promote advancement of open embedded Linux systems for consumer electronics devices. Another industry organization, the Embedded Linux Consortium (ELC), is focused on general embedded Linux standard development.

Due to the recent development of Linux smart phones and wireless network equipment, the future of Linux in mobile computing seems quite exciting. For Linux to reach its full potential in the mobile sector, it must address "fear, uncertainty, and doubt (FUD)" with regard to how Linux will perform as a reliable, robust, and yet low-cost platform, let alone the recent litigation addressing UNIX intellectual property between SCO Group Inc. and some UNIX and Linux providers.

Application development on open source or commercial embedded Linux systems differs significantly from system to system. Aside from vendor-specific tools, standard Linux toolchains coupled with specific libraries are often used. For example, QT/Embedded (http://www.trolltech.com/products/embedded/) is a popular GUI development framework optimized for Linux mobile devices. Widely used embedded application development tools include GnuPro Toolkit (http://www.redhat.com/software/gnupro/) and the general Gnu toolkit.

4.7 Java 2 Micro Edition (J2ME)

Java is not a mobile operating system, but it is mentioned here because other mobile operating systems can leverage Java platforms to enable code portability and enforced security. In this sense, Java, in particular J2ME, performs as a common middle layer between a specific mobile operating system and valued-added services and applications offered by wireless service providers.

4.7.1 Java Primer

Java by definition is a cross-platform, object-oriented, interpretive programming language developed by Sun Microsystems. At the heart of Java is a set of Java virtual machines (JVMs) for different platforms that interpret compiled Java bytecode. Here, the notion of *platform* refers to a computer system's processor and its operating system. For example, Linux on an x86 processor and Linux on a Sun SPARC processor are considered two distinct platforms. Popular platforms that have Java support are Windows on x86 processors, Linux on x86 processors, Solaris on SPARC processors, Solaris on x86 processors, Linux on AMD processors, and Windows CE on ARM processors, among others. On each of those platforms a Java runtime environment (JRE) is required to run Java applications. A JRE consists of a JVM for the underlying platform in question, core classes in standard packages, and some supporting files. The latest Java version is Java 2 version 1.5, code-named Tiger.

A Java platform is the combination of a JRE and its software development kit. A Java source file is first compiled into a bytecode file by a compiler. Bytecode is platform independent and is not machine code. Java was designed to allow the same bytecode to be interpreted by JVMs on different platforms, a feature referred to as "compile once, run everywhere." In reality, this is not always the case due to the inconsistency of JVM interpretation across different platforms. As an interpretation language, Java has long been criticized for being considerably slower than those languages that generate native code for a platform. To improve the performance of Java programs, today's JVMs offer an option to leverage the just-in-time (JIT) technology, which loads bytecode and compiles it into native machine code on the fly before a program is executed. Compared to traditional interpretation of bytecode, native machine code runs significantly faster. As a side note, recall that similar JIT technique is also used in Microsoft .Net Framework.

Interestingly, Java was initially created to allow code to be easily ported across a variety of embedded devices. This design goal fell short as few embedded device manufacturers considered code portability a very important feature of their offerings. Along with the explosive growth of the Internet, Java found its way to the World Wide Web. In addition to static HTML web pages, a new scheme was needed to allow the delivery of a web-based dynamic user interface, usually wrapped into some software components over different platforms. For some time Java seemed to be fitting well into the web environment; as long as a JVM for the underlying web browser was available for each targeted platform, the same web-based Java application (Applet classes) could be downloaded and interpreted by the JVM. This idea of Java applets served to promote Java technology in the software development community, but this momentum did not last for long as more and more people began to complain about the poor performance of client-side applet executions.

Key characteristics of Java language include:

- Java bytecode is platform independent.
- Java is an object-oriented language with a large set of standard packages and third-party packages.
- Java has multithreading and thread synchronization support.
- Java employs automatic memory management and garbage collection; pointers are not used in Java.
- Java comes with internationalization support.

The following is a list of official Java platforms available from Sun Microsystems, each targeting one type of computer system:

- Java 2 Standard Edition (J2SE), for desktop and server systems
- Java 2 Enterprise Edition (J2EE), for enterprise application systems

- Java 2 Micro Edition (J2ME), for embedded systems
- Java Card, for smart card applications

Java 2 Standard Edition is a complete software development solution of Java applications for desktop and server systems, whereas J2EE subsumes J2SE and provides a set of tools for developing enterprise multiple-tier component-based services that implement specific business logic across heterogeneous application servers. It was J2EE that eventually contributed to Java's highly anticipated success — not on embedded devices or the web, but on enterprise server systems, which nobody ever expected. J2ME is aimed at the ever-expanding embedded devices market. Due to the diversity of devices in this segment, J2ME has been furthered packaged into two distinct configurations: Connected Limited Device Configuration (CLDC) for mobile devices and Connected Device Configuration (CDC) for consumer and embedded devices. Java Card is specifically designed for smart card application development.

4.7.2 J2ME Configurations

A J2ME configuration consists of a set of fundamental requirements for JVMs and supportive Java classes and APIs that as a whole represent Java runtime environment for a collection of embedded devices with similar hardware and network capabilities. J2ME has provided CLDC and CDC configurations. The characteristics of CLDC devices can be summarized as follows:

- A 16-bit or 32-bit processor with a clock speed of 16 MHz or higher
- Low memory budget: 160 to 512 Kbyte
- Lower power consumption and limited power supply (mostly battery)
- Simple user interface or no interface at all
- Low bandwidth and possibly intermittent network connection

Examples of CLDC devices include cell phones, low-end PDAs, pagers, wireless sensors, radio frequency identification (RFID) devices, etc. CLDC devices are resource constrained in terms of power, processing capability, memory capacity, and network capability; therefore, standard JVM cannot be used. Instead, a stripped-down JVM, the kilobyte virtual machine (KVM), is highly desirable for CLDC configuration. Sun has supplied a reference implementation (RI) of KVM along with the CLDC configuration. Other KVM implementations are also available, such as IBM's J9 virtual machine, which has been selected for the BREW platform by Qualcomm. KVMs for different operating systems such as Symbian OS, Palm OS, Embedded Linux, and Windows Mobile enable Java applications to run on these systems.

The following are characteristics of CDC devices:

- A 32-bit processor
- Large memory budget: 2 MB RAM and 2.5 MB ROM available to the Java application environment
- Wired power supply or long battery time
- Stable network connection
- Various user interface, from sophisticated GUIs to no UI

Examples include high-end PDAs, television set-top boxes, and RFID readers. Unlike CLDC, the CDC configuration uses standard JVM. J2ME configurations are defined based on hardware and network specifications rather than vertical classification of application scenarios of various devices. The functionality provided by devices of the same J2ME configuration may vary greatly, requiring additional APIs that are not provided by the underlying common-ground J2ME configuration. For this reason, some J2ME profiles have been defined for each configuration. A profile is a set of standard Java APIs for a specific narrower class of devices within a J2ME configuration. For the CDC configuration, a foundation profile, a personal basis profile, and a personal profile are defined; the foundation profile acts

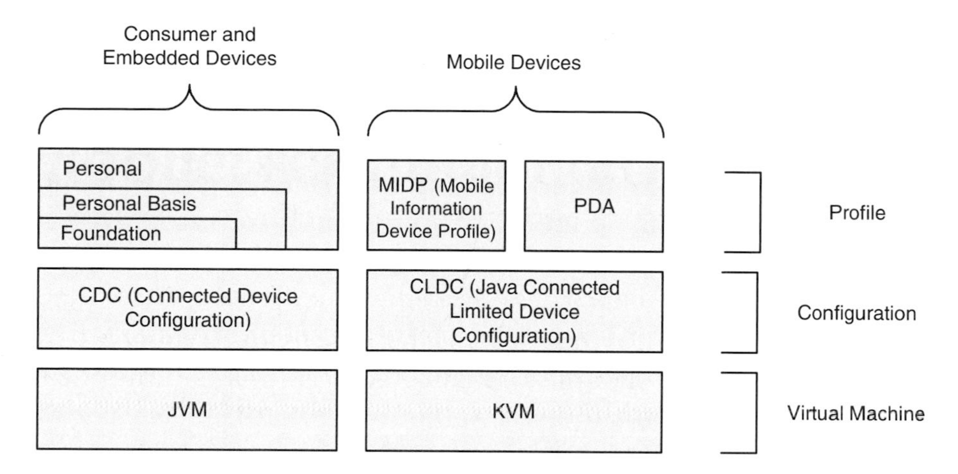

Figure 4.8 J2ME configurations.

as the core to the other two profiles. For the CLDC configuration, mobile information device profiles (MIDPs) and PDA profiles have been devised. For cell phones and smart phones, the CLDC configuration and the MIDP profile made up a standard Java platform, as shown in Figure 4.8. Similar to the Java applet scheme for the web, MIDP applets can be downloaded and executed in a CLDC configuration on a mobile device. For each configuration, optional packages are also provided for specific purposes such as database access and remote method invocation.

Java 2 Micro Edition provides inherent security mechanisms such as sandbox and crypto APIs. MIDP 2.0 of the CLDC configuration introduced digital signatures and code signing to ensure bytecode security.

To date, J2ME has been widely used in the development of mobile applications on cell phones, PDAs, pagers, handheld PCs, set-top boxes, and many consumer electronics devices. An assortment of J2ME-based mobile applications is being used worldwide, ranging from cell phone games [5] to multimedia download and playback to enterprise applications.

Unlike desktop computers, the operating system of a mobile device is often greatly customized to leverage underlying mobile hardware.

Sometimes installing a new operating system on a device designed to run a different operating system can be extremely difficult and dangerous, because it can render the device unusable due to hardware failure. On the other hand, in order to improve developer productivity and mobile code security, people are free to choose a virtual-machine-based, managed software development platform such as J2ME and .Net Compact Framework. (Managed code is executed within a software execution environment, which is responsible for enforcing type-safe execution, memory management, and code security.) A comparison between J2ME and .Net Compact Framework is presented in the last section of this chapter.

4.8 BREW

Binary Runtime Environment for Wireless (BREW) is a wireless software platform solution for CDMA cell phones developed by Qualcomm, the company that created CDMA technology. According to Qualcomm, as of November 2004, more than 40 million BREW units were being used worldwide, and more than 150 device models were utilizing BREW as the runtime environment and application development platform. Contrary to popular perception, BREW is independent of the air interface of the underlying cellular technology, meaning that GSM and UMTS/WCDMA cell phones are also supported by BREW.

BREW consists of three components: a binary runtime environment, an application development environment, and a distribution system. The binary runtime environment allows native BREW applications to operate regardless of the air interface used by the cell phone. The application development environment provides a set of platform APIs and supporting tools for BREW software development. The BREW distribution system (BDS) allows wireless data service providers to deliver content to subscribers. Like Microsoft's .Net Compact Framework, any mobile applications that are about to be deployed on a mobile device must be signed using a digital key and certified. BREW is generally regarded as a hardware-independent wireless solution for mobile devices. In this sense, BREW is competing with J2ME. However, these two technologies are not entirely competitors. For example, Qualcomm has selected IBM's J9 Java virtual machine for BREW, thereby making it possible to execute Java applications.

4.9 Comparisons of Mobile Software Platforms

Symbian OS and Palm OS are historically dedicated mobile operating systems for cell phones and PDAs, respectively. Both of them are tightly connected to a group of mobile device manufacturers: Symbian is in effect funded by a number of cell phone manufacturers, and Palm OS is dedicated to Palm devices. The trend of convergence of these mobile devices opens up enormous opportunities for Symbian and Palm, potentially allowing each to enter the other's stronghold with smart phone offerings. The challenge is to provide a wide range of mobile telephony and mobile data services with very limited resource on a small form factor. In contrast, Microsoft Windows Mobile for Smartphone is rooted in the general-purpose Windows CE operating system and is known to require much more in the way of computing resources due to the generalized architecture of Windows CE even though it has been tailored for smart phones.

As an open-source operating system, Linux has been embraced by a number of mobile device manufacturers as a low-cost and highly customizable solution. Both Windows Mobile and Linux are general-purpose operating systems and must be tailored for use on mobile devices. J2ME CLDC is platform independent and can be used on any mobile operating systems supporting J2ME. Among proprietary mobile platforms, Qualcomm's BREW managed to gain considerable acceptance, primarily by CDMA cell phone manufacturers.

Palm OS, Microsoft Smartphone, and BREW are proprietary systems in that they only provide largely restricted APIs but no mechanism to tweak the system itself. On the other hand, Symbian, Linux, and J2ME are completely open systems in this sense. Table 4.2 provides a comparison of these mobile platforms.

Table 4.2 Mobile platform comparisons

	Symbian	Palm OS	Microsoft Smartphone (with .Net Compact Framework)	Linux	J2ME	BREW
Multitasking	Yes	Yes	Yes	Yes	Depending on OS	Yes
Supported processors	TI OMAP (ARM based) and Intel XScale	Intel XScale, Motorola DragonBall, TI OMAP (ARM based)	Intel StrongARM and XScale	Intel StrongARM and XScale	Depending on devices	Qualcomm processors
Footprint	Small	Small	Large	Large	Medium	Small
Power	Excellent power management	Excellent power management	Lacks power management	Shows potential for power management	Relies on the operating system	Chipset for power management
Mobile telephony	Inherent support	Enhanced feature for high-end Palm devices	Limited support	Limited support	Extensive support for mobile operating systems	Strong support
Wireless data connection support	Bluetooth, IrDA	Bluetooth, Wi-Fi, IrDA	Bluetooth, Wi-Fi, IrDA	Bluetooth, Wi-Fi, IrDA	Bluetooth	Bluetooth and IrDA
Open or not?	Open	Proprietary	Proprietary	Open	Open	Proprietary
Programming language support	Native C and C++	C	Native C, C++, and .Net runtime support	Native C and C++	Java	Native C/C++
Industry support	Major cell phone manufacturers	PalmOne	Several wireless service providers and some Palm smartphones	Asia cell phone manufacturers	Licensees include major cell phone manufacturers	CDMA cell phones

Due to its supreme portability and established support in the mobile application developer community, J2ME is well positioned for next-generation mobile application development. The fact that every mobile software platform has some kind of Java environment (even Windows Mobile may be equipped with a third-party KVM) is a clear indication of J2ME's advantage over other application platform solutions. On the other hand, .Net Compact Framework along with Windows Mobile does have a chance to gain some traction, provided mobile device manufacturers continue to embrace Microsoft Windows Mobile in their future offerings. Considering the heterogeneity of mobile devices in terms of hardware and software platforms, J2ME and .Net Compact Framework are likely to flourish, while other mobile software platforms dedicated to specific operating systems may experience only limited acceptance, which depends on the popularity of mobile devices running the operating system in question.

4.10 Supporting Tools

In addition to the operating system and its development interfaces, a complete mobile software platform also includes a range of middleware and mobile development supporting tools that can be used to address the issue of device and network heterogeneity, as well as to facilitate efficient cross-platform development. In this section, we will look at three major elements in this arena: mobile web browsers and markup languages, and device emulators.

4.10.1 Mobile Web and Markup Languages

Heterogeneity is everywhere in the mobile wireless world. In the case of mobile markup language, several HTML-like languages are being used on a variety of mobile devices. Table 4.3 provides a summary of these markup languages.

Table 4.3 Wireless Markup Languages

Language	Description	Example
Hypertext markup language (HTML)	The markup language of web pages displayed on web browsers	This is a paragraph: <h2><b>example</b></h2>
Extensible markup language (XML)	The markup language that allows users to create their own tags for the definition, validation, and interpretation of structured data	<Team> <Employee> <Name>John Doe</Name> <EmployeeID> 1234 </EmployeeID> </Employee> <Employee> <Name>Joe Smith</Name> <EmployeeID> 1235 </EmployeeID> </Employee> </Team>
Wireless markup language (WML)	The markup language optimized for text-based WAP (1.0) devices; WML code is compiled into binaries before being sent out	<card id="mycard" title="card example"> text example</p> </card>
Extensible hypertext markup language (xHTML)	A transition between HTML 4.0 and XML; compared to HTML, xHTML imposes stricter rules with tagging (*e.g.*, elements must be properly nested and closed); used in WAP 2.0	This is a paragraph </p> <h2><b>text example</b></h2>
Compact HTML (cHTML)	A subset of HTML optimized for small information devices; primarily used by Japan's iMode service	See HTML example
Standard generalized markup language (SGML)	A meta-language that defines how to define a specific markup language; HTML and XML are based on SGML	N/A

Mobile web platforms such as WAP and i-Mode (as discussed in Chapter 3) use one or two markup languages in the above-mentioned list. Both of them require a special browser application on mobile software platforms to render web pages designed using selected markup languages. A more general mobile platform is to use a stripped-down version of web browsers to access both normal web pages designed for desktop computer displays and mobile web pages for small mobile displays. Major "micro web browsers" include Opera Mobile (http://www.opera.com/products/mobile/), Microsoft Pocket Internet Explorer (http://www.microsoft.com/windowsmobile/about/tours/sp/pocketie.mspx), and Mozilla Minimo (http://www.mozilla.org/projects/minimo/). Opera Mobile is by far the most widely used mobile web browser on cell phones. Of course, mobile website designers face the challenge of authoring mobile web pages optimized for small display rendering. To convert a normal web page into a format suitable for small displays, some "page transformation" techniques are often applied for mobile web servers or wireless application gateways. This issue is discussed further in Chapter 7 (Mobile Application Challenges).

4.10.2 Mobile Device Emulators and Simulators

Each software platform discussed above provides a device emulator or a simulator program that can be used to debug and test code without a physical target mobile device. A mobile device emulator is a software-based execution environment that emulates a complete set of hardware as well as the mobile operating system on a target mobile device. In contrast, a mobile device simulator is an instance of the target mobile operating system or software platforms compiled natively for the hardware platform of the development computer hosting the simulator software. For Palm OS mobile application development, both an emulator and a simulator are provided. Other platforms only provide an emulator or a simulator. The emulators and simulators provide a virtualized execution environment supporting target CPU instruction set emulation/simulation and memory and I/O device emulation/simulation. Figure 4.9 shows screenshots of these emulators.

Symbian emulator for Nokia Series 60

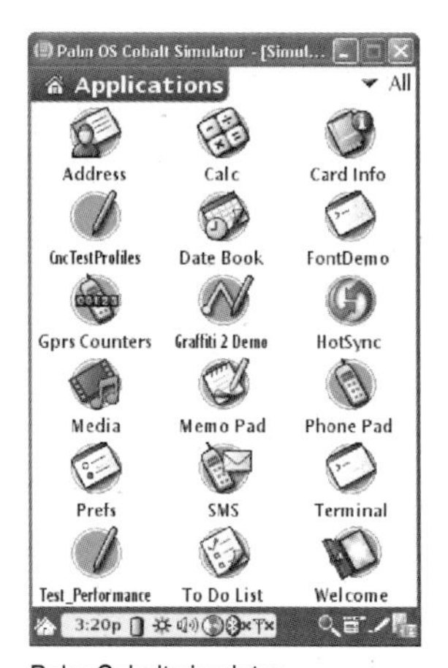

Palm Cobalt simulator

Microsoft Smartphone emulator

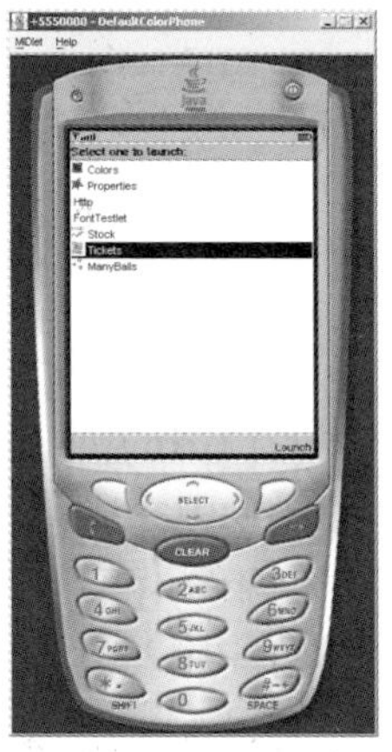

J2ME MIDP emulator

BREW simulator

Figure 4.9 Device emulators. (Courtesy of Nokia Communications; Palm Source; Microsoft; Sun Microsystems, and Qualcomm.)

4.11 Summary

All wireless services and applications eventually rely on a mobile terminal to deliver to users. Terminal platforms of next-generation mobile devices are undergoing significant changes in both hardware and software. On the hardware side, mobile processors continue to achieve faster speed. We can expect dramatic improvement of mobile hardware capabilities primarily in the areas of wireless connectivity and storage. A broad range of wireless interfaces have been developed as built-in features or extension interfaces pluggable to a mobile device. Radio frequency functionality, digital signal processing, multimedia functionality, and security mechanisms will appear in the system-on-chip form and will be integrated into various mobile devices. Memory and persistent storage of mobile devices will not be a problem for future mobile applications. More nonvolatile memory is being used at a lower cost, and secondary storage for mobile devices is beginning to appear, although low-power computing and power management will remain key challenges. The power consumption of each hardware component and the overall system has been greatly reduced. In addition, system-wide power management schemes based on hardware control chips have been introduced to some mobile devices. Flexible display technologies may provide a solution to the issue of mobile display.

On the software side, multiple mobile software platforms such as Symbian, Palm OS, Microsoft Smartphone, BREW, and J2ME are supplied by various industry consortiums targeting the same market of next-generation mobile devices. Because of this remarkable diversity, it is quite clear that in the foreseeable future no single platform will dominate the entire segment as is the case in desktop computer systems. It is also evident that all the mobile platforms are moving quickly toward converged mobile devices, and smart phones are among the top priorities to target. Another trend is that support for a wide range of wireless technologies is becoming standard.

This chapter presented coverage of existing mobile terminals that are likely to leverage those wireless technologies discussed in Chapter 3 with respect to providing building blocks for next generation mobile services and applications. Needless to say, a number of issues in this regard remain unsolved. For one thing, mobile convergence calls for seamless integration and interoperation among mobile terminals and network infrastructures, but the diversity of those non-interoperable mobile platforms would seem to be a barrier to this trend. In the next several chapters, we will discuss these challenges and solutions surrounding smart phones in the domain of next-generation mobile computing.

Further Reading

A. Wigley, S. Wheelwright, R. Burbidge, R. MacLeod *et al.*, *Microsoft .Net Compact Framework (Core Reference)*, Microsoft Press, Redmond, CA, 2003.

ARM processors core documentation, http://www.arm.com/documentation/ARMProcessor_Cores/index.html.

For a good reference on mobile software platforms and tools, see M. Mallick, *Mobile and Wireless Design Essentials*, John Wiley & Sons, New York, 2003.

For a good source on J2ME, see C. E. Ortiz, E. Giguere, and C. E. Ortiz, *Mobile Information Device Profile for Java 2 Micro Edition (J2ME): Professional Developer's Guide*, John Wiley & Sons, New York, 2001.

GPL license, http://www.gnu.org/copyleft/gpl.html.

Intel XScale technology, http://www.intel.com/design/intelxscale/.

J2ME, http://java.sun.com/j2me/; J2ME CLDC configuration, http://java.sun.com/products/cldc/overview.html.

Linux on PDAs, http://www.handhelds.org/; Linux devices, http://www.linuxdevices.com/ (website about the latest embedded devices using Linux).

Microsoft Windows Mobile, http://www.microsoft.com/windowsmobile/default.mspx; Windows Mobile for Smartphone, http://www.microsoft.com/windowsmobile/smartphone/default.mspx.

Palm OS at Palm Source, http://www.palmsource.com/palmos/; overview of Palm OS Cobalt, version 6.1, http://www.palmsource.com/palmos/cobalt.html.

Qualcomm Brew, http://brew.qualcomm.com/.

Symbian OS White Papers: http://www.symbian.com/technology/whitepapers.html.

References

[1] F. Koushanfar, V. Prabhu, M. Potkonjak, and J. M. Rabaey, Processors for mobile applications, in *Proc. 2000 IEEE Int. Conf. Computer Design: VLSI in Computers and Processors*, Austin, TX, 2000, page 603–608.

[2] M. Fleischmann, *Dynamic Power Management for Crusoe Processors*, White Paper, Transmeta Corp., Santa Clara, CA, 2001.

[3] Symbian OS v8.1 technical specifications, 2004 (http://www.symbian.com/technology/symbos-v81-det.html).

[4] The Mono Project website: http://www.mono-project.com/.

[5] J. Krikke, Samurai Romanesque: J2ME, and the battle for mobile cyberspace, *IEEE Comput. Graph. Appl.*, 23(1): 2004, page 16–23.

5

Mobile Networking Challenges

The trend of convergence in the domain of mobile computing encompasses a multitude of wireless technologies that must coexist and interoperate seamlessly. These technologies have opened up enormous opportunities for applications that can provide ubiquitous wireless data access and real-time communication over the forthcoming, fully integrated, high-speed networks. This chapter concerns a number of open issues with regard to the heterogeneous mobile wireless networks that smart phones or other mobile devices will access. A significant amount of effort by the network research community has been made to address these problems. The following issues represent the most important problems that must be addressed as 3G cellular networks and a variety of new wireless technologies are being rolled out:

- Mobile IP and mobile IPv6 (the foundation of future mobile wireless networks)
- Wireless transmission control protocol (TCP; enhancement for wireless environment)
- Convergence of heterogeneous wireless networks (issues and solutions in the integration of different wireless technologies)
- Wireless local area networks (LANs) and Bluetooth coexistence (schemes for wireless LANs and Bluetooth coexistence)
- Mobile next-generation networks (mobility support in next-generation networks)

- Mobile *ad hoc* networks (major issues such as topology control, routing, and service discovery)
- Quality of service (QoS) in mobile computing (QoS support in various wireless technologies)

5.1 Mobile IP

Recall that cdma2000 offers two types of services: simple IP and mobile IP. Simple IP service only allows a mobile station to roam seamlessly within areas under the umbrella of the same packet data service node (PDSN), because the IP address assigned to the mobile terminal is valid across these areas. If the mobile terminal moves to an area controlled by another PDSN, it has to obtain a new IP address to replace the old one, thereby disconnecting the current connection. In order to maintain a steady packet stream on a mobile device, transparent handoff is needed, which in turn implies transparent IP mobility. Of course, theoretically, mobility can also be achieved on layers other than the IP layer, such as on layer 2, using some kind of hardware address, or on an application layer. These approaches, however, do not scale well.

Mobile IP was initially proposed to solve the problem of IP mobility in a cellular network. Figure 5.1 shows how mobile IP works. A mobile device will always be identified by its home address, maintained by its home agent (HA). When the mobile device moves to another area, it receives a care-of address (CoA) from a foreign agent (FA). The foreign agent and home agent will coordinate to provide a connection as though the mobile device can still be reached via its home address. In fact, the home agent, most likely an access router, maintains a mapping between the mobile device's home address and foreign address. Any packet sent to the home address will be put into an IP-in-IP tunnel by the HA, and forwarded to the foreign address. At the end of this tunnel is the FA, whose job is to decapsulate the packet and forward the original data to the mobile device. To make the tunnel go through a network address translation (NAT) box, IP-in-IP

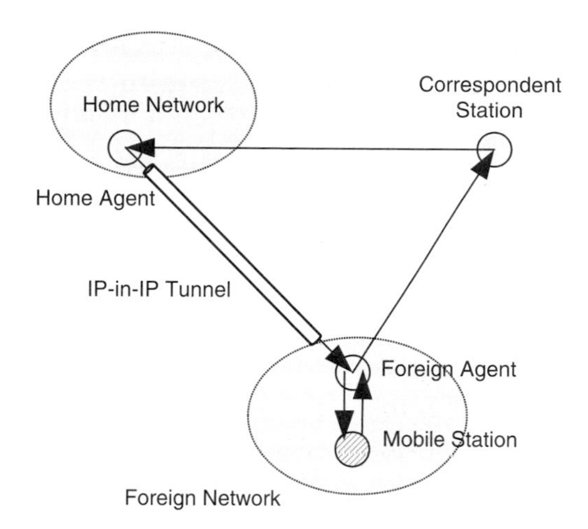

Figure 5.1 Mobile IP.

has been replaced by IP-in-UDP (User Datagram Protocol). Details of the mobile IP protocol can be found in RFC 3220 (All RFCs can be obtained from IETF's website: http://www.ietf.org/rfc/)[2][1]. Mobile IP essentially offers a way to identify a mobile device using the same address when it moves to another location and cannot be serviced by its original home agent. It does not require any modification to the application layer. Its drawback is the overhead of signaling and IP tunneling. To reduce the delay of signaling between the FA and HA, some techniques have been proposed to offer uninterrupted service during roaming. For example, one approach tries to create a temporary tunnel *before* a handoff begins (make-before-break). As a result, the packet sent to the mobile device will always be forwarded without interruption.

5.1.1 Mobile IPv6

It is commonly agreed that next-generation mobile networks will be based on IP. Ideally, in the foreseeable future, every networked computing or communication device will have an IP address allowing for some kind of network communication. This means that your land-line telephone, television, portable music players, cell phones, and even your watch will all have IP addresses, and the list of devices is growing longer everyday. Given the IP address space (version 4) of 32 bit, the total number of IPv4 addresses is about 4 billion, less than the number of human beings on Earth. A short-term solution to the IP address shortage problem is network address translation (NAT), as described in RFC 1631. NAT leverages ranges of globally nonroutable, private IP addresses[1] for internal use. NAT relies on a gateway known as the NAT box to translate source IP address (*e.g.*, 192.168.1.10) to a global IP address when an internal host is about to access the Internet. Outside the private network, hosts will only see the global IP address of the NAT box. For TCP and UDP connections, the original source port of a packet must be replaced as well. To deliver incoming TCP and UDP traffic corresponding to packets sent by an internal host, the NAT box must maintain a mapping between its serving source port and a tuple (internal host IP address, its original source port). NAT effectively enables many hosts to access the Internet with a single global IP address. Because of this unique advantage, NAT has been widely used in enterprise networks, cellular data networks, home networks, and wireless LAN hotspots; however, NAT does have some obvious problems. For one thing, a NAT extension is required for each application layer protocol using the IP address in the TCP or UDP payload (such as FTP and some instant messaging protocols), complicating the configuration and upgrade of NAT boxes because these applications will not pass through the NAT box otherwise. Furthermore, some argue that NAT violates the fundamental rule of protocol layering, because NAT in effect involves data from both the network layer and the transport layer. RFC 2993 describes the major drawbacks of NAT. Considering the staggering increase in the number of mobile wireless devices, it is not difficult to realize that the days of

---

1 As defined in RFC 1918, private IP addresses are: 10.0.0.0 -10.255.255.255, 172.16.0.0 - 172.31.255.255, 192.168.0.0-192.168.255.255. Additionally, an APIPA (Automatic Private IP Address) range is defined as 169.254.0.0 - 169.254.255.255. The APIPA is not reserved but often used by a host in case no DHCP server is available.

IPv4 are numbered. IPv6, the long-awaited upgrade to the foundation of today's Internet protocol, is finally making its way to reality, thanks to the convergence of computing, communication, and consumer electronics. Details of IPv6 can be found in RFC 2460, 2461, 2462, and 1881.

5.2 Wireless TCP

The TCP/IP protocol suite effectively dominates the Internet. TCP provides connection-oriented service over packet switching networks. Acknowledgment schemes are used to ensure TCP connection reliability. Round-trip time (RTT) estimates allow dynamic adjustment of the sliding window in order to achieve maximum throughput. The flow control and congestion control mechanisms of TCP help the transport adapt to network conditions and endpoint characteristics.

TCP assumes all packet losses are caused by congestion at links where buffers overrun and packets are dropped. Congestion control using slow-start, congestion avoidance, fast retransmission, and fast recovery are commonplace among many TCP implementations. Slow-start works by allowing exponential growth of the sender's congestion window for each received acknowledgment until it reaches a threshold, when the congestion window will enlarge linearly (congestion avoidance) until it reaches the receiver's advised window. The threshold is an estimate by the sender of how many packets can be transmitted at one time without causing congestion. Any packet loss detected by a timeout or duplicate acknowledgment will result in a threshold that is half the current congestion window size and a congestion avoidance procedure starting from a congestion window of 1 (or 2 in the Linux kernel). A receiver will send an immediate "duplicate ACK" message with an indication of the lost sequence number when an out-of-order segment is received. After receiving three "duplicate ACK" messages, the sender initiates fast retransmission by retransmitting the lost segment immediately without waiting for that segment's timeout. After that, the sender will begin fast recovery by conducting congestion avoidance rather than slow-start. Details of TCP congestion control can be found at RFC 2001 and Jacobson[13].

5.2.1 Wireless TCP Challenges

It should be noted that all the TCP schemes mentioned above are based on characteristics of wired networks. With regard to TCP for wireless communication, however, one must consider whether these TCP schemes will still perform well, as the underlying communication infrastructure is completely different. Wireless communications have imposed a number of challenges on TCP, including:

- *High packet drop rate* — In contrast to wired networks, wireless channels are known to have a quite high bit error rate, resulting in a high packet drop rate over IP and TCP layers. Handoff of mobile terminals across base stations also causes packet losses. Common TCP does not distinguish between packet drop due to network congestion or transmission error because the latter is extremely rare; however, wireless TCP must deal with both cases.
- *Unreliable wireless link* — The physical connection of wireless networks is subject to open-air interference from any radios in the proximity. The connection may break at any time, interrupting the TCP connection.
- *Mobility* — Mobile devices connecting to a wireless network move within a base station, across base stations, across administrative domains, or across heterogeneous networks, making it difficult to maintain a stable TCP connection.
- *Resource-limited mobile device* — Mobile devices usually have quite limited computing power, memory and storage, and power supply. Wireless TCP on mobile devices must take these into account.
- *Wireless link bandwidth vs. wired link bandwidth* — Wireless link bandwidths are generally much lower than wired link bandwidths.

5.2.2 Wireless TCP Protocols

Wireless TCP can be enhanced in several ways, including TCP-split, link-layer-assisted TCP, and regular TCP modifications (see Table 5.1). For wireless communication between a fixed host and a mobile host, TCP-split separates the wired connection between a fixed host and a base station or wireless access point with the wireless connection being made between a mobile terminal and the base station or wireless access point. As a result, the involved wired link does not require any modification because it is isolated from the wireless part where specific TCP modules are needed. Link-layer- and network-layer-assisted TCP aims to address packet loss at the link layer, instead of leaving it to the transport layer to detect and recover. Regular TCP modifications include enhancements to TCP mechanisms for wireless communication.

Below is a brief summary of some of the well-known solutions within each category.

- *TCP-split (I-TCP)* — In I-TCP [14], every connection from a mobile host (MH) to a fixed host (FH) through a mobility support router (MSR) is decomposed into two distinct portions: one wireless connection between the MH and MSR and one wired connection between the MSR and FH. The MSR maintains the association

Table 5.1 Wireless TCP Protocols

Category	Examples
TCP-split	I-TCP
	M-TCP
Link-layer-assisted TCP	Snoop/delayed duplicated acknowledgment (DDA)
	AIRMAIL
	RLP
TCP modification	Eifel
	TCP SACK
	Fast start

between the two links. If the MH moves to an area under the control of another MSR, related links will handoff to the new MSR. TCP-split solutions leverage the isolation of wired and wireless links, thus the FH does not require any modification for wireless communication. A drawback is that the TCP semantics of end-to-end connections is broken because the MSR serves as a proxy of MHs for the FH, and ACKs to the FH actually come from the MSR, not the MH. For applications that are unaware of the underlying environment, TCP semantics should not be altered. M-TCP [15] is another TCP-split scheme that was mainly proposed to work in wireless environment with frequent disconnections. M-TCP does not alter TCP semantics.

- *Link-layer-assisted wireless TCP (Snoop)* — Snoop [16] is a network-layer solution that utilizes a traffic interception module at the base station that resides between the fixed host sender and the mobile host receiver. In some sense, it can be considered a TCP-split solution, with the exception that Snoop does not change the semantics of TCP for either the sender and receiver. The Snoop module caches every unacknowledged TCP packet sent by the FH, which can be used for fast retransmission to the MH without notifying the FH in case of packet losses. Only if the base station receives ACKs from the MH does the Snoop module clear the buffered messages and propagate these ACKs back to the FH. For TCP packets sent from the MH to the FH, the Snoop module maintains those unacknowledged messages in case of packet loss in the wired network. It also generates negative acknowledges of lost packets in the wireless link (from the MH to the base station) and sends them to the MH for fast retransmission. Duplicate ACKs are suppressed by the Snoop module in either case. Delayed duplicated acknowledgment (DDA) is an extension of Snoop. It differs from Snoop in that the third and subsequent duplicate ACKs will be delayed for some time to avoid immediate retransmission to the sender.
- *TCP modifications (Eifel algorithm)* — For a long end-to-end, high-delay path that is composed of a number of wireless links, wireless

TCP suffers two major problems: spurious timeout and spurious retransmits. Spurious timeout occurs when the TCP sender does not wait for a sufficiently long time before timeout of the packet. Spurious fast retransmit happens because IP out-of-order delivery causes the sender to retransmit the packet when the acknowledgment duplicate threshold is reached. These two issues are largely due to the intrinsic TCP characteristics of ambiguous acknowledgments that may be caused by either the original TCP packet or the retransmitted ones. Aiming to address these issues, the Eifel algorithm [17] introduces the TCP time-stamp option in the TCP header. The time stamp consists of the sender clock, slow-start threshold, and congestion window size. Any ACKs will contain the time stamp of the corresponding packets, allowing the sender to restore the slow-start threshold and congestion window size in case the original packet is being acknowledged. As a result, spurious timeouts and fast retransmits are eliminated. TCP selective ACK (SACK) uses two TCP option fields to carry information of noncontiguous blocks of data in the ACKs from the receiver to the sender. TCP SACK is not specially designed for wireless networks, thus it cannot handle mobile handoff.

TCP modifications and link-layer-assisted protocols generally satisfy TCP semantics; however, TCP modifications require an upgrade of TCP modules on FHs, which does not seem to be feasible. Link-layer-assisted protocols are most suitable for a wireless environment that shows a higher bit error rate (BER), as the base station effectively reduces link-layer overhead by suppressing duplicate ACKs from the MH. TCP-split does not require any modification of the fixed network side, but it breaks TCP end-to-end semantics. Some TCP-split schemes such as I-TCP do not handle handoff disconnections. For a detailed review of wireless TCP protocols and comparisons, please refer to Hu and Sharma [18] and Elaarag [19]. Another major issue, mobile security and privacy, is discussed in the next chapter.

5.3 Convergence of Heterogeneous Wireless Networks

Wireless is everywhere. A large array of wireless technologies is being used in our everyday life, not to mention those used in industry, military, government agencies and organization, and scientific research; consequently, there are many types of communication networks, each comprised of intelligent mobile devices and fixed network infrastructures. In this broad domain, convergence is bound to happen in many aspects. Below is a list of issues that have been extensively explored by researchers from both academia and the wireless industry:

- Integration of cellular networks and wireless LAN hotspots
- Integration of wireless LAN and Bluetooth
- Integration of wireless LAN and cellular for enterprises
- Integration of wireless LAN and corporate IP private branch exchange (PBX)

An imminent trend in the domain is integration of cellular networks and wireless LAN hotspots. Integration of wireless LAN and Bluetooth implies coexistence with reduced interference rather than full integration. Integration of wireless LAN and cellular for enterprise emphasizes using wireless LAN as the preferred access network whenever available, whereas the problem of integration of wireless LAN with existing corporate IP PBX is primarily concerned with providing voice over wireless IP for mobile users.

5.3.1 Integration of Cellular Network and Wireless LANs

The integration of cellular networks and wireless LANs will benefit both wireless service providers who operate 2G and 3G cellular systems and wireless Internet service providers (WISPs) who control and administer a large number of wireless LAN hotspots. In many ways, these two services complement each other. Roaming between wireless LANs has been standardized by IEEE Std 802.11f (inter-access point protocol, or IAPP) and 802.11r (fast roaming eliminating perceptible disconnections). The wireless LAN section of Chapter 3 provides more details regarding these two standards. Cellular services have historically been facilitated by wireless wide area networks (WANs), in which each base station covers an area of a few square miles and a large number of such stations and backend mobile service controllers have been deployed to allow roaming across cities and even countries. Cellular networks are a well-established infrastructure in most populated areas. Designed for voice communication, cellular services cannot provide a very high data rate; even the 3G cdma2000 and universal mobile telecommunications system (UMTS)/wideband code-division multiple access (WCDMA) networks can only offer a peak data rate of up to 2 or 3 Mbps. Wireless LANs such as 802.11b, 802.11a, and 802.11g, on the other hand, are local area network technologies covering a distance of only a few hundred feet from the access point, but wireless LANs can deliver significantly higher performance in terms of data rate, which usually ranges from 11 to 54 Mbps. Because of the high data rate and the easy, no-wire setup, wireless LANs are becoming the most popular network access technology for home networks, enterprise networks, and networks for airports, hotels, conferences, parks, sport stadiums, school campuses, and so on. The marriage of wireless LANs and cellular networks will allow wireless operators to offer high data rate for mobile data access, while enabling WISPs to reach the huge user base of cellular service subscribers. End-to-end voice communication can be achieved via pure cellular networks or a combination of cellular and wireless LANs using voice over IP, and mobile data access can be optimized as well, depending on the coverage and traffic load of these two networks.

To access both cellular networks and wireless LANs, mobile terminals must be equipped with these two radio interfaces. More importantly, the mobile terminal must be smart enough to determine which access method will deliver better performance at minimal cost dynamically at a physical location. With regard to the convergence of cellular networks and wireless LANs, some relevant guidelines are listed here:

- The core of both networks should not be altered. This means that both networks should still work independently for their own business. In particular, radio interfaces and functional components of both networks need not be changed; software upgrades to these components are preferred.
- Both roaming and handoff across various types of networks should be supported in an integrated architecture. Roaming enables the use of a mobile access method in a new system or domain, while handoff ensures that network connection will not drop while a mobile terminal is moving across different networks or domains. Within the same type of cellular networks, roaming and handoff are already in place. These problems are particularly interesting when different types of networks are involved.
- Different scenarios may require different integrated network architectures. For example, for consumers who roam across wireless LAN hot spots and cellular covered areas, the key to the integration solution is to allow seamless roaming for both voice and data sessions. Such a solution is often provided by a union of wireless operators. For mobile workers who have a constant need to access an enterprise network, cellular-based mobile access can be augmented by wireless LAN based mobile access via some specialized components in the enterprise network. In either case, authentication, authorization, and accounting (AAA) mechanisms in both networks have to been coordinated.
- QoS policies and guarantees in one network should be maintained in the other.

5.3.2 Internetworking Systems Architecture

Let's begin with a most general case. A cellular network and a number of well-maintained wireless LANs operated by a WISP or a wireless operator are about to integrate to meet the needs of the above-mentioned list. The cellular network could be a global system for mobile (GSM)/general packet radio service (GPRS), a UMTS/WCDMA, or a type of cdma2000, each of which was discussed in the previous chapter. Whatever the cellular network, it is apparent that one or more "gateway"-like components are necessary to connect the cellular network and many wireless LANs. Depending on the way in which these components operate, two types of integration schemes have been devised [1]:

- Tightly coupled internetworking
- Loosely coupled internetworking

In the tightly coupled integration architecture, a wireless LAN gateway is necessary to act as an agent for a group of access points. It appears to the core cellular network as in a normal wireless access network. Data access to the Internet from a wireless LAN has to go through the cellular network. In the loosely coupled internetworking scheme, wireless LANs directly connect to the Internet and do not have any links to cellular core elements. High-speed data access from a mobile terminal within a wireless LAN is completely separated from that originating from cellular networks. Figure 5.2 shows a conceptual architectural comparison of these two schemes.

Recall that in a cdma2000 network, the packet data service node (PDSN) is responsible for controlling packet data traffic between mobile stations and the Internet. In GRPS and UMTS/WCDMA, it is the gateway GPRS node (GGSN) that performs the same task. According to the tight coupling scheme, a wireless LAN has a direct link to a PDSN in cdma2000 network or a GGSN in a UMTS network via different interfaces. By utilizing the wireless LAN air interface, high-speed data rate for a mobile station within the range of that wireless LAN can be achieved. The mobile station must support both types of air interfaces and implement both signaling and transport protocols. The wireless gateway that controls the connection from the wireless LAN to the PDSN works is a packet control function (PCF) in cdma2000 or a serving GPRS service node (SGSN) to the

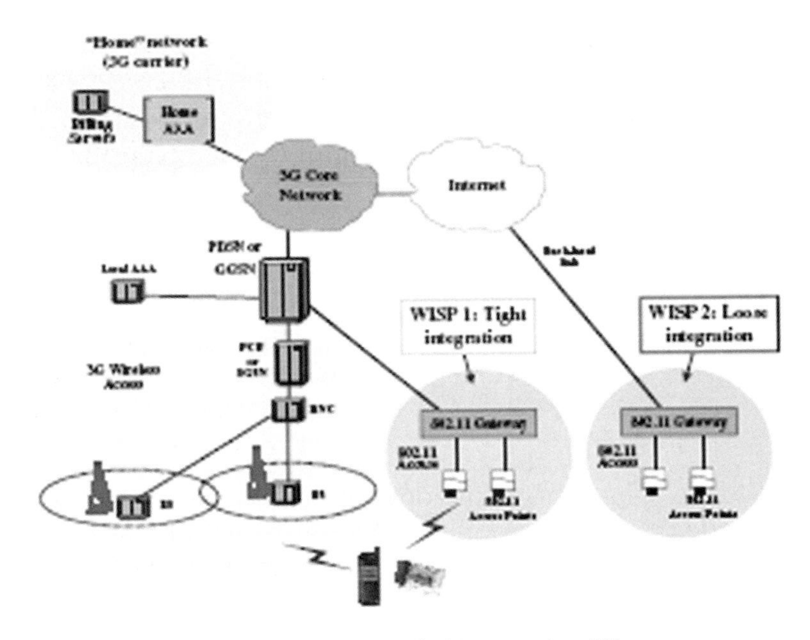

Figure 5.2 Tightly coupled integration and loose coupled integration [1].

PDSN in charge. Consequently, cellular-based authentication, authorization, and accounting (AAA) services can be extended to those wireless LANs. Handoffs between a wireless LAN and cellular access network (UMTS terrestrial radio access network [UTRAN] in UMTS and radio access network [RAN] in cdma2000) are carried out the same as handoffs within a cellular network. The prime disadvantage of the tightly coupled scheme is that a wireless operator's cellular network components are exposed to coupled wireless LANs that may not be under the same administration of the operator. Integrating wireless LANs deeply into the infrastructure of a cellular network necessitates reengineering the traffic load to avoid degradation of voice and data communication performance, because the wireless LAN brings a possibly heavy load into the cellular network.

The loosely coupled integration takes another approach to solve the problem. Here, the wireless LAN docs not have a direct link to PDSN or GGSN; instead, as usual, it connects to the Internet via backhaul links. A mobile station equipped with a wireless LAN air interface will no longer go through the cellular network for packet data access. When the mobile terminal moves within the coverage area of a cellular base station where no wireless LAN hotspot is available, data access falls back to cellular-based packet switching service. Because the cellular network and the wireless LAN are physically separated, a gateway component is required on the wireless LAN side to enable the interoperability of these two types of networks. To allow a unified per-user policing, the gateway component must collect identification information from mobile stations and AAA data from AAA servers in cellular networks. Specifically, when the cellular network is a cdma2000, the wireless LAN can take advantage of the mobile IP service provided by cdma2000 to enable roaming and handoff.

The existing 802.11 wireless LANs do not implement any quality of services (QoS) mechanism. Like many IP networks, they provide best-effort packet switching services with no guarantee of performance. No specialized signaling protocol has been accepted as the predominant mechanism for resource reservation for data access; however, cellular networks are based on connection-oriented circuit switching for both voice and data communication. QoS in cellular networks is generally implemented by employing signaling protocols such as packet data protocol (PDP) signaling in GRPS and UMTS networks. In an integrated network, wireless LANs are required to impose QoS such that the user's service level can be ensured in both types of networks. Expanding on the idea of QoS in wired IP networks, some researchers have proposed the use of differentiated services, integrated services, or multiple protocol label switching (MPLS) for the signaling of resource reservation in the core heterogeneous network and access network.

5.3.3 Integration of UMTS Networks and Wireless LANs

In UMTS networks coupled with wireless LANs, the GGSNs and SGSNs handle mobility across different types of networks. Because mobile IP has not officially been adopted as the protocol, this is an open issue that must be addressed. The general idea is to have handoff control functionality implemented at some selected components in the integrated network, with or without an implementation of mobile IP service. Figures 5.3 and 5.4 show two examples of this general approach.

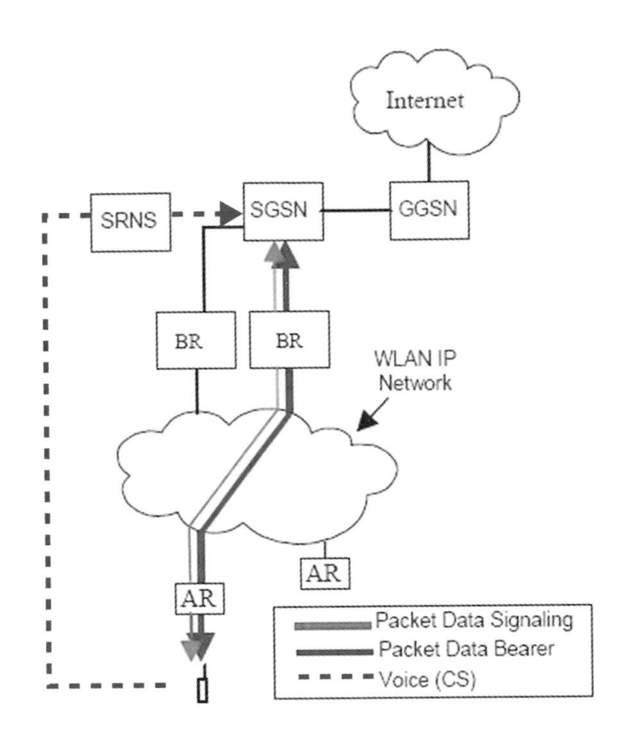

Figure 5.3 A Client-centric UMTS and WLAN integration [3].

A client-centric approach [3] requires the mobile station to maintain two simultaneous connections: a data connection (PS) to a wireless LAN and a voice connection (CS) through UMTS, in parallel. The UMTS connection is used only for voice service, while the wireless LAN connection provides both PS signaling and PS data transmission. This is effectively a tightly coupled solution in that a wireless LAN in the integrated network relies on a GGSN to connect to the Internet. This implies that a mobile station is able to use the IP address acquired from a serving GGSN of the GRPS/UMTS network, even if it is served by a wireless LAN for data access to the Internet. In addition, the mobile station and some access routers in the wireless LAN maintain the wireless LAN mobility context, whereas SGSNs and the mobile station both utilize the GPRS mobility context. Intersystem handover is performed by the mobile station, an access router (AR), an SGSN, and the serving radio network controller (SRNC),

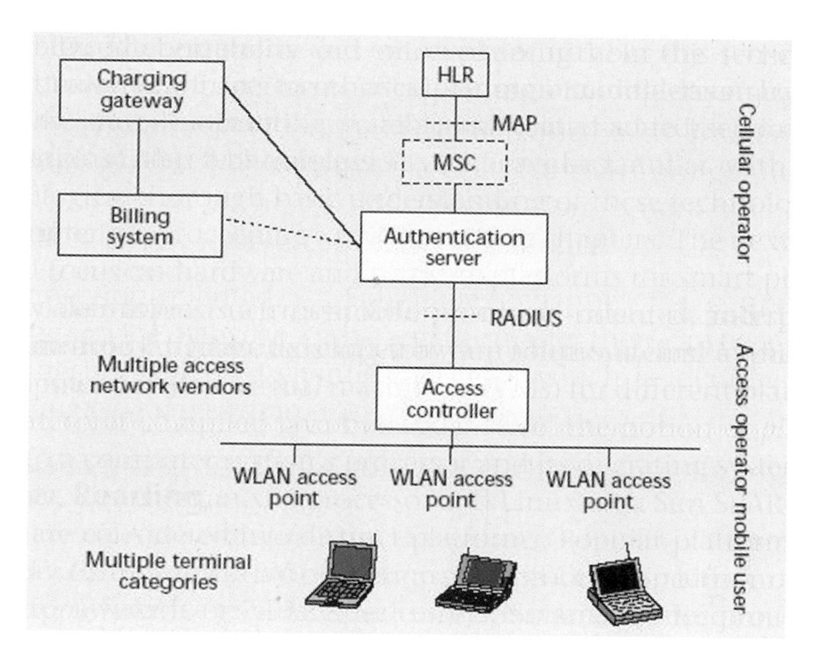

Figure 5.4 OWLAN elements [4].

on layer 2 and layer 3 sequentially. The mobile station detects the handover condition by constantly monitoring the wireless LAN beacon messages. PDP contexts are sent from SGSN to the AR if the mobile station is roaming from a cellular network to a wireless LAN.

Ala-Laurila et al. [4] presented a loosely coupled integration scheme for wireless operators to complete their cellular-based offering with operator wireless LANs (OWLANs). An OWLAN operates the same way as regular wireless LANs, the only exception being that control signaling of mobile station authentication is done in association with the GSM/GPRS/UMTS core network. To allow for subscriber identification module (SIM) authentication within an OWLAN, the proposed architecture must have four key components — authentication server, access controller, access point, and mobile terminal, each conceptually corresponding to a counterpart in a GPRS network. An authentication server acts as the gateway between the cellular core network and the operator IP core network. Like the SGSN of a GPRS network, the authentication server is in charge of authentication and billing signaling to the home location register (HLR) and GPRS charging server via Signaling System No. 7 (SS7). It may communicate with many access controllers according to the remote authentication dial-in user service (RADIUS) protocol. An access controller is the gateway on the wireless LAN side which gathers accounting information regarding mobile terminals within the range of those access points under its control. An access control is like a GGSN in a regular GRPS network. An access point provides physical wireless links to serving mobile terminals. It interfaces with the fixed operator IP core network through an access controller. A mobile terminal in this architecture is not necessarily a cell phone; instead, it could be any mobile device with a wireless LAN interface, a SIM reader, and a SIM authentication software module.

The OWLAN architecture employs a network access authentication and accounting protocol (NAAP) for signaling between a mobile terminal and an access controller. NNAP is a UDP/IP-based transport protocol but is enhanced with retransmission mechanism in case a packet is lost. GSM authentication messages are encapsulated into NNAP packets when they are sent to an access controller. The OWLAN approach can only provide SIM-based mobile roaming, not seamless handover, as uninterrupted dual-mode (wireless LAN and UMTS) mobile access is not considered in the architectural design; however, the OWLAN approach can be regarded as a near-term integration scheme for wireless operators to quickly incorporate the widespread wireless LAN technology into their cellular core, taking advantage of the high data rate offered by wireless LANs without affecting the core business of cellular voice services.

In practice, mobile network operators and cell phone manufactures have begun to design integration protocols for GSM/GPRS and wireless LANs. Unlicensed Mobile Access (UMA; http://www.umatechnology.org/) is such a specification supported by mobile network operators such as AT&T Wireless (now Cingular), T-Mobile US, and cell phone manufacturers such as Nokia and Ericsson, as well as wireless network equipment providers such as Alcatel and Nortel Networks. As a loosely coupled integration architecture, UMA enables a mobile device to access GSM/GPRS cellular backhaul via a local unlicensed wireless LAN and Bluetooth connections. According to the UMA specification, a key component in the UMA architecture is the UMA network controller (UNC), which interfaces with a mobile switching center (MSC), SGSN, and AAA server (or proxy) via an A-interface between an MSC and a base station controller (BSC)/UNC for circuit switch service, a Gb interface between SGSN and a BSC/UNC for packet switch service, and a Wm interface between an AAA server and a BSC or a secure gateway in UNC for AAA handling, respectively. A Wd interface is defined for communication between the AAA server in a home mobile network and the AAA proxy server in a visiting mobile network. A mobile station (MS) connects to a UNC via one or more wireless LAN/Bluetooth access points, which in turn connect to a broadband IP network. Figure 5.5 depicts the UMA functional architecture.

The MS in the UMA architecture must be able to operate in dual mode (GSM and unlicensed) and switch between them transparently. MS roaming and transaction control are provided by network elements in the core network such as MSC, visitor location register (VLR)/home location register (HLR), and SGSN/GGSN; a UNC is not involved but simply passes location updates and transaction data to

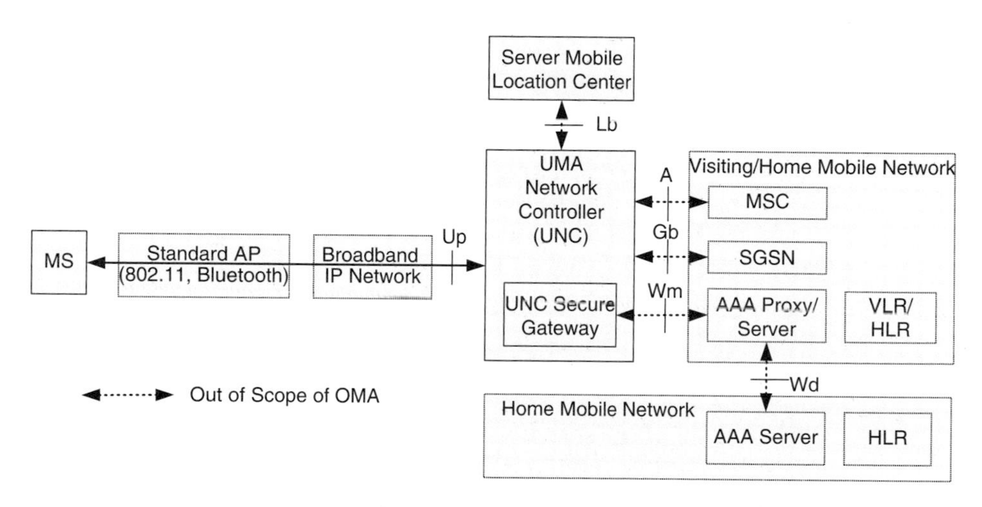

Figure 5.5 Unlicensed Mobile Access (UMA) functional architecture [5].

corresponding elements in the core network. UMA specifies the use of a number of GSM/GPRS and wireless LAN protocols without modification. Particularly, a UNC includes a security gateway (SGW) that terminates secure remote access tunnels from the MS, interfacing to an AAA server and providing mutual authentication, encryption, and data integrity for signaling, voice, and data traffic. UMA employs a layered security solution for voice, data, and control signaling security, including radio interface security between an MS and an AP, GSM/GPRS "Up" interface between an MS and a UNC, GSM/GPRS CN authentication and encryption between MS and the core network, and application security such as SSL between an MS and an application server. Overall, we consider UMA to be a solution likely to win the favor of mobile network operators, as it basically allows any public or private wireless LANs or Bluetooth networks to be a part of a cellular system. UMA could be a temporary solution to wireless LAN and cellular integration, but it may not prevail in the long run because it does not support the session initialization protocol (SIP; see Chapter 7) and handoff between wireless access points. In the communication industry, despite increasing debate over the idea of voice over wireless IP, voice over IP service providers have begun to offer voice over Wi-Fi servers along with Wi-Fi phones. For example, Vonage launched such services in the United Kingdom in early 2005.

5.3.4 Integration of cdma2000 Networks and Wireless LANs

For a cdma2000 network to support mobile IP, PDSNs must implement FA functionality to deal with visiting mobile stations, and the carrier's network must maintain an HA for its users. To integrate loosely with a wireless LAN, the wireless LAN gateway must also implement FA functionality so as to establish a tunnel between the visiting mobile station and its home agent back in the cellular network. The wireless gateway also must be able to interface with the AAA server of the mobile station to retrieve user account policy and accounting information, as well as tunneling parameters and home IP address. An example of cdma2000 cellular network and wireless LAN integration is the integration of two access (IOTA) technologies gateway proposed and implemented by researchers at Bell Labs, Lucent Technologies [1]. The IOTA gateway is deployed in each wireless LAN to support mobile IP functions for the mobile devices served by the wireless LAN. In addition to mobility management, the IOTA gateway also has a component of AAA server that can either relay 802.11-based information between the underlying wireless LAN's access points and the mobile device's home AAA server or interact with the gateway's mobile IP agent to authenticate a mobile device with its home AAA server. Seamless QoS implementation can also be achieved on the IOTA gateway.

The IOTA gateway is software system consisting of the following components:

- *Remote authentication dial-in user service* — A RADIUS server is an AAA server that relays authentication packets between a mobile station's home AAA server and the 802.11 wireless LAN access points.
- *Mobile IP agent* — The IOTA implements both the foreign agent and home agent, thereby bringing mobility support to wireless LANs. The FA will handle new moving-in mobile devices by issuing a care-of address (CoA) and notifying the HA, just like a foreign agent works in cellular networks. The HA is somewhat special in that it will be used to support dynamic home agent allocation, which allows a mobile device to have a non-fixed home agent (and home address) close to the foreign agent.
- *Dynamic firewall* — Per-user firewall polices and packet mangling rules obtained from home AAA servers can be installed dynamically.
- *QoS module* — The IOTA gateway obtains user service class from home AAA servers and imposes individual bandwidth control over wireless LAN access points and backhaul links to the Internet. The QoS guarantee is only valid within the wireless LAN access network. The provision of true end-to-end QoS service requires the entire cdma2000 network infrastructure to implement a QoS mechanism.

- *Accounting module* — The IOTA gateway maintains an accounting table that can be queried by authentication entities such as foreign agents, local AAA servers, or web authenticators used in the simple IP service.
- *Integrated web cache* — The IOTA employs a web cache that utilizes IP packet mangling techniques to avoid unnecessary round-trips to a mobile station's home network for mobile IP services.
- *Simple IP module* — This module enables roaming of mobile access but not seamless handoff. It combines dynamic IP address assignment and IP packet filtering to provide browser redirection service, in which the user must be authenticated when a web browser is launched.

The IOTA approach can also be applied to the integration of UMTS/WCDMA and wireless LANs. More precisely, mobile IP and IEEE AAA can be introduced as an overlay on top of UMTS networks. GGSN in UMTS will handle the tasks of a FA, and authentication can be done by standard mobile IP procedures using SIM cards. No matter what underlying cellular network is being used, the client software of the IOTA architecture remains the same, within which a complete mobile IP stack and an intelligent multi-interface selector have been provided. The client software has an interface abstraction layer that supplies a unified virtual interface to the underlying mobile station system.

5.3.5 Integration of Wireless LANs and Cellular Networks for Enterprise

The above-mentioned cellular and wireless LAN integration architectures are operator oriented in that the wireless service providers (wireless operators or WISPs, or both) must build an integrated or interoperable wireless WAN and negotiate service and billing agreements. Another scenario in this domain must be addressed, though: integrated data access to enterprise networks. Very often mobile professionals must frequently access enterprise networks via secured connections, such as virtual private networks (VPNs). Cellular data service (*i.e.*, using a GSM or CDMA modem on a laptop computer or a handheld mobile device) used to be the prime access method for this purpose. With the proliferation of public and private wireless LANs, it would be quite compelling for mobile users to be able to access enterprise networks via a secure, integrated environment within which cellular data service and wireless LAN data service are interoperable and roaming between them does not break the ongoing connection. To achieve this goal, the enterprise network must be enhanced to realize functionality for secure IP mobility.

An solution to this issue is Internet roaming [6,7], which provides seamless internetworking across office wireless LANs, residential wireless LANs, public wireless LANs, and cellular networks for corporate mobile users. Internet roaming is a mobile access system built into the enterprise network and mobile devices, rather than a set of add-on entities to existing cellular networks or wireless LAN networks; therefore it can be applied to mobile access for different cellular data networks. It is comprised of four building blocks:

- Virtual single account (VSA) server, which is a backend user authentication server. It stores various security credentials for users and provides a single sign-on procedure for users who may use various wireless access networks to log onto the enterprise network.

- Secure mobility gateway (SMG), which acts as the gateway between an enterprise network and the Internet to enable secure data transmission between the mobile station and an enterprise computer. Depending on the wireless access method the mobile station is using, packets relayed by SMG may be encrypted using IPSec.

- Internet roaming client (IRC), which is the client (hardware or software) installed on the mobile station. The major task of the IRC is to create and maintain a mobile IPSec tunnel (mobile IP with IPSec encryption) to the SMG interface. In this case, the SMG is the home agent, and the IRC acts as the foreign agent. Mobile IP packets in the tunnel are encrypted and encapsulated while being transmitted in the tunnel in an IP-in-UDP fashion.

When the mobile station is using an office wireless LAN to access the enterprise network, IPSec encryption over the mobile IP tunnel is not necessary, assuming that the office wireless LAN is well secured. Similar to OWLAN, the Internet roaming architecture requires mobile stations to detect the optimal (not necessarily faster) wireless network access method and to dynamically and properly configure itself without user intervention. Researchers at AT&T who proposed the Internet roaming solution have implemented a hardware-based IRC called iCard. More details can be found at the AT&T website (provided in the Further Reading section at the end of this chapter).

Virtual private networks are widely used to secure enterprise data access at public and residential wireless LANs. A VPN essentially provides secured communication over public networks by encrypting traffic in a tunnel. Traditional VPNs are designed for wired network access; roaming is not supported. VPNs supporting station roaming are called mobile VPNs and include NetMotion and Ecutel.

5.3.6 Integration of Wireless LANs and Corporate IP PBX

We have talked about dual-mode (cellular and wireless LAN) mobile devices for corporate professionals allowing access to wireless LANs and wide area cellular networks, but what about replacing aged corporate PBXs with an IP PBX supporting voice over IP using a LAN? Not only do IP PBX-based solutions reduce operating cost, but they also make it easier to integrate the communication infrastructure and applications with the computing infrastructure and applications. To this end, companies such as Cisco and Nortel have developed IP PBX systems and associated IP telephones and mobile phones. For example, Cisco's 7900 series IP phones include both Ethernet-based telephones and 802.11-based wireless phones.

While wireless IP phones may work well within a corporate network, roaming between an enterprise VoIP system and a wide area cellular system remains a big challenge. UMA is not ideal in this scenario because it lacks support for SIP, a seemingly prevalent multimedia session management protocol in all-IP networks. 3GPP proposed an alternative solution called IP Multimedia Subsystem (IMS). Unlike UMA, IMS supports SIP and largely utilizes the Internet for call routing and presence services. Compared to UMA, IMS is a more fully fledged all-IP solution for 3G systems. The IMS architecture supports multiple application servers providing both traditional telephony services and enhanced data services such as instant messaging (IM), push-to-talk (PTT), audio and video streaming, or multimedia messaging service (MMS). Readers interested in reading more about IMS should refer to the following 3GPP technical specifications: 3GPP TS 22.228 (Service Requirements for the IP Multimedia Core Network Subsystem, Stage 1) and 3GPP TS 23.228 (IP Multimedia Subsystem, Stage 2).

5.4 Wireless LAN and Bluetooth Coexistence

Wireless LANs and Bluetooth are the two most widely used wireless technologies aside from cellular networks. Bluetooth is mainly used for wireless personal area networks (PANs) as a cable replacement extending up to 10 m, whereas wireless LANs are able to cover up to several hundred meters. In this sense, they complement each other rather than competing head to head. In fact, in many cases, mobile devices are equipped with both interfaces.

5.4.1 Frequency Overlapping

As discussed in Chapter 3, wireless LAN and Bluetooth both operate at the 2.4-GHz unlicensed frequency band. Recall that wireless LANs (*e.g.*, IEEE 802.11b and g), employ the spread spectrum over the 2.4-GHz band, which greatly reduces interference with other radios operating in the same band. Bluetooth radio uses frequency hopping at a rate of 1600 times per second over 79 channels shared by a master device and up to 7 slave devices. Nevertheless, it has been observed that when these two radios operate in very close proximity (for example, on a PDA with both air interfaces), interference will occur and the performance of both will drop significantly. To address this issue, IEEE 802.15.2 and Bluetooth SIG working groups have been formed. The IEEE 802.15.2-2003 standard was completed in August 2003. The basic approaches can be divided into two categories: collaborative and noncollaborative. The collaborative approach aims to coordinate frequency use between two types of radios, thus requiring interprotocol communication that must be implemented on the same device. The noncollaborative approach looks at only one side; that is, it tries to mitigate interference by reducing the chances of frequency overlapping, but it requires the acting device to detect interference or conduct interference estimation. We introduce packet traffic arbitration (PTA) [8] as an example of the collaborative approach and adaptive frequency hopping (APH) [9] as an example of the noncollaborative approach.

5.4.2 PTA and Adaptive Frequency Hopping

Packet traffic arbitration is a coexistence scheme recommended by the IEEE 802.15.2 group. As shown in Figure 5.6, the PTA control entity acts as an arbitrator between wireless LAN MAC and Bluetooth MAC in a time-division multiplexing (TDM) fashion. Communication between them is accomplished using a handshake protocol. The PTA control entity constantly monitors the status of each MAC and ensures that transmissions from each side are first authorized before they can take place. To do so, the wireless LAN interface and the Bluetooth interface have to be located on the same board or device where the PTA resides. Alternating wireless medium access (AWMA) is a similar approach that divides time into Bluetooth time slots and wireless LAN time slots, thereby avoiding any overlapping of frequency use.

Adaptive frequency hopping is a noncollaborative approaching aimed at addressing the coexistence issue by dynamically adjusting the frequency hopping sequence of Bluetooth. The master device selects 32 frequencies out of a total number of 79 available frequencies within the 2.402- to 2.480-GHz range. After all the 32 frequencies have been visited, 16 of them and another 16 new ones from the

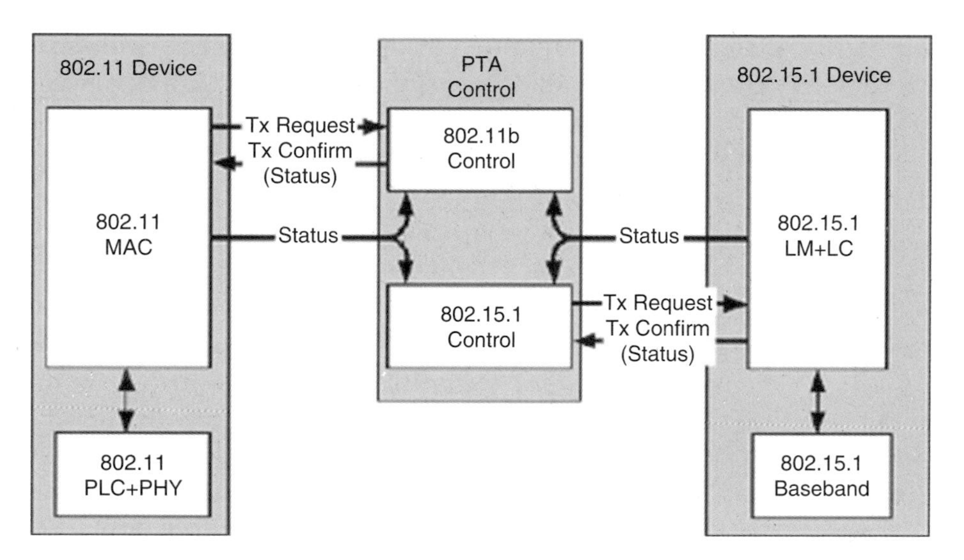

Figure 5.6 Packet traffic arbitration (PTA) collaborativemechanism for coexistence of 802.11b and Bluetooth technology (courtesy of Intel Corp. © 2005 Intel Corporation).

original list are selected for the next hopping sequence. The new hopping sequence among devices is broadcast within a link management protocol (LMP) message. Thus, the first problem that must be addressed is to let the master device be aware of "bad" frequencies so it can skip them to avoid overlapping. This is often done by both the master device and the slaves to conduct measurement based on received data. Metrics frequently used are received signal strength indication (RSSI), packet error rate (PER), signal-to-interference ratio (SIR), and packet loss rate, on a frequency basis. For each metric a threshold can be used to identify occupied frequencies. To further improve resilience against frequency overlapping interferences, AFH requires the master and the slave device to communicate within the same channel, unlike regular Bluetooth specifications in which slaves respond using a frequency channel other than that used by the master. This method eliminates the possibility that one of the two channels used by either the master or the slave is bad. Aside from AFH, another idea is to postpone data transmission on those bad frequencies, rather than changing hopping sequence. Bluetooth interference aware scheduling (BIAS) is such a mechanism. It is worth noting that the collaborative approach requires hardware modification to the Bluetooth radio interface. A detailed review of solutions to Bluetooth and wireless LAN coexistence problems can be found in Golmie *et al.* [10].

5.5 Mobile Next-Generation Networks

The notion of mobile next-generation networks (NGNs) refers to futuristic computer and communication networks that are characterized by a dazzling diversity of wireless and wired technologies as well as a myriad of mobile applications and services. Some view NGNs as a merger of the Internet and intranets with mobile networks and with media and broadcast technologies [11]. 3G cellular systems such as cdma2000 and UMTS are simply technological steps toward mobile NGNs, which essentially enable pervasive and ubiquitous computing on a variety of mobile terminal devices. The integration of cellular networks and wireless LANs and coexistence problems of wireless LAN and Bluetooth are subsets within the scope of mobile NGNs.

Integrating and coordinating various wireless and wired technologies in mobile NGNs have generated several challenges, such as mobility management, security, and terminal convergence. In addressing those issues, Internet protocol (IP) has been generally recognized as the universal network layer protocol in the interconnected infrastructure. Notice that the cdma2000 standard has shown a clear trend toward an all-IP solution with the introduction of mobile IP services. UMTS has also introduced IP Multimedia Subsystem (IMS), an IPv6-based architecture for real-time voice and video data services in a packet switch domain. In this section, we focus on IP-based mobility management in mobile NGNs, leaving security to the next chapter and mobile application issues to Chapter 7.

Issues of mobility management in mobile NGNs can be divided into two categories: roaming and handoff. Here, roaming refers to mobile access availability in a new system (or domain), whereas handoff refers to network connections staying active when the mobile station moves to a new system (or domain). No wonder handoff is more seamless but more difficult to achieve. In particular, within the context of mobile NGNs, vertical handoff (*i.e.*, handover between different networks, systems, service provides, or administrative domains) is the key issue, as horizontal handoff (*i.e.*, handover between the same types of networks) can be resolved relatively easily due to homogeneity of underlying networks. Mobility management of mobile NGNs can be performed at different layers; for example, mobile IP, as described in Section 5.1, is a network layer solution for both roaming and handoff (although enhancement is needed to achieve true seamless mobility), and cdma2000 provides support for mobility at the link layer.

5.5.1 Mobile IP for Macromobility

Mobile IP is able to facilitate both macromobility and micromobility. In the case of macromobility, where a mobile station is moving across two administrative domains (thus, two access routers) of the same or different wireless networks, mobile IP requires both networks to implement home agents and foreign agents on some network elements. Recall that packets sent by a correspondent station from a home network to the mobile station are intercepted by the home agent and then tunneled to the mobile station in the form of an encapsulated packet headed for the care-of address (CoA) via the foreign agent. Because the mobile station is able to directly communicate with the correspondent station, a triangular routing path is formed which, on the one hand, effectively allows communication from and to the mobile station but, on the other hand, introduces IP routing overhead for traffic from correspondent station to the mobile station, as shown in Figure 5.7a. This problem can be solved by allowing the home agent to notify the correspondent station the CoA of the mobile station in question using a "binding update" message. The corresponding station maintains what is called a "binding cache" for the latest known CoA of the mobile station. Hence, direct communication between the two parties can be performed within the home agent's relay. Figure 5.7b depicts the routing optimization scheme

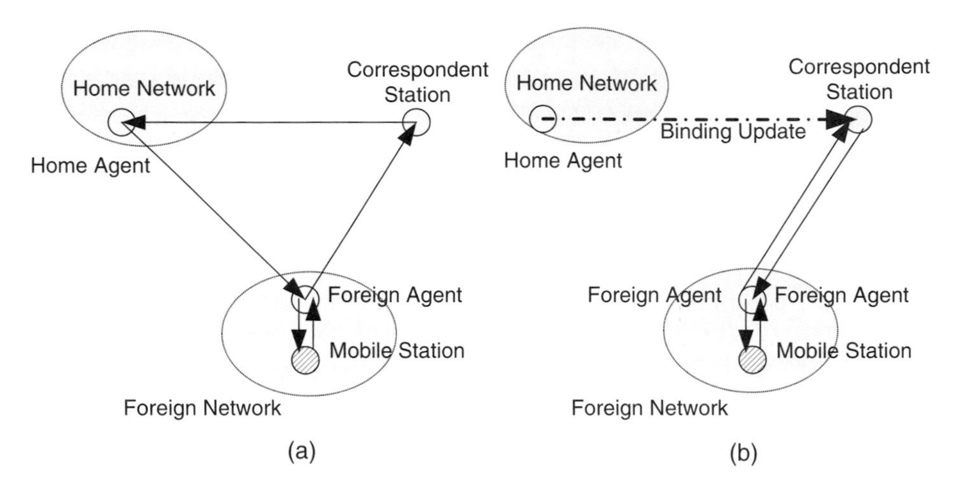

Figure 5.7 Mobile IP enhancements.

for mobile IP. Whenever the mobile station moves from one foreign agent to another, the new foreign agent notifies the old one of this change, and corresponding stations will be informed as well.

5.5.2 Mobile IP for Micromobility

Within one domain a mobile station may frequently move across different subnets. It is inefficient to notify the home agent every time such intradomain roaming occurs, as signaling of the CoA change and binding update may impose a large overhead on communication between a correspondent station and the mobile station. The solution, as in many other computer or network scenarios, is to leverage locality in some sense. In this case, a basic idea is to build a hierarchical architecture based on mobile IP such that mobility within a domain will not result in any CoA updates on the home agent of a mobile station. In its simplest form, a two-layer scheme called mobile IP regional registration, shown in Figure 5.8a, introduces a gateway foreign agent (GFA) in a domain along the path between the home agent in the home network and the foreign agents. The home agent simply maintains the public routable address of the GFA, rather than the CoA of the mobile station. CoA updates from

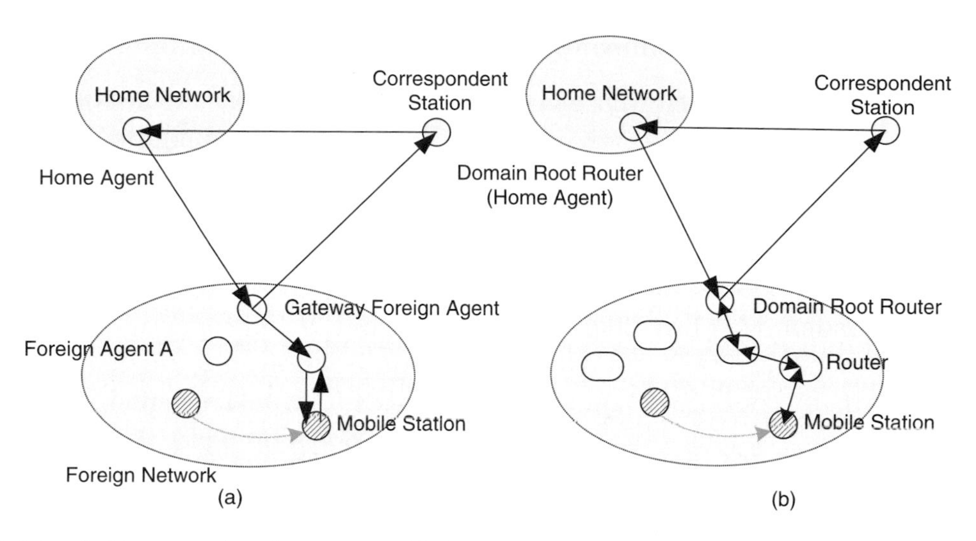

Figure 5.8 Mobile IP regional registration and HAWAII.

a foreign agent only reach the underlying GFA; consequently, the home agent has been relieved of micromobility update messaging. For data traffic from corresponding stations to the mobile station, the GFA is responsible for forwarding those packets to the foreign agent at that point of time, which further relays traffic to the mobile station. If the mobile station roams beyond the control of the GFA, regular mobile IP operation that involved signaling to the home agent will be performed. For large domains with many subnets, multiple levels of foreign agents can be formed to reduce the overhead of signaling for mobility management.

Another scheme, called handoff aware wireless access internet infrastructure (HAWAII), aims to build host-specific paths among routers in a domain for visiting mobile stations [12]. A "domain root router" for each domain is designated to handle intradomain and interdomain mobility, as shown in Figure 5.8b. When the mobile station is in its home domain, it initiates a path discovery by sending out a power-up message to serving base stations toward the domain root router. Routers on the way to the domain root router create a soft-state routing entry for the mobile station. In addition, the mobile station will send path refresh messages periodically to indicate its presence, thereby refreshing relevant routing caches. When the mobile station moves within the home domain, it performs a path update operation so as to build an updated path to the domain root router. When it moves to another domain (foreign domain), the domain root router of the new domain will act as the foreign agent to the mobile station, allowing intramobility within the foreign domain. It achieves this by assigning a co-located care-of address (CCoA) to the visiting mobile station and notifying the home agent (the domain root router of the home domain) of this change. The mobile station also performs path update and path refresh operations in order to build a host-specific path toward the domain root router in the foreign domain, as it does in its home domain. The path is made up of a sequence of routers that relay packets destined to the mobile station in a hop-by-hop manner. As in common mobile IP architecture, packets destined to the mobile station from correspondent stations will still be tunneled by the home agent to the foreign agent. Intradomain mobility in the foreign domain results in routing path updates over those intermediate routers and the domain root router of the foreign domain, but the CCoA of the mobile station will not change, thus no signaling to the home agent is needed. Notice that host-specific routing based on specific mobile addresses rather than IP routing is used in both the home domain and foreign domain.

The major advantage of host-specific routing in HAWAII is that it avoids the overhead of IP address changes and additional layers of gateways. The disadvantage is the high cost of building and maintaining paths within a domain, which considerably limits its scalability in serving many mobile stations in the domain. A similar scheme is used in cell IP.

5.5.3 Link Layer Mobility Management

Mobility management for heterogeneous next-generation networks can be performed at the data link layer as well as the network layer. Roaming and seamless handoff of mobile stations can be achieved by translating different formats of signaling messages, data packets, and QoS specs among heterogeneous wireless systems. This often implies the introduction of some gateway entities such as the wireless LAN gateway used for integration of wireless LAN and cellular networks discussed in Section 5.3.1. The newly added gateway entities not only perform mobility management but also deal with location management across a variety of wireless systems.

The UMTS standard defines an optional network entity called gateway location register (GLR) for intersystem roaming. In UMTS, the service area is partitioned into gateway location areas (G-LAs), which are further divided into location areas (LAs), which are comprised of a number of cell sites. Depending on the extent of mobility, three scenarios can be identified:

- *G-LA* — An HLR location update is performed when a mobile station crosses the boundary of a G-LA.
- *LA* — A GLR location update is performed when a mobile station crosses the boundary of an LA.
- *Cell* — A VLR location update is performed when a mobile station completes d movements between cells, where d is the movement threshold

As shown in Figure 5.9, the gateway location register (GLR) is a node between the VLR and the HLR which may be used to optimize the handling of subscriber profile data across network boundaries. In the visited network, the GLR plays the role of the HLR toward the VLR, whereas in the home network the GLR plays the role of VLR toward the HLR. Because of the layered location management used in UMTS, the GLR handles any location changes between various VLR service areas (LAs) in the visited network without involving the HLR. The G-LAs and LAs can represent of different types of wireless networks such as UMTS and cdma2000.

Link layer mobility management usually involves MAC layer communication across interconnected system. Due to the diversity of wireless technologies, it is rather difficult to devise a unified

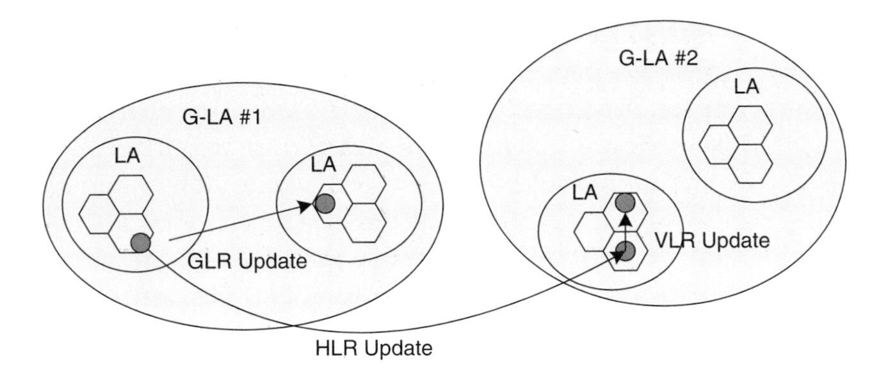

Figure 5.9 Gateway location register (GLR) in UMTS.

link layer handoff mechanism for mobile next-generation networks. Instead, depending on the overlapping or adjacent wireless systems, specialized MAC-based entities are added to one or both of the interconnected networks; for example, the PCA for Bluetooth and wireless LAN coexistence monitors and controls the MAC operations of both networks.

5.6 Mobile *Ad Hoc* Networks

Mobile *ad hoc* networks (MANETs) have been the subject of mobile computing research for years. Unlike infrastructure-based networks, an *ad hoc* network does not have any preexisting or fixed centralized server nodes to perform packet routing, directory assistance, service discovery, resource management, and signaling. Nodes in an *ad hoc* network must self-organize to form a fully distributed, cooperative environment within which a variety of computing systems can be built. A MANET is an *ad hoc* network of mobile nodes that communicate with each other over wireless links, thereby creating a highly a dynamic network environment in terms of topology, packet routing, and service provision and consumption. Communication between two nonadjacent nodes in a MANET is conducted in a self-discovered multihop fashion. In many circumstances, a hybrid *ad hoc* network of a fixed network infrastructure and collaborative mobile nodes is often desired. For simplicity, MANET is defined as including both fully mobile multihop *ad hoc* networks and hybrid *ad hoc* networks.

The intrinsic flexible network architecture of MANET, along with the advancement of mobile wireless technologies, has spawned a variety of potential applications. Some of the examples are listed as follows:

- *Rescue operations and disaster recovery* — A MANET of mobile devices such as laptop computers, PDAs, smart phones, and handheld industrial devices can be formed autonomously onsite to supply fast and reliable data communication.
- *Military operations in a battlefield* — Vehicles and soldiers equipped with wireless communication capabilities and embedded sensory devices may utilize the underlying *ad hoc* network for real-time communication in a remote battlefield. In conjunction with military satellites and global positioning system (GPS) facilities, the mobile *ad hoc* network can be further empowered to allow accurate positioning, quick response, and commanding.
- *Conferences and events* — Attendees in a conference room, a sports stadium, or anywhere an event takes place can form temporary *ad hoc* networks that permit efficient mobile access to the Internet or other resources, as well as one-to-one and group-based instant collaboration. For example, smart phone users in a sports stadium can leverage a MANET of cellular *ad hoc* network for mobile access.
- *Remote monitoring and surveillance system* — Wireless-enabled cameras, sensors, and computing devices can be used to build a pervasive *ad hoc* surveillance system that does not require preconfigured central control.
- *Vehicular communication system* — Cars on a highway can temporarily form an *ad hoc* network to provide Internet access, real-time traffic information, location tracking, and location-based searches. In conjunction with the increasing use of telematics,

vehicular *ad hoc* networks can be used to link data access to in-car sensory functionality.

5.6.1 MANET Categories

A mobile *ad hoc* network (MANET) essentially encompasses several other networking applications involving various wireless technologies. Wireless sensor networks (WSNs), for example, are MANETs in the sense that sensor nodes, either fixed or moving, are very often used as an *ad hoc* network environment for various purposes such as habitat monitoring, industrial monitoring and control, and home networking, among others. A wireless mesh network is another example of a MANET. In a wireless mesh network, wireless LAN or cellular base stations placed across a geographical area form an *ad hoc* network to provide fixed wireless Internet access to mobile devices. Because of the diversity of specific MANET application scenarios, this section concentrates on general MANET.

The challenge of providing reliable and highly efficient computing and communication services in a fully decentralized architecture has spurred a vast amount of research effort in the domain of mobile computing.

- *Topology control* — A mobile node in a MANET must establish a localized view of the network within which neighboring nodes can be identified. Because the network topology of a MANET is subject to rapid changes, it has to be adaptive for performance with minimal communication overhead.
- *Routing* — Not only are the network dynamics of a MANET (such as topology changes) considered with regard to packet routing, but also power consumption and location-based data.
- *Service discovery and access* — A key issue in a MANET is to discover provisioned services and consume them in a highly efficient and reliable manner. This involves node addressing, data caching, and service proxying.

- *Data acquisition and aggregation* — In WSNs, a predominant application scenario is a sensory network database, in which sensory operations are tightly coupled with network-based data acquisition and aggregation over the *ad hoc* transport.
- *Security and privacy* — Because mobile nodes in a MANET rely on others to work properly, a number of security and privacy issues such as node authentication, integrity security of data communication, and bilateral trust must be carefully addressed.

In the following subsections, we focus on the first four issues, leaving the last security issue to the next chapter.

5.6.2 Topology Control in a MANET

Two key issues of topology control in a MANET are *neighbor discovery* and *network organization*. The neighbor discovery problem is concerned with controlling transmission power and antenna of a mobile node with the objective of constructing an optimal topology. The network organization problem aims to adjust MANET topology in response to node mobility and wireless network dynamics. In either case, network throughput is a key factor in choosing the appropriate topology for a MANET. In addition, a topology control scheme has to be energy efficient; that is, the construction of a desired topology must consume only a minimal amount of power. The fundamental challenge of topology control in MANET lies in the fact that a mobile node can only possess a local knowledge of the network connectivity, thereby making a globally optimal topology very difficult to achieve.

Many topology control algorithms have been proposed to solve these two problems. The basic idea is described as follows: Given a *transmission graph* that models node connectivity of a MANET, find a distributed algorithm that generates a subgraph comprised of optimal paths between any nodes with respect to network throughput and power consumption. Figure 5.10 illustrates an example of MANET topology control. There are seven mobile nodes, N1 to N7. Some nodes, such as N1 and N3, are capable of reaching everyone else

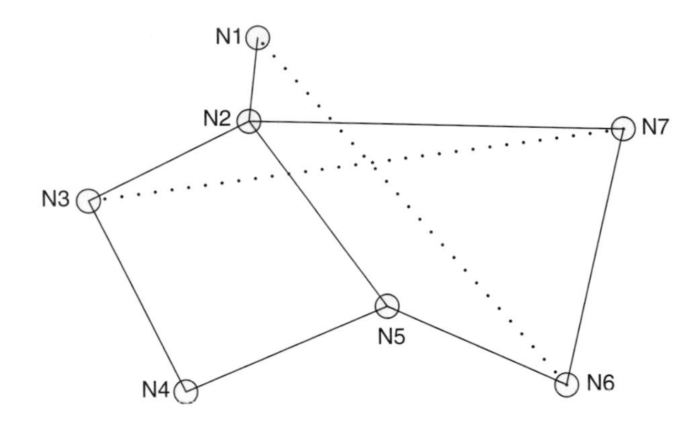

Figure 5.10 An example of MANET topology control.

in the network by adjusting their transmission power. Others can only directly communicate with some adjacent nodes. Solid lines are used to denote direct wireless links between two nodes. The length of a line in the figure represents the distance between two neighbor nodes. As a simple example of topology control, let's look at N1. For N1 to communicate with N6, it can simply increase transmission power to discover N6 as a neighbor node; hence, as long as the received signal power is above some threshold at N6 after path loss, the unidirectional link between N1 and N6 works. However, N1 may opt for a path of N1–N2–N5–N6 because it is highly possible that the cumulative power required for N1, N2, and N5 to transmit over direct links is lower than the power needed to transmit directly between N1 and N6 (the reason for this is discussed in the following paragraph), and the performance (*e.g.*, throughput, delay) of this transmission still satisfies the needs of applications. Such a justification is based on the objective of topology control in this example: overall power efficiency for network longevity assuming every node possesses the same amount of power. Similar topology control occurs with N3 and N7, as well. A slightly different example of traffic control is N4, which has three nodes, N3, N5, and N6, within its signal transmission range. Suppose N4 uses a directional antenna for wireless communication; it can either link to both N3 and N5 in one direction or N6 in another direction. If N4 can detect that, by adjusting the antenna, it can reach anyone else through N3 and N5 with moderate power consumption and acceptable performance, it may not have to establish a direct link to N6 because unnecessary wireless links tend to result in contention and interference, as well as hidden terminal and exposed terminal problems.

These examples illustrate that the basic idea of topology control in MANET is to leverage controllable parameters such as transmission power and directional antenna to build a network topology for power-efficient operations, network connectivity, and good performance. The challenges are also evident: How does a node know, based on its local knowledge, that some nodes should be selected as neighbors and some should not? How does a node know another reachable node has to be a neighbor node in order to maintain network connectivity?

Among almost all topology control schemes, it is often assumed that mobile nodes in a MANET are denoted by points on a Euclidean plane, and there is a line between two points if and only if the two nodes can communicate directly. A unit disk graph is thus applied to model a MANET, in which two nodes are connected if and only if their Euclidean distance is at most 1, indicating a normalized transmission range. Generally, the power loss ratio along a wireless link in a MANET is determined by d^r, where d is the distance between the two nodes and r is a path loss exponent between 2 and 4. For one node to reach another, its transmission power must be sufficiently higher than the product of d^r and the receive power threshold of the destination node. If the receive power threshold is the same across all the nodes in a network, then the "cost" of each wireless link in the network can be modeled by d^r. A path (route) that consists of a number of links is characterized by the sum of the costs of all links in the path. Thus, it is quite clear that using a route with several intermediate links is more efficient in terms of power consumption than the use of a long, single, direct link. The minimum power route is a route that has the smallest link cost among all routes between two nodes. The total energy consumption of the network is the sum of the transmission power of all nodes. Furthermore, minimum energy network connectivity (MENC) [20] requires a strongly connected *ad hoc* network topology (the graph) and minimized total energy consumption of the network.

Topology control in a MANET is often addressed by applying computational geometry structures, such as Delaunay triangulation, spanning trees, Gabriel graphs, and three-dimensional cones. Delaunay triangulation is a computational geometric method that, given a set of points on a graph, produces a set of edges connecting each point to some neighbors for triangulation, such that the smallest angle in each triangle is maximized [21]. An approach based on Delaunay triangulation [22] employs a few heuristic guidelines, such as not exceeding an upper bound of node degree and choosing links with an eye toward the creation of regular and uniform graph, but it does not take power consumption into account.

An approach based on the spanning tree [23] aims to minimize transmission power under the constraints of connectivity and biconnectivity, using a centralized graph structure for static networks. For mobile networks, two distributed heuristics for topology control are introduced: local information, no topology (LINT) and local information, link-state topology (LILT). As the names of these two heuristics implies, LINT depends solely on local neighbor information, while LILT exploits both local neighbor information and global topology snapshot information obtained from some routing protocols such as link-state protocols. They both try to minimize power consumption by limiting node degrees in an incremental manner. As a pioneering work toward achieving power-efficient topology control, this approach opens up many opportunities. The power model used in this approach is rather simplified in that the power consumption of a node is denoted by the number of its neighbors (*i.e.*, node degree in the graph). Because LINT and LILT are heuristics, they do not guarantee network connectivity.

The idea of using the spanning tree for topology control has been extended to a graph structure known as the local minimum spanning tree (LMST) [24]. In this case, each node in a MANET intermittently uses its maximum transmission power to conduct a broadcast-based neighbor discovery process by sending out "hello" messages. Any node receiving "hello" messages replies with its identification and position. In this way, the initiator node is able to determine its distance from all reachable nodes. It then uses that information to build a local minimum spanning tree, the root of which is the initiator node. The initiator node selects as neighbors any nodes that are one hop away on the local spanning tree.

Using the minimum spanning tree to model a subgraph of reachable nodes at each node offers some very nice features. First, it can be proved that in a minimum spanning tree node degree is bounded up to 6, which effectively limits the number of neighbors for each node, thereby reducing power consumption. Second, the network topology derived from LMST is connected. Finally, the network topology can be transformed into a bidirectional topology. Another spanning-tree-based topology control algorithm utilizes an important observation: The optimal solution of total power consumption of a MANET is upper bounded by the sum of the costs (power consumption) in a spanning tree and the edges in its critical paths [20]. A critical path is the longest (highest cost) path from internal nodes to leaves in the spanning tree. It is quite straightforward to devise centralized and distributed algorithms to identify these critical paths and reduce their costs as much as possible. Two centralized heuristics, one based on the minimal spanning tree and one based on the minimum incremental power tree, have been proposed.

A Gabriel graph of point set S contains an edge between point u and point v if and only if the disk with diameter (u, v) does not enclose any other point from S. A Gabriel graph is planar; that is, no two edges intersect. The Gabriel-graph-based approach [25] is a distributed topology control algorithm that guarantees strong connectivity. It introduces the notion of relay region and enclosure so as to build a Gabriel graph. This approach generates the global minimal energy solution for static networks and suboptimal solutions for mobile networks. For any node i that intends to transmit to node j using r as the relay node, a relay region of pair (i, r) is defined as the area within which j may reside and node i will consume less power when it chooses to relay through node r to transmit to j instead of transmitting directly. The enclosure of node i is then defined as the union of the complement of relay regions of all the nodes that node i can reach by using its maximal transmission power. The proposed approach aims at topology control for a MANET in which a master-site node is the data sink of all other nodes. A two-phase distributed protocol has been devised to find the minimum power topology for a static network. In the first phase, each node i begins to perform a broadcast-based discovery process using maximum power and to detect relay regions of every node pair. After that, the enclosure graph of node i can be computed, which in turn will be used in the second phase to compute the shortest path to the master-site node, with power consumption as the cost of the edges. For mobile networks, the same two-phase operations are conducted periodically to cope with node mobility.

A cone-based topology control algorithm [26] has been proposed to address the problem of facilitating power efficient operations. It achieves this goal by permitting each node to find its minimal operational power based solely on local information, while still satisfying global network connectivity requirements. As a consequence, each node will be able to control transmission power in favor of both energy efficiency and network connectivity. This distributed algorithm is performed in two stages. In the first stage, each node on a two-dimensional space continues to broadcast neighbor-discovery messages at an increasing transmission radius until it finds at least one neighbor in every cone of α degree centered at the node or maximum transmission power is reached. Each receiving node acknowledges all neighbor-discovery messages. The original sender node then collects those acknowledgments and derives spatial directions (cone) of those receivers using directional antennas or angle of arrival (AOA) techniques in place in many wireless systems. In the second stage, redundant links of each node are pruned in an edge removal procedure, which results in a low-degree network topology and thus low transmission interference among nodes. This cone based algorithm has been proved to always yield the maximum connected set of nodes for the network when $\alpha \leq 2\pi/3$. For smaller angels, the algorithm can guarantee minimum power routes as well.

In a more general sense, the topology control problem in MANET can be characterized by a triple of the form <**M**, **P**, **O**>, where **M** denotes the graph model (directional or unidirectional), **P** denotes the desired graph property such as the *k*-edge connected and bounded maximum node degree, and **O** represents the power minimization objective (*i.e.*, maximum power and total power) [27]. The problem of minimizing the maximum power is NP-complete (A decision problem is NP-complete if and only if it is in NP and it is NP-hard. Readers who are interested in computational complexity can refer to M. Garey and D. Johnson's book "Computers and Intractability: A Guide to the Theory of NP-Completeness".). Furthermore, the problem of minimizing the number of nodes that use maximum power is also NP-complete. Total power minimization has been proved to be NP-hard, and only approximation and heuristics such as those approaches described above may be applied.

5.6.3 Routing in MANET

A major advantage of a MANET is its configuration-free deployment, which means that nodes will evolve to form a fully distributed network for data communication without any routing configuration on each node. Furthermore, a MANET must adapt itself to dynamically changing network topology. In contrast to topology control, *ad hoc* routing aims to react to topology changes for power-efficient and high-performance operations rather than generate a good topology. The prime problem of *ad hoc* routing is to maintain an up-to-date routing table at each node against a number of factors such as overhead and scalability, power consumption, fast convergence, and stability.

Routing in a MANET is done on a multihop basis, as there is no preexisting or elected dedicated router in a MANET. Unlike a host in a wired network, a node in MANET cannot have a preconfigured default router specified in the routing table. Nodes collaborate with each other in maintaining routing tables, thus each may serve as a router. As shown in Figure 5.11, at its initial location node N1 relies on nodes N2 and N3 to forward a packet to node N4. When it moves

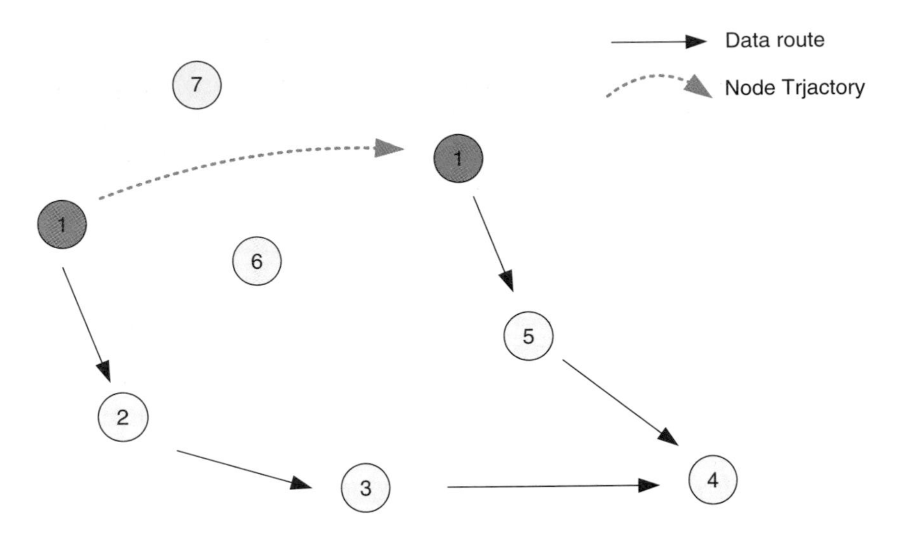

Figure 5.11 MANET routing.

to the second location, it then detects another path (N5–N4) and uses it. Note that every node in this example may move around, including those intermediate "router" nodes and the destination node. The problem will be far more complicated in that case.

Routing protocols in MANET can be broadly classified into the following categories:

- *Unicast routing protocols* — (1) Topology-based routing, including proactive routing protocols, on-demand routing protocols, and hybrid routing protocols, and (2) location-based routing protocols
- *Multicast routing protocols*
- *Broadcast routing protocols*

Some routing protocols emphasize low-power operations, but security and QoS are also considered in some protocols.

5.6.3.1 Routing Basics: Link State and Distance Vector Algorithms

Intuitively, one might think that traditional routing protocols in the wired network could be employed in wireless *ad hoc* networks. These protocols use either link state or distance vector routing algorithms [3], such as Dijkstra's and Bellman–Form's algorithms. These algorithms will find the shortest paths from one node to all the other nodes it can reach in the network. The metric used by the Internet can be the number of hops, physical distance, or route trip delay. In link state protocols such as OSPF (Open Shortest Path First)[28], each router discovers and maintains a list of link costs to its neighbors (a local view) and broadcasts the list to all other routers whenever a link cost in the list changes in the form of link state packets. Based on the received link state packets, each router computes the shortest path to every other router (a global view). In distance vector protocols such as RIP (Routing Information Protocol)[29], each router maintains a distance vector of link cost for each router in the network. A node also monitors the link cost to each of its neighbors. When some of these link costs change, the updated vector is reported to all its neighbors and a router computes an updated routing table. In addition to the routing-loop problem and counting-to-infinity problem, the major drawbacks of these traditional routing protocols are long convergence times and the highly complex messages, which may create too much overhead for the CPU [30].

Both link state and distance vector algorithms are proactive routing protocols in the sense that they require a router to periodically broadcast network change. Because topology changes are far more frequent in *ad hoc* networks than in fixed networks, simply porting proactive routing protocols to a MANET may not work well because of the resulting significant communication overhead.

5.6.3.2 Proactive Routing: DSDV and OLSR

A destination sequenced distance vector (DSDV) [31] is a modification of the distributed Bellman–Ford algorithm in that a tag of sequence number for each destination node is used in the vector maintained at each node. The sequence number can be used to distinguish between stale routes and fresh ones because a router will increment the sequence number when sending a route update message to its neighbors. To reduce the amount of information in these packets two types of update messages are defined: full dump and incremental dump. The full dump carries all available routing information, and the incremental dump only carries the information that has changed since the last dump. As based on the Bellman–Form algorithm, DSDV is dependent on periodic broadcasts of route update packets. It requires some time to converge before a route can be used. This convergence time can probably be considered negligible in a fixed wired network, where the topology does not change frequently; however, in an *ad hoc* network, where the network topology is highly dynamic, the long convergence time may result in a large number of dropped packets before a valid route is detected. The periodic broadcast also adds a large amount of overhead into the network.

Optimized link state routing (OLSR) aims at providing low-overhead routing for large, dense MANETs using a combination of a link state algorithm and a multipoint relay (MPR) technique. MPR is an improvement over the plain broadcasting method. With MPR, a node chooses a subset of its neighbors discovered by the schemes, discussed in the preceding Topology Control section, and sends link state "hello" messages to them. As a result, only those selected nodes, namely MPR nodes, will further forward those messages from the node in question (the selector). The MPR nodes also periodically broadcast their selector lists using MPR, and every node learns the route to every other node by collecting these selector lists. OLSR avoids flooding of "hello" messages and reduces communication overhead while maintaining link state.

5.6.3.3 On-Demand Routing: DSR

DSR (Dynamic Source Routing) [32] uses source routing rather than hop-by-hop routing. Each packet to be routed carries in its header the complete, ordered list of nodes through which the packet must pass to its destination. The key advantage of source routing is that intermediate nodes do not need to maintain up-to-date routing information in order to route the packets, as the packets themselves already contain all the routing decision. DSR works in an on-demand fashion, in which a route will be discovered by a node only when necessary, eliminating the need for periodic route advertisement and neighbor-detection packets in those proactive routing protocols; therefore, DSR dramatically reduces routing overhead. Additionally, battery power is also saved for the same reason, as a node could choose to switch into sleep mode while it is not involved in packet forwarding. In DSR, a node uses a route cache to temporarily store entries of source route, which is a complete, hop-by-hop path to a destination. If no source route to a destination is present in the route cache, the node must perform route discovery. It does this by broadcasting route query packets throughput the network. Each query packet records the hop sequence it traverses. Thus, when the query packet reaches its destination, a route reply packet will be sent back that contains the entire hop sequence of that query packet. All the intermediate nodes between the source and destination may also benefit by collecting route information from those route reply packets.

The downside of DSR is twofold: First, each packet must carry the complete sequence of the route; this overhead grows when the packet has to go through more hops to reach the destination. Second, the route discovery imposes some delay when a packet is being routed and the route cache does not contain the source route for it.

5.6.3.4 On-Demand Routing: AODV

AODV (Ad-hoc On-demand Distance Vector) is essentially a combination of both DSR and DSDV. It borrows the basic on-demand mechanism of route discovery and route maintenance from DSR and the use of hop-by-hop routing, sequence numbers, and periodic beacons from DSDV. It enables multihop routing between participating mobile nodes seeking to establish and maintain an *ad hoc* network. As an on-demand routing protocol, AODV only requests a route when needed and does not require nodes to maintain routes to destinations that are not actively used in the communications. As long as the endpoints of a communication connection have valid routes to each other, AODV does not take any action. Features of this protocol include loop freedom and the fact that link breakages cause immediate notifications to be sent to the affected set of nodes, but only that set. The use of destination sequence numbers guarantees that a route is "fresh."

5.6.3.5 TORA

TORA (Temporally-Ordered Routing Algorithm) is a distributed routing protocol based on a link-reversal algorithm. It can be separated into three basic functions: creating routes, maintaining routes, and erasing routes. It is designed to discover routes on demand, provide multiple routes to a destination, establish routes quickly, and minimize communication overhead by localizing algorithmic reaction to topological changes when possible. Route optimality (shortest-path routing) is considered of secondary importance, and longer routes are often used to avoid the overhead of discovering newer routes. TORA does not guarantee the shortest path. The actions taken by TORA can be described in terms of water flowing downhill toward a destination node through a network of tubes that models the routing state of the real network. The tubes represent links between nodes in the network, the junctions of tubes represent the nodes, and the water in the tubes represents the packets flowing toward the destination. Each node has a height with respect to the destination that is computed by the routing protocol. If a tube between nodes A and B becomes blocked such that water can no longer flow through it, the height of A is set to a height greater than that of any of its remaining neighbors, such that water will now flow back out of A (and toward the other nodes that had been routing packets to the destination via A). A comparison of unicast *ad hoc* routing protocols is presented in Table 5.2.

5.6.3.6 Hybrid Routing Protocols (Gradient, Clustering)

Hybrid routing protocols aim at leveraging advantages of both proactive and on-demand routing protocols at the same time. Zone routing protocol (ZRP) is an example of hybrid routing protocol. In ZRP, a MANET is divided into zones, which are defined by a set of all the nodes within a distance of n hops away. Within a zone, a proactive intrazone routing technique is used; across zones, an on-demand interzone routing technique is employed. Consequently,

Table 5.2 Unicast *Ad Hoc* Routing Protocols

	DSDV	OLSR	DSR	AODV	TORA
Routing method	Proactive	Proactive	On-demand	On-demand	On-demand
Neighbor discovery	Periodical broadcast of "hello" messages	Multipoint relay of link state messages (selector list)	N/A	Periodical broadcast of "hello" messages	N/A
Communication overhead	High	High	Low	Medium	Medium
Routing delay	No delay	No delay	No delay if a route cache is available	No delay	No delay if a route cache is available

communication overhead incurred by proactive routing advertisement is largely reduced by the scope of the zone. Within a zone, a route is always available whenever it is needed. Link cost changes in a zone do not affect other zones.

Cluster-based routing protocol (CBRP) [33] takes a similar approach. A cluster in CBRP has a diameter of two hops. Nodes in a cluster elect the one with the smallest ID (e.g., IP address) as the "head" of the cluster. The head knows everyone in the cluster; therefore, routing is performed between cluster heads rather than all the nodes. A node in the network periodically sends "hello" messages to its neighbors, and a head uses "hello" messages to notify members of its existence. A regular node uses "hello" messages to give each neighbor a list of all neighbors and adjacent cluster heads. Thus, a node is able to obtain a view of nodes that are two hops away. It can also collect "hello" messages from its neighbors (nodes from the same cluster or other clusters) so as to maintain a *cluster adjacency table*, which contains a list of neighboring cluster heads and the next-hop nodes to reach them. To detect other neighboring heads, a head can inspect the "hello" messages of its members that contain cluster adjacency tables. As a result, a two-tier hierarchy of network topology can be established. Similar to DSR, CBRP uses on-demand source routing but only among heads across clusters. When a source has to send data to a destination, it sends route request packets to the neighboring cluster heads. Upon receiving the request a cluster head checks to see if the destination is in its cluster. If yes, then it sends the request directly to the destination; otherwise, it forwards the request to all its adjacent cluster heads, and so on. Due to the two-hop cluster size, CBRP is more suitable for a dense MANET with slow movement, where topology updates are not very frequent.

5.6.3.7 Location-Based Routing Protocols

Outdoor localization technologies such as GPS and cellular triangulation are becoming standard on mobile devices, enabling a broad range of location based networks, services, and applications. For example, routing in a MANET can leverage location information of mobile nodes rather then network topology to reduce communication and computing overhead associated with topology updates. Location information is obtained via proactive flooding of location change packets or reactive location query packets. A packet is routed toward its destination based on the location of the destination node; thus, design challenges of location-based routing in a MANET have been reduced to: (1) obtain location information at each node with low overhead, and (2) provide a location-based routing decision for the destination without using flooding.

Location-aided routing (LAR) [34] is an example of location-based routing protocols for MANET. In LAR, a node is assumed to know the location of the destination node at some point of time. Using this information, a source node is able to discover an *expected zone*, a region within which the destination node may fall during the routing process. Route discovery is done in a similar way as in DSR but with one exception — only neighbor nodes in the *request zone* will forward the route request. The request zone of a source node is of limited scope defined as a rectangle that includes the location of the source node and the expected zone. The source node explicitly puts the request zone into its route request message; hence, only those nodes falling into the request zone will forward these messages toward the destination node. Other nodes will simply discard these messages. LAR also provides another scheme to guide the route request messages toward the destination node. In this scheme, the route request message contains the distance between the source node and the destination node at some point D_S. Each intermediate node, upon receiving this message, calculates its distance to the destination node (D_i) and compares the result with D_S. The node forwards the message only if D_i is smaller than D_S. In this case, D_i will replace D_s in the message for the next hop node.

Location-aided routing effectively reduces routing overhead but lacks a mechanism to distribute location information throughput the network. In fact, in location-based routing, aside from location message flooding, it is quite difficult to obtain up-to-date location information for every node in a MANET. Intuitively, a hierarchical location system could be a solution to this problem. For example, the Landmark hierarchy [35] employs a scheme that leverages well-known root nodes (the landmarks) to distribute location information. Each node in the Landmark system has a location server that is responsible for providing locations of the nodes in question. Nodes in a Landmark network have unique permanent IDs that are not directly useful for routing; instead, the node ID is hashed to obtain the address of the location server. As in source routing, the location information of a node maintained at its location server is identified by a list of intermediate nodes along the path to the root landmark. To find the location of a destination node, a source node simply applies the same hashing algorithm to obtain the location server's address of the destination node and then queries the location server for the location of the destination node. After receiving this information, the source node can compute a path that runs through the hierarchy. The Landmark approach works well when the nodes close to the root landmark in the hierarchy are quite stable; otherwise, maintenance overhead becomes a significant issue. Grid [36] is another hierarchical location-based routing protocol. The area of a MANET in Grid is divided into small squares, four of which compose a bigger square, and so on. Location updates for a mobile node are made among all other nodes within the 1-order square and some selected nodes in upper-level squares. Table 5.3 provides a brief comparison of these location-based *ad hoc* routing protocols.

Table 5.3 Location Based *Ad Hoc* Routing Protocols

	LAR	Landmark	Grid
Network organization	Flat	Hierarchy based on well-known root nodes	Hierarchical grid
Location discovery	N/A	Path to the root nodes	Position in the grid
Location maintenance	On-demand	Location server for each node	Selected nodes in the grid hierarchy
Communication overhead	Low	Medium	High
Scalability	Low	Medium	High

5.6.3.8 Multicast Routing

Multicast is a point-to-multipoint communication method in which a single message can be received by a group of designated mobile nodes. Compared to broadcast or multi-unicast, multicast is much more efficient with respect to communication overhead, which is a crucial problem in mobile environment where network bandwidth is limited and connections are not quite reliable. For multicast routing protocols in a MANET, the problem is to build a multicast group against node mobility. Tree-based multicast routing protocols focus on building an on-demand spanning tree for a source node that wishes to send a message to a group of nodes. Mesh-based multicast routing protocols use mesh topology for group management, thus improving protocol resilience against network dynamics. By definition, multicast is carried out on the network layer (IP), meaning that it requires the participating routers to support IP multicast; however, application layer, or overlay, multicast is also used in some cases to allow end-to-end deployment of multicast at the absence of multicast support on mobile nodes.

5.6.3.9 Broadcasting

Broadcasting refers to the communication means of delivering a single message to every node within the range of a network or subnet. In a wired network, the broadcast "storm" is restricted within the subnet by routers at the border of a network. In a MANET, to broadcast a message, a node initiates a wireless transmission to a specially defined broadcast address. Hence, all neighbors of the node can receive the message. Notice that wireless signal broadcasting is subject to contentions and collisions within the physical channel among the neighbor nodes. Network-wide broadcast allows a note receiving a message to forward the message to its neighbors, resulting in a "flooding" or "blind broadcasting" of the message. It is a simple and effective way to reach every node in the network. Because of this, flooding serves as a building block in a number of unicast and multicast routing protocols. Flooding is used in the following circumstances: route updates in proactive routing protocols such as DSDV and OLSR, route discovery in on-demand routing protocols, location updates in location-based routing protocols, and group establishment in multicast routing protocols.

A major problem with flooding is the redundant transmission of the same message. This problem is particularly important in wireless networks where the channels are shared media and bandwidth is largely limited. To cope with this problem, flooding is very often coupled with tagging a message identifier on the message to eliminate redundant retransmission and loops. Some consolidation techniques for flooding messages can also be used to reduce the number of messages to be transmitted at a node. Instead of simply forwarding a message, a node may choose to aggregate several messages to be forwarded into a single message that it then transmits. The aggregation operation often involves application-layer logic, an example of cross-layer design in a MANET.

Aside from simple flooding, other broadcast methods are location-based broadcast, topology-based broadcast, and probability-based broadcast. Similar to location-based unicast routing (discussed earlier), the idea of a location-based broadcast is to let a node rebroadcast only if the expanded coverage area obtained via rebroadcast is sufficiently large compared with the area where the sender and the node reside.

Topology-based broadcast schemes utilize a local topology to decide if a rebroadcast is necessary. Multipoint relay (MPR) and scalable broadcast algorithms (SBAs) are two examples of this category. Schemes falling into this category require a node to periodically broadcast a list of its one-hop neighbor nodes to all of its one-hop neighbors; thus, a node is able to maintain a local topology within the range of two hops. In MPR flooding [37], a node selects some of its one-hop neighbor nodes, called MPRs, and asks them to forward the message being broadcast. The selected MPRs are those that can reach all the two-hop neighbor nodes. Unlike MPR, where the source node makes the decision as to which neighbors will rebroadcast a message, SBA [38] shifts this decision-making procedure to those neighbor nodes. More precisely, upon receiving a message, a node will check to see if its own neighbor list contains those nodes that are one-hop neighbors of the message sender. Only those that have not been covered by the sender will receive the message. Probability-based broadcast shifts this decision-making procedure to those neighbors that can simply rebroadcast with a fixed or adjustable probability in response to local traffic statistics and topology change. All of the schemes introduced here require a buffering mechanism at each node to track messages received in the last short period of time. This information will be used by a node to decide whether or not to rebroadcast and who should receive the message according to the specific algorithm used by each protocol.

5.6.3.10 Power-Aware Routing Protocols

Power-aware computing is a common design goal in the domain of mobile computing. In the case of routing protocols in a MANET, power-aware computing imposes a new dimension of constraint with regard to route discovery, route selection, and routing table maintenance. There are several approaches to power-aware unicast routing in a MANET. One is to find the optimal, energy-efficient path at each node when a message is being forwarded to the destination step by step; therefore, route updates among nodes will contain power-related information that can be used to compute the cost of a path. The energy metric usually consists of two interrelated parts: the power necessary to transmit and receive over the link and residual power of the sender normalized to the initial energy.

The second approach aims at keeping the entire network connected and operating as long as possible, rather than looking for and using an individual energy-efficient path that may result in a disconnected network due to uneven power consumption. Here, multiple good paths will be discovered for a connection, and one of them will be selected probabilistically. Another way to reduce power consumption for a routing protocol in a MANET is to make use of the idle listening time; for example, a protocol may choose to turn off some nodes with little energy for a certain duration to forcibly direct messages to other nodes. For multicast and broadcast in a MANET, power consumption could be considered as a factor when choosing the set of nodes to rebroadcast. In addition to topology-related factors such as coverage area or node connectivity, the transmission power required to reach a specific node may be considered as well so as to reduce total transmission power needed for rebroadcast. For spanning-tree-based multicast protocols, the cost of the spanning tree may be denoted by the total transmission power of all the direct links in the tree. Thus, while building a spanning tree, an energy metric can be factored in.

5.6.3.11 Security in *Ad Hoc* Routing

Security is an important issue for MANETs. The *ad hoc* nature of a MANET fosters a variety of challenging security issues, such as message confidentiality and integrity, node authentication, certificate management, and trust association. Because each node in a MANET works as a router for others, mobile privacy becomes another key issue. Ideally a node should be completely unaware of messages being forwarded, and the system should also guarantee that sensitive information regarding the node, such as identity and location, is revealed only to designated nodes, not anyone else.

General security-related mechanisms of a MANET are discussed in detail in the next chapter. It should be quite obvious that secure *ad hoc* routing protocols are based on the routing protocols described above. Here, we briefly introduce security issues and solutions in the domain of *ad hoc* routing. First, a simple approach for secure routing would be directing messages to highly trusted nodes for routing; aside from network topology, a node's level of credibility can also be used when choosing a path. Second, a symmetric key or PKI (Public Key Infrastructure) infrastructure (PKI, digital signitures, and digital certificates are introduced in the next chapter) can be introduced to a MANET such that message integrity can be ensured and the source node can be authenticated before a message is received. A node authenticates a sender by checking its digital signature enclosed in the route discovery message using the sender's public key. The route discovery message is also encrypted by the receiver's public key so only the designated receiver is able to decrypt the message. Ariadne [39] is an example of such protocols.

Ariadne is a DSR-based, on-demand routing protocol. In Ariadne, each route request message is appended with a message authentication code (MAC) computed with a shared secret key (K_{SD}) between the initiator (S) and the target (D) over a time stamp. This allows the target to verify the authenticity and freshness of the route request. Recall that in DSR a route request will contain a list of intermediate nodes when it finally reaches the target. The target then generates a request reply message and sends it back to the initiator following the node list. Two security problems arise during this procedure: (1) The target has to authenticate the nodes in the node list, and (2) the initiator must also authenticate those nodes. Three techniques have been devised to tackle these two problems: digital signature, pairwise secret key, and a broadcast authentication protocol called TELSA (Timed Efficient Stream Loss-Tolerant Authentication):

- *Digital signature* requires each node to have a private key and a public key certified by a certificate authority (CA). The public keys are either preinstalled at each node or are distributed in a PKI so the nodes can authenticate each other's public key by checking its digital certificate using the CA's public key.
- *Pairwise secret key* assumes that a shared secret key is made known to any pair of nodes.

- *TELSA* requires loose clock synchronization among nodes and a hash key chain that is disclosed at a certain interval by the sender. The one-way hash key chain is initialized by an initial random key but published by the sender in a reverse order according to a well-known schedule. Because of clock synchronization, the receiver is able to detect, at a specific time point upon receiving a message, whether a key in the chain has been published by the sender. The receiver simply discards the message if that round of hash key has been published by the sender, indicating that the message may be forged; otherwise, the receiver stores the message and authenticates it later when it receives the key. TELSA is described in RFC 4082.

The digital signature technique and the pairwise secret key technique require a trusted third party to provide preexisting keys, whereas TELSA does not require any such servers. The downside of TELSA is the mechanism by which a key is authenticated: clock synchronization and delay estimate in the underlying MANET. In highly dynamic networks, such a mechanism may become impractical. It is also difficult to select a keychain publish interval to balance network performance and security control message (key publishing messages) overhead.

Aside from message integrity and node authentication, another problem in this domain is anonymous misbehaving or uncooperative nodes in a MANET. Three types of nodes fall into this category: malicious nodes controlled by attackers, faulty nodes that do not function correctly, and selfish nodes (leeches) that consume services but do not contribute enough. These problems will become even more important when MANETs begin to find their way into consumer applications where no single administrative authority is available. Again, these issues are discussed under the general topic of mobile computing security in the next chapter. For a routing protocol in a MANET to take advantage of all the cooperative nodes, the second and third categories of nodes that do not contribute enough to packet routing must be identified and penalized or eliminated from the network. One approach is to build a distributed credit system in which mobile nodes receive credit by forwarding the messages of others and consume credit by sending out their own messages. For this purpose, there are two types of credit systems that address the issues in different perspectives:

- A *self-claim credit system* relies on the node itself to maintain its credit using either tamper-proof hardware or incentive techniques supplied by a central credit server.
- A *neighbor-claim credit system* provides neighbor-based ratings on a node's behavior with respect to forwarding messages.

In both systems, countermeasures to forged claims and collaborative cheating are required, such as coupling distributed evaluation techniques with security mechanisms used in node authentication and security associations. This topic is beyond the scope of *ad hoc* routing.

5.6.4 Service Discovery

To a mobile node, a MANET is a virtual server that provides a number of services. Those services can be broadly divided into the following three classes:

- Network services, such as point-to-point or group communication, mobile access proxying, network performance monitoring, or power alert
- Computational services, such as distributed job scheduling, distributed storage, or distributed caching
- Application services, such as printing, web caching, or web services

A mobile node must have an efficient way to determine what services are offered by a MANET and how to consume those services. This problem (service discovery) raises another key issue in MANETs.

Sometimes the notion of *service discovery problem* encompasses two other issues: network self-configuration and service delivery. Network self-configuration refers to the process where mobile devices identify themselves and others using addresses and thus form into a network with necessary configurations on each device. Service delivery specifies how a mobile device uses a service provided by another device after the location of that service is discovered.

In a wired network, a centralized directory server is often used to provide a mapping between a service and a pointer to the service. For example, a Windows active directory is responsible for providing pointers to a number of services ranging from domain management and application sharing. In a MANET, where such a centralized infrastructure is not available, mobile nodes must self-organize into a distributed structure to advertise services. In a sense, the service directory establishment process can be considered one stage toward the establishment of the network; other stages are addressing, localization, and service provision.

Several service discovery protocols are utilized in a general networking environment, such as:

- Universal plug and play (uPnP)
- Jini (Java Naming and Directory Interface)
- Rendezvous
- IETF's service location protocol (SLP)

Universal plug and play is an open peer-to-peer service discovery and delivery framework promoted by the UPnP Forum led by Microsoft. It mainly targets small *ad hoc* networking environments such as home networks and business office networks. No centralized directory server is needed to register and advertise services for a network. In UPnP, a consumer of a service is called a *control point*, a software module operating on a device. All devices in a UPnP network use IP addressing. If a DHCP server is available in the vicinity, a device will simply contact the DHCP server to obtain a dynamic IP address. If no DHCP server is detected, the device chooses an unused private IP address in the range of 169.254.0.0. A CP is able to retrieve UPnP service and device descriptions from a device that offer some services, send actions (commands) to services, and receive events from services. After addressing is done, a CP initiates the service discovery process in the network. The communication protocol between a CP and a service provider is the simple service discovery protocol (SSDP). SSDP utilizes HTTP multicast (HTTPMU) and HTTP unicast (HTTPU) to provide the service search and service announcement functions. HTTPU is a protocol using UDP for message exchange, whereas HTTPMU is a protocol that sends HTTP packets to a multicast group over UDP. In the "description" stage, the URL sent from the device to the CP points to a network location that provides description of available services, as well as a list of URLs that will be used to access those services. In the "control" stage, the CP will in turn use those URLs obtained in the "description" stage to gather additional information about service commands and parameters. In the "eventing" stage, the CP subscribes to some services and the services notify the CP of changes in device status. If the device supports a web interface, a presentation URL can be used to control the service via a web interface. All the messages exchanged between a service provider and a CP through URLs can be formatted in the form of XML and encapsulated into simple object access protocol (SOAP) messages to allow simple HTTP-based, machine-to-machine communication over the TCP.

Jini is a Java-based device coordination solution from Sun Microsystems. Jini uses the term *federation* to define a set of autonomous Java-compatible devices forming an importune network and providing some *services* by *resources* in the federation. The services may include communication, storage, and computation alike. Unlike UPnP, Jini requires centralized directory entities called *lookup services* to work closely with every participating device in the underlying federation. If the location of a lookup service is not known, a device can use multicast to discover one. The lookup service then serves as a directory for those devices to maintain service registration and answer service queries. Jini employs a unique Java technology, Java's remote method invocation (RMI), to provide convenient service interface distribution among consumers. A distributed messaging mechanism like RPC (Remote Procedure Call), Java RMI allows a Java object to communicate remotely with other Java objects residing on different computers. During the service registration process, a device is able to upload some Java stub code to the lookup service. The code is a complete description of the interface on the serving device, which is similar to the Web Service Definition Language (WSDL) used by web services. When a querying device (the potential consumer) contacts the lookup service, it can automatically download the stub code and use Java RMI to call the service of the serving device. The lookup services can even form a hierarchy for service query resolution. Security in Jini is provided in two aspects: user authentication based on access control list and Java virtual machine (JVM) security mechanisms. Only authenticated users can make requests to a service. A service can also be "leased" to a user for a specific duration of time. Jini does not specify a low-level device addressing scheme.

Rendezvous is a zero-configuration networking solution from Apple that is supported by every network application in the Mac world. Although mainly designed for Macintosh computers in wired IP networks, Rendezvous can also be ported to other software platforms in a wireless environment. Rendezvous requires that devices do three essential things: allocate IP addresses without a DHCP server, translate between names and IP addresses without a DNS server, and locate or advertise services without using a directory server. The addressing is achieved by a protocol called Dynamic Configuration of IPv4 Link-Local Addresses, which is very similar to auto IP. Rendezvous conducts service or device name resolution in the absence of a traditional DNS server utilizing the multicast DNS service discovery (mDNS-SD) protocol. In essence, each participating device advertises its services via a multicast tree of devices. Each device also has a lightweight DNS server to store caches of service notifications from other devices. For a device to look up a service, it first consults its local cache for such a service before sending out requests to other devices in a multicast fashion. Any devices that can fulfill the request can send a multicast reply back to the requester; therefore, mDNS-SD solves the problem of service location as well. The downside of Rendezvous is that it does not support service discovery in different subnets separated by router interfaces.

IETF's service location protocol (SLP) is another service discovery protocol. In SLP, a device can assume one or more of the following three roles at the same time: *user agents* (UAs), which are devices that search for services in a local area network; *service agents* (SAs), which are devices that announce one or more services, and optional *directory agents* (DAs), which are devices that act as directory servers to collect and cache service advertisements and answer service queries. Each service offered by a device is uniquely identified by a URL. Service-specific name/value pairs (called *attributes*) can also be used. When a device joins a network, it first searches for a DA using multicast. If a DA is present, subsequent service registration and query will be done with the DA. If not, the device will simply send a UDP service query, again using multicast. All SAs that can fulfill the query will send a UDP reply back to the UA. In addition, the device will periodically send heartbeat multicast messages to discover DAs. SLP can operate on both UDP and TCP. As a general service location protocol, SLP assumes that devices in a local area network have been properly addressed prior to service discovery or advertisement. SLP can be used by other protocols as a component for service location. For example, the session initialization protocol (SIP) can use SLP to locate multimedia services over a distributed, self-organized network.

Table 5.4 presents a comparison of service discovery protocols discussed in this section. It can be seen that multicast is very often used in an *ad hoc* network environment for service advertisement and look up. For service discovery in a MANET, a protocol that requires a centralized directory is not practical. To this end, distributed service discovery protocols have shown clear advantages over those using a directory server. Ideally, in a pervasive computing environment, additional mobile-device-specific information such as power, location, and service load, should also be considered in the service discovery process such that a MANET is able to operate in a globally optimal fashion.

Table 5.4 Service Discovery Protocols

	Universal Plug and Play (uPnP)	Jini	Rendezvous	Service Location Protocol (SLP)
Directory	No	Yes	Yes	Optional directory
Addressing	Dynamic host configuration protocol (DHCP) or auto IP	Not specified	Dynamic configuration of IPv4 link-local address	Not specified
Service discovery	Simple service discovery protocol (SSDP)	Directory-assisted remote method invocation (RMI)	Multicast DNS service discovery (mDNS-SD)	Multicast lookup
Service locator and attributes	Standard	Standard	Standard	Standard
Service invocation	Simple object access protocol (SOAP)	Java's RMI	Not specified	Not specified
Communication method	Unicast and multicast	Unicast and multicast	Multicast	Multicast
Service event and status notification	Eventing (notification) and polling	Notification	Notification and polling	Not specified

5.6.5 Mobile *Ad Hoc* Networks in the Real World

A major application of mobile *ad hoc* networks in consumer electronic market is *wireless gaming*. Portable gaming devices such as Nintendo's Game Boy are mostly single-player game devices, where a gamer usually plays against the device. Because many computer games allow remote gamers to play together via a network such as a LAN or the Internet, it would be natural to facilitate multiplayer wireless gaming on portal game devices, cell phones, and any other mobile devices. Like mobile nodes in an 802.11 wireless LAN, gaming devices can work in two modes: infrastructure mode and *ad hoc* mode. For example, numerous portable gaming devices may connect to a GPRS data network or some access points of a wireless LAN for communication. In *ad hoc* mode, no access points or cellular connections are needed; instead, gaming devices may self-organize into an *ad hoc* network within which gamers can discover others and join one or more shared gaming sessions. In addition, gamers holding gaming devices are free to move, thus creating a mobile *ad hoc* network supporting a rich set of applications beyond gaming, as gamers could engage in text chat, voice chat, video conferencing, and so on. Compared with infrastructure-based wireless gaming, mobile *ad hoc* gaming does not suffer from single point failure and incurs no communication overhead for the wireless network infrastructure. Moreover, because gamers would be physically close to each other, the game could be extended to incorporate the gamers' locations, thus leading to a marriage of location-based service and wireless gaming.

Sony's Play Station Portable (PSP) is arguably the first portal gaming device to support *ad hoc* wireless gaming. The PSP device features a 4.3-inch, 16:9 widescreen, thin-film transistor (TFT), liquid-crystal display (LCD) with 24-bit colors on a 480 × 272 pixel, high-resolution screen; a USB 2.0 interface; and an 802.11b wireless LAN interface. The PSP supports both online gaming at a Wi-Fi hotspot or *ad hoc* gaming with up to 16 gamers within a range of 100 feet. Game titles leveraging the *ad hoc* mode of PSP are already available in the market. PSP's major competition, Nintendo DS, has similar functionality. As multiplayer gaming continues to gain momentum among gamers, we will see more portable gaming devices equipped with wireless network interfaces. These devices will support infrastructure-based networking, *ad hoc* networking, and a hybrid of them.

On the other hand, from the perspective of digital convergence, we expect future PDAs, music players (*e.g.*, Apple iPod), cell phones, and smart phones to be "*ad hoc*-able," enabling a broad set of mobile application in addition to wireless gaming. For example, an interesting application of mobile *ad hoc* networks is multimedia streaming among mobile devices. Users may choose to share their MP3 collections, pictures, books, and any other resources with other mobile devices users within the mobile *ad hoc* network. An example of enabling tools is Pocketster (http://www.simeda.com/pocketster.html), which is a wireless personal web server for Windows Pocket PCs that advertises itself to other Pocket PCs in the neighborhood in a wireless LAN. Its resource advertisement-and-discovery protocol is Rendezvous. Green Packet's SONbuddy (http://www.sonbuddy.com) is yet another Wi-Fi-based mobile *ad hoc* networking tool. SONbuddy allows mobile device users to communicate securely via a self-organized private community network. Applications provided by SONbuddy include chat rooms, virtual whiteboards, application sharing, screen sharing, voIP, file sharing, and gaming. A feature worth noting is Internet sharing, which effectively allows multiple users to share a single Internet connection offered by another user acting as a BuddyGateway.

5.7 Quality of Service in Mobile Computing

Quality of service (QoS) in a network refers to the capability of the network to prioritize traffic and guarantee performance of data communication. The principle reason for QoS provisioning is that the Internet is inherently a best-effort network, meaning that packets are delivered from the source to the destination without guarantee of performance such as bandwidth, packet delay, delay jitter, packet drop rate, bit packet error rate, and in-order delivery, among others. However, in many cases, the traffic of some applications and services, due to their intrinsic characteristics, must have higher priority over others in order to achieve desired performance. For example, IP telephony applications may require strict limit on packet delay and delay jitter, while video streaming applications may demand guaranteed bandwidth. Because the Internet or IP networks do not offer QoS guarantees, it is very common for the data link layer and network layer of some protocols to rely on higher layers for QoS provisioning on an end-to-end basis. For the network to provide QoS, special traffic engineering mechanisms must be implemented. In the mobile computing arena, the need for QoS of mobile applications becomes even more evident in resource-constrained settings where overprovisioning is not possible. Additionally, mobility support in a mobile wireless network imposes another challenge on QoS provisioning. This section discusses these QoS issues within the context of mobile computing.

5.7.1 An Overview of QoS

Key elements of a QoS provisioning mechanism are listed as follows:

- *Traffic classification and marking* (the process of classifying traffic into a set of classes)
- *Traffic policing* (the process of discarding traffic that exceeds a profile)
- *Traffic scheduling and shaping* (the process of controlling multiple queues and possibly delaying or dropping some packets according to some rules)
- *Resource reservation and signaling protocols* (the process of explicitly reserving and allocating network resources such as bandwidth for specific traffic)
- *Admission control* (the process of deciding if a packet stream should be allowed to enter a network)

The following is a list of three QoS mechanisms for wired networks:

- Differentiated services (DiffServ)
- Integrated services (IntServ) with resource reservation protocol (RSVP)
- Multiprotocol label switching (MPLS)

DiffServ uses a portion of the type of service (TOS) byte, namely the DiffServ CodePoint (DSCP), in the IPv4 header to define a set of traffic classes that will be treated differently by DS-compatible routers on a per-hop basis. An autonomous domain consisting of DS-compatible edge routers and core routers is a DS domain. Traffic coming into a network is classified, evaluated, and possibly shaped at the boundaries of the network and assigned to different behavior aggregates identified by a single DSCP. The mapping rule between a user's traffic and a DSCP ID is derived from the service-level agreement between the user and the network service provider. Class-based queuing technique and various scheduling techniques can be used to implement these sophisticated operations. Notice that these operations only take place at the boundary of a network. Within the core of the network, packets are forwarded according to the per-hop behavior (PHB) associated with the DSCP.

A number of traffic classes have been defined in two PHB groups: expedited forwarding (EF) and assured forwarding (AF). With a single DSCP pattern, The EF PHB group primarily targets applications requiring low delay and low delay jitter of traffic. User traffic that complies with a profile will be guaranteed to enjoy relatively better performance than others. Out-of-profile traffic will simply be discarded. The AF PHB group uses a total of 12 DSCP patterns, each indicating a combination of traffic class and a drop precedence. Out-of-profile traffic may be delayed or have a lower probability of delivery than that of in-profile traffic but may not necessarily be dropped. The AF PHB group can be used for applications that require better reliability than plain best-effort service. For example, the AF PHB group could be used to implement so-called Olympic service, which consists of three service classes: bronze, silver, and gold, with increasing quality. Packets classified into different services are treated by corresponding PHBs such that high-priority packets have a greater probability for timely forwarding than low-priority packets.

DiffServ does not use any signaling mechanism for resource reservation. QoS provisioning of DiffServ is based on relative priority of traffic classification, rather then real performance metrics. It operates in the granularity of traffic class instead of packet flow, making it scalable in providing QoS for a large number of user connections. On the downside, end-to-end QoS is difficult to achieve as a result of the coarse service differentiation. The problem becomes more difficult when multiple DS domains under different administrations engage in QoS provisioning.

IntServ is a finely grained QoS mechanism that operates at individual traffic flows instead of traffic classes. It utilizes RSVP to signal network resource reservations for specific packet streams along a routing path. The flow specs of IntServ define both traffic specification and service request specifications. Two service classes in addition to the best effort service have been defined: guaranteed service for delay-sensitive applications and controlled-load service for reliable best-effort services.

According to the RSVP, to be able to utilize these two services a sender must first send a PATH message to the receiver before data transmission. The PATH message contains characteristics of the upcoming traffic flow. Intermediate routers receiving the PATH message will forward it to the next hop determined by regular routing protocols. When the PATH message reaches the receiver, it first determines the appropriate resource needed for the flow, such as dedicated bandwidth and buffer space, and then sends an RESV message containing the flow spec to the sender. Depending on the admission control routine, intermediate routers may either grant requested resources to the flow and forward the RESV message to the next hop or reject the request and send an error message to the receiver, which in turns terminates the resource reservation process. Clearly, an IntServ router must maintain the flow state for the duration of the flow.

IntServ essentially provides a simple resource reservation mechanism for the Internet, but it conflicts with a fundamental design objective of the Internet: The network should operate at the packet level rather than flow level. Some flow aggregation techniques have been proposed to reduce the flow states that must be maintained at each router, but this again adds overhead to the underlying traffic and significant complexity to participating routers. As a result, it is generally agreed that IntServ works on a small-scale network but not on a large scale.

Multiprotocol label switching (MPLS) was initially proposed as a fast packet forwarding mechanism for wide area networks. MPLS uses a label stack, a 32-byte field called MPLS SHIM between the IP header and link layer header, to identify the path — namely, the label switching path (LSP) — by which the packet will be forwarded to a MPLS domain, as well as the QoS specifications of the packet. Within the MPLS domain, traditional packet routing is replaced by much faster label switching on a hop-by-hop basis, so MPLS does not provide an end-to-end service guarantee.

Routers in an MPLS domain can be classified into label edge routers (LERs) and label switching routers (LSRs) in the core of a network. The LER operates at the edge of the access network and MPLS network, classifies a packet into a new or existing forward equivalence class (FEC, not to be confused with forward error checking) according to some service level agreement, and assigns a label to the packet. Label distribution can be done by static configuration or by employing dynamic signal protocols. In the latter case, LERs and LSRs work together to distribute new labels in the MPLS using the label distribution protocol (LDP), RSVP, CR–LDP (constraint-based routing), protocol-independent multicast (PIM), or border gateway protocol (BGP). LSRs only perform label switching of the data traffic based on a locally maintained label table and the label stack of the packet. Depending on the content of the topmost label in the packet's label stack, the LSR may perform one of three operations: swap, push, or pop. The swap operation replaces the topmost label with a new one. The push operation adds a new label to the label stack, and the pop operation removes the topmost label from the label stack. After that, the packet is forwarded to the appropriate next hop based on the label table of the LSR. Notice that, during the process, the contents of the packet, including source and destination addresses, are not examined at all; instead, the MPLS label is used as an efficient index to a particular LSP.

It has been debated as to whether an ISP should begin to deploy a particular QoS mechanism and supply such services to users in conjunction with a service-level agreement. A principal barrier to QoS deployment is that the Internet in itself is a collection of self-administrative networks which makes interdomain service-level agreement quite difficult to achieve; hence, large ISPs do not see an incentive to offer QoS. In reality, many ISPs that operate an IP backbone network choose to simply provide more network capacity (as known as overprovision) to avoid any packet loss and traffic degradation, largely due to the fact that the cost to deploy fibers has dropped in recent years. Some access network ISPs leverage QoS techniques to impose resource limitations instead of providing QoS. The situation of slow QoS deployment may change when more resource-intensive applications begin to surface and eventually dominate network traffic.

5.7.2 End-to-End QoS Support in Mobile Computing

Providing QoS in wireless networks is significantly different and far more challenging than in wired networks. QoS mechanisms designated for wired networks are not suitable for wireless networks, as most of those solutions are based on some assumptions that cannot hold in wireless networks. As in wired networks, QoS provision in wireless networks is performed in one of two ways: resource reservation *before* traffic begins to flow or per-hop differentiated treatment *while* the traffic is flowing in the network. In the first case, a signaling protocol over wireless links is required to build a path based on the QoS specifications for the application in question. In the latter case, the wireless data routing protocol must be augmented with QoS support. In both cases, some fundamental differences between wireless and wired networks have created an array of challenging issues in providing QoS in wireless networks and interconnected wireless systems. Below is a brief summary of these issues:

- *Link bandwidth on QoS* — Wireless link bandwidth is much lower than wired network bandwidth. In a cellular network, it is extremely costly and even impossible to acquire more bandwidth for data communication; hence, bandwidth overprovision is not possible.
- *Link quality on QoS* — Wireless links, in contrast to wired network links, are strongly affected by environmental factors such as rain

and lightening, as well as obstructions and other radio signals in space. A low signal-to-noise ratio and high bit error rate are commonplace in wireless networks. Even worse, wireless link quality is highly variable due to various changes in radio propagation, mobility, and power supply. This in turns makes resource reservation difficult to guarantee. This fundamental difference between wireless links and wired links results in a natural principle of QoS in wireless networks: The wireless link layer must be taken into account.

- *Mobility on QoS* — Movement of a mobile device has a huge impact on performing reliable and adaptive operations across several layers. During a handover, dynamic QoS management mechanisms can be applied to discover the most appropriate base station to which a mobile station should be linked. Mobility also brings about topology changes and route advertisements in *ad hoc* networks, suggesting that QoS provision in such scenarios must be adaptable to these frequent changes.
- *Mobile devices on QoS* — Due to the intrinsic limitation of mobile devices, particularly battery life and form factor, QoS provision over wireless networks of mobile devices must ensure low power and highly effective operations on mobile devices.

Quality of service mechanisms in wireless networks address the above-mentioned problems at different layers and even employ a cross-layer design to add QoS support for various wireless circumstances. For example, because wireless links play a significant role in QoS provision in wireless networks, some proposed mechanisms have focused on enhancing the link layers of different wireless network protocol stacks to achieve QoS adaptation. As mentioned in the previous chapter, the IEEE 802.11e standard introduced the enhanced distribution coordination function (EDCF), which leverages eight traffic categories or service levels to provide link-layer probabilistic service differentiation. The Maser device in a Bluetooth piconet can take a similar approach to manage QoS across slave devices in the piconet. As it turns out, link-layer QoS mechanisms primarily target wireless access networks. The impact of mobility and mobile device capability on QoS provision are being intensively investigated with regard to routing protocols, group communication methods, and topology control schemes, mainly targeting MANETs as well as hybrid wireless systems.

Figure 5.12 shows an overall view of QoS provision in the general context of mobile computing. Three scenarios are identified on the figure. First, link-layer QoS provision mechanisms in a wireless access network provide service differentiation that is transparent to applications. Thus, when the wireless access network is connected to a wired backbone network via an access router, it can be easily integrated into the QoS framework of the wired network. Second, in a multi-hop *ad hoc* network, QoS routing protocols are able to find a path that satisfies a given end-to-end QoS requirement. Finally, fixed wireless mesh networks, as part of the mobile computing paradigm, also present interesting QoS issues. In providing a complete end-to-end solution for QoS, QoS

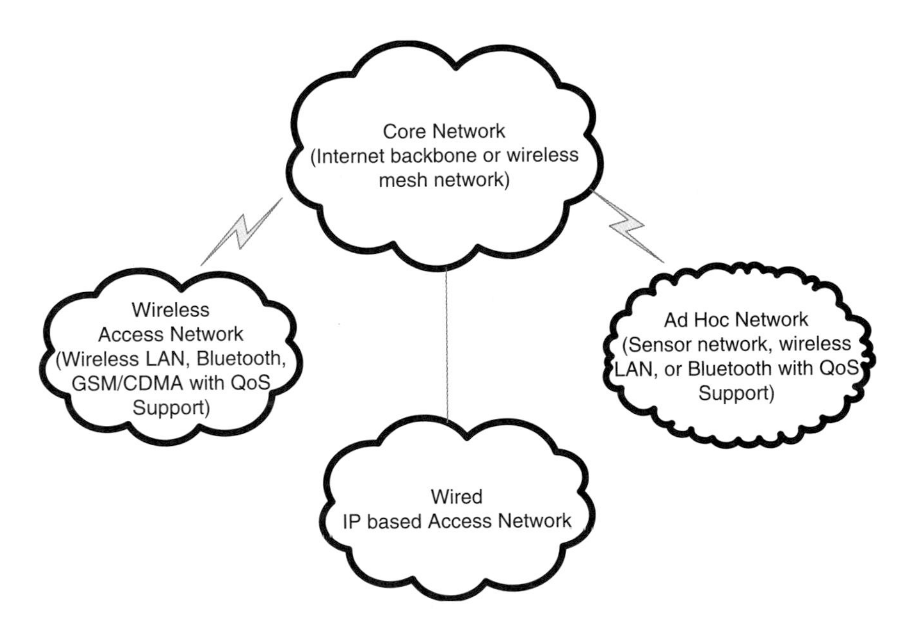

Figure 5.12 QoS Provision in a heterogeneous wireless environment.

mechanisms for different scenarios must coordinate. Not surprisingly, this is sometime an administrative issue rather than a technological issue. The following discussion addresses QoS mechanisms in these three scenarios.

5.7.2.1 QoS in Wireless Access Networks

For wireless access networks (a cell controlled by a base station or access point) to provide QoS, it is quite intuitive to adopt DiffServ to the wireless and wired combined environment because DiffServ does not require stateful signaling as is the case in IntServ, which not only results in communication overhead and delay but also may become ineffective when wireless link quality is largely affected by environmental factors. Furthermore, in many cases it is meaningless to absolutely guarantee a performance metric in a wireless network. A statistical differentiation of wireless traffic coupled with some per-hop treatment seems a better approach. Therefore, a basic idea is to use DiffServ in the core wired network that provides abundant network resources and design a QoS mechanism for the wireless access network that can be integrated into the overall DiffServ architecture.

As a component in the DiffServ network, the access router between a wireless access network and the core network acts as the gateway for all mobile nodes within the access network and performs QoS management for the wireless access network. The access router may implement an admission control or a signaling protocol that allows mobile nodes to request network resources such as bandwidth for egress traffic. The access router then has the choice to determine whether or not to grant the resource based on the SLA of the source. Network level traffic classification, marking, policing, and shaping can be applied to the access router according to the service level provision for an individual packet stream. (A dedicated router device or a Linux-based host supporting advanced traffic control can work as the access router for QoS provision.) The layer 3 QoS approach can be further augmented with link-layer scheduling schemes (also called "over-the-air" QoS such as those defined in the IEEE 802.11e standard) to reduce contentions and retransmission over wireless links, a requirement for voice over wireless LAN applications. Because of the nature of voice communication, voice traffic in a wireless access network has to be differentiated from data traffic and treated differently at layer 2 before any layer 3 QoS mechanisms take effect. To this end, wireless LAN chipmakers, such as Atheros and ViXS, and access point manufacturers, such as AutoCell and Meru Networks, have developed proprietary QoS solutions to support multimedia streaming over wireless LAN.

Another function of the access router is to support QoS-aware handoff with other access routers. When a mobile node moves from one access network to another, the access router in the new access network receiving the handoff request must check to see if sufficient resource can be granted to the incoming mobile node, based on the mobile node's QoS spec obtained from the access router of its original network. If the handoff request is granted, the corresponding QoS configuration and state on the original access router must be transferred to the new access router in charge. If the handoff request is denied, the mobile node may either resort to another wireless network, if any, or simply accept degraded service from the requested one.

5.7.2.2 QoS in Mobile *Ad Hoc* Networks

Node mobility and mobile node capability essentially make wired QoS mechanisms inapplicable to MANETs. Constrained resources on mobile nodes in MANETs generally cannot handle the processing, storage, and communication overhead brought about by stateful QoS mechanisms such as IntServ/RSVP. The mobility of MANET further complicates these solutions by incurring higher communication overhead. On the other hand, DiffServ seems a better solution, but it still must be modified because a MANET does not distinguish between ingress nodes, core nodes, and edge nodes. Each node in a MANET may assume one of three roles at any time for a particular packet stream and may assume more than one role for different packet streams simultaneously. An example of this idea is the flexible QoS model for MANETs (FQMM) [40]. FQMM combines both IntServ and DiffServ but with major enhancements for MANETs. As in normal DiffServ, packets are classified into classes that require different service levels. The role of a mobile node in this model varies significantly over time depending on its operation for a packet stream: transmission (ingress node), forwarding (code node), and reception (egress node). Per-flow QoS provisioning is only performed for traffic of the highest service level, thus flow state management will be limited to a small scale.

Node mobility also has a profound impact on QoS aware routing protocols for MANETs. Not only route reachability but also link quality become highly variable and both of them must be taken into consideration when selecting one or more paths if multi-path routing is used. For a routing protocol to support QoS, a set of QoS parameters should be encapsulated into route update messages as characteristics of a wireless link. For example, the QoS version of AODV adds a QoS extension of four parameters to Route Request (RREQ) messages before a route discovery is performed. The four parameters are maximum delay, minimum available bandwidth, list of sources requesting delay guarantees, and list of sources requesting bandwidth guarantees. Along a path from the original source to the destination, these fields are updated on a per-hop basis to reflect the current QoS request. Only those nodes that meet the requirement of these parameters will take further action with the updated RREQ message. If a local route cache (again, with updated QoS parameters) is available, the node will send a Route Reply (RREP) back to the original source; otherwise, the node will rebroadcast the updated RREQ message. The QoS AODV also copes with link dynamics as a result of node mobility or battery drain. A node periodically broadcasts perceive QoS parameters of its links in a "hello" message to direct neighbors. Any time a node notices a change of link capacity or latency of its direct links, it will notify all the nodes that may be affected by this change with respect to QoS guarantee. The list of nodes can be obtained from the two fields in cached RREQ messages: (1) list of sources requesting delay guarantees and (2) list of sources requesting bandwidth guarantees.

A number of QoS routing protocols for MANETs make use of QoS parameters in conjunction with the routing algorithms discussed in Section 5.6.3. The fundamental design choice is to achieve a balance between communication overhead incurred by the augmentation of QoS support and the need for QoS provision toward building a globally optimized MANET. In some sense, power-aware *ad hoc* routing can also be considered as a QoS routing problem.

5.7.2.3 QoS in Fixed Wireless Mesh Networks

A wireless mesh network is a special type of *ad hoc* networks. In a wireless mesh network, base stations are wirelessly connected to form a mesh to provide wireless connections for mobile nodes. Each base station acts as an access router in the wireless access network but may have multiple wireless links to other base stations; however, the backbone network is a fixed wireless network rather than a wired network. The problem of QoS provision in wireless mesh networks can be decomposed into two related issues: QoS management at a base station (*e.g.*, the wireless broadband "last-mile") and QoS management in the backend mesh.

The first issue can be addressed by adopting QoS solutions to the wireless access networks described above (*i.e.*, MAC layer service differentiation). For instance, IEEE 802.16, a standard for broadband wireless mesh networks (WiMax), specifies a MAC protocol that defines a variety of QoS parameters for packets flows, including traffic priority, maximum and sustained rate, minimum reserved rate, maximum latency, and tolerated jitter. These parameters are used for QoS scheduling in the granularity of individual frames.

The second issue is not that easy to tackle because the backend mesh is not as reliable as a wired network, and it is not as dynamic as an *ad hoc* network. Although mobility is not a problem in this scenario, those base stations have the capability to accommodate newly deployed base stations and self-organize into a new mesh. As in a MANET, QoS provision in a mesh network must include some signaling scheme that performs network resource reservation for mobile nodes served by a base station. In addition, routing protocols in a wireless mesh network must also support QoS. In discovering an appropriate path for a packet stream or a traffic class, a base station may choose to maintain a routing table with QoS extensions or preconfigured with a set of per-class packet forwarding functions. It is understandable that QoS provision in wireless mesh networks does not seem to be viewed as a high-priority issue, as upstarts are more focused on robustness, performance, and cost.

5.8 Summary

This chapter has examined a broad set of challenging issues in the domain of mobile computing. Our emphasis is on the design issues of the underlying heterogeneous networks. Mobile terminals such as smart phones are also considered when a network service is delivered to the end unit. With regard to wireless networks, this chapter has discussed network organization issues (heterogeneous network integration, mobile next-generation networks, and *ad hoc* networks), network service issues (QoS provision and routing protocols), and network transport (wireless TCP). As wireless technologies continue to emerge with new applications and services, it is undoubtedly impossible to cover every aspect of the entire wireless arena. This chapter has merely presented a glimpse into the latest development of wireless technologies and has provided in-depth discussions of selected topics regarding the utilization of fundamental computer networking techniques and methodologies.

Major issues not addressed in this chapter are mobile security and privacy. Indeed, as more and more mobile devices begin to invade our daily life and people become increasingly dependent on them, security and privacy issues are receiving tremendous attention from both academic researchers and industry practitioners. The next chapter talks about wireless data communication security, mobile network security, and mobile privacy issues in a variety of wireless environments.

Further Reading

3GPP IMS specifications, 3GPP TS 22.228 service requirements for the Internet protocol (IP) multimedia core network subsystem (IMS), stage 1, http://www.3gpp.org/ftp/Specs/html-info/22228.htm.

Dynamic configuration of IPv4 link-local addresses (Internet draft), http://files.zeroconf.org/draft-ietf-zeroconf-ipv4-linklocal.txt.

For a comprehensive discussion of mobile *ad hoc* networks, refer to G. Aggelou, *Mobile Ad Hoc Networking*, McGraw-Hill Professional, New York, 2004.

For general QoS issues and solutions, see Z. Wang, *Internet QoS: Architectures and Mechanisms for Quality of Service*, Morgan Kaufmann, San Francisco, CA, 2001.

For Linux's QoS support, see *Linux Advanced Traffic Control*, http://lartc.org.

For QoS in IEEE 802.16/WiMax networks, see IEEE 802.16 MAC and service provisioning, *Intel Technical Journal*, http://www.intel.com/technology/itj/2004/volume08issue03/art04_ieee80216mac/vol8_art04.pdf.

For wireless sensor networks, refer to F. Zhao, *Wireless Sensor Networks: An Information Processing Approach*, Morgan Kaufmann, San Francisco, CA, 2004.

iCard and Secure Mobility Management at AT&T Labs, http://www.research.att.com/areas/wireless/Mobile_Interdomain_Roaming/Mobility_Management/internet_roaming.html.

IETF Mobile IP Charter, http://www.ietf.org/html.charters/mip4-charter.html; RFC 3344, IP mobility support for IPv4, http://www.ietf.org/rfc/rfc3344.txt; mobile IP regional registration, http://www.ietf.org/internet-drafts/draft-ietf-mip4-reg-tunnel-00.txt.

IETF SLP (Service Discovery Protocol), http://www.ietf.org/rfc/rfc2608.txt.

Jini specification, http://java.sun.com/products/jini/.

Multicast DNS, http://files.multicastdns.org/draft-cheshire-dnsext-multicastdns.txt.

UPnP device architecture, http://www.upnp.org/resources/documents.asp; SSDP, http://www.upnp.org/download/draft_cai_ssdp_v1_03.txt.

References

[1] M. Buddhikot, G. Chandranmenon, S. Han, Y. W. Lee, S. Miller, and L. Salgarelli, Integration of 802.11 and Third-Generation Wireless Data Networks, in *Proc. INFOCOM '03*, 2003.

[2] Network Working Group, *RFC 3220: IP Mobility Support for IPv4*, 2003 (http://www.ietf.org/rfc/rfc3220.txt).

[3] M. Jaseemuddin, An architecture for integrating UMTS and 802.11 WLAN networks, in *Proc. Eighth IEEE International Symposium on Computers and Communications*, July 2003.

[4] J. Ala-Laurila, J. Mikkonen, and J. Rinnemaa, Wireless LAN access network architecture for mobile operators, *IEEE Commun.*, 39:11, November 2001, pp. 82–89.

[5] Unlicensed Mobile Access, UMA Stage 2 Specification R1.0.3, 2005.

[6] H. Luo, Z. Jiang, B.-J. Kim, N. K. Shankaranarayanan, and P. Henry, Integrating wireless LAN and cellular data for the enterprise, *IEEE Internet Comput.*, 7:2, March 2003, pp. 25–33.

[7] H. Luo, Z. Jiang, B. J. Kim, N. K. Shankar, and P. Henry, Internet roaming: A WLAN/3G integration system for enterprise, in *Proc. Asia–Pacific Optical and Wireless Communications: Wireless and Mobile Communications II*, Shanghai, China, 2002.

[8] O. Bar-Shalom, K. Chinn, and U. Gadamsetty, On the union of WPAN and WLAN mobile computers and hand-held devices, *Intel Technol. J.*, 7:2003, pp. 20–36.

[9] N. Golmie, Bluetooth adaptive frequency hopping and scheduling, in *Proc. MILCOM '03, Military Communications Conference*, October 2003.

[10] N. Golmie, N. Chevrollier, and O. Rebala, Bluetooth and WLAN Coexistence: Challenges and Solutions, *IEEE Wireless Commun.*, 10:6, December 2003, pp. 22–29.

[11] J. F. Huber, Mobile next-generation networks, *IEEE Multimedia*, 11:72–83, 2004.

[12] R. Ramjee, K. Varadhan, L. Salgarelli, S. R. Thuel, S.-Y. Wang, and T. L. Porta, HAWAII: a domain-based approach for supporting mobility in wide area wireless networks, *IEEE/ACM Trans. Networking*, 10:396–410, 2002.

[13] V. Jacobson, Congestion avoidance and control, in *Proc. ACM SIGCOMM '88*, 1988, pp. 314–329.

[14] A. Bakre and B. R. Badrinath, I-TCP: indirect TCP for mobile hosts, in *Proc. 15th International Conference on Distributed Computing Systems*, 1995.

[15] K. Brown and S. Singh, M-TCP: TCP for mobile cellular networks, *ACM SIGCOMM Computer Commun. Rev.*, 27:19–43, 1997.

[16] H. Balarishnan, V. N. Padmanabhan, S. Seshan, and R. H. Katz, A comparison of mechanisms for improving TCP performance over wireless links, *IEEE/ACM Trans. Networking*, 5:756–769, 1997.

[17] R. Ludwig and R. H. Katz, The Eifel algorithm: making TCP robust against spurious retransmissions, *Computer Commun. Rev.*, 30:30–36, 2000.

[18] F. Hu and N. K. Sharma, Enhancing wireless Internet performance, *IEEE Commun. Surv. Tutorials*, 4:1, 2002.

[19] H. Elaarag, Improving TCP performance over mobile networks, *ACM Comput. Surv.*, 34:357–374, 2002.

[20] M. X. Cheng, M. Cardei, X. Cheng, L. Wang, Y. Xu, and D.-Z. Du, Topology control of *ad hoc* wireless networks for energy efficiency, *IEEE Trans. Computer*, 53:12, 2004, pp.1629–1635.

[21] S. Meguerdichian, F. Koushanfar, M. Potkonjak, and M. B. Srivastava, Coverage problems in wireless *ad hoc* sensor networks, in *Proc. IEEE INFOCOM '01*, 2001.

[22] L. Hu, Topology control for multihop packet radio networks, in *Proc. IEEE INFOCOM '91*, 1991.

[23] R. Ramanathan and R. Rosales-Hain, Topology control of multihop wireless networks using transmit power adjustment, in *Proc. IEEE INFOCOM '00*, 2000.

[24] N. Li, J. Hou, and L. Sha, Design and analysis of an MST-based topology control algorithm, in *Proc. IEEE INFOCOM '03*, 2003.

[25] V. Rodoplu and T. H. Meng, Minimum energy mobile wireless networks, *IEEE J. Selected Areas Commun.*, 17:8, 1999, pp.1333–1344.

[26] R. Wattenhofer, L. Li, P. Bahl, and Y.-M. Wang, Distributed topology control for power efficient operation in multihop wireless *ad hoc* networks, in *Proc. IEEE INFOCOM '01*, 2001.

[27] E. L. Lloyd, R. Liu, and M. V. Marathe, Algorithmic aspects of topology control problems for *ad hoc* networks, in *Proc. MobiHoc '02*, 2002.

[28] N. B. Priyantha, A. Chakraborty, and H. Balakrishnan, The Cricket location-support system, in *Proc. Annual ACM/IEEE Int. Conf. on Mobile Computing and Networking (MOBICOM '00)*, 2000.

[29] L. M. Ni, Y. Liu, Y. C. Lau, and A. P. Patil, LANDMARC: indoor location sensing using active RFID, in *Proc. IEEE Int. Conf. on Pervasive Computing and Communication (PERCOM '03)*, 2003.

[30] D. B. Johnson, Routing in *ad hoc* networks of mobile hosts, in *Proc. IEEE Workshop on Mobile Computing Systems and Applications*, 1994.

[31] C. E. Perkins and P. Bhagwat, Highly dynamic destination-sequenced distance-vector (DSDV) routing for mobile computers, in *Proc. SIGCOMM '94 Conference on Communications Architectures, Protocols, and Applications*, 1994.

[32] D. B. Johnson and D. A. Maltz, Dynamic source routing in *ad hoc* wireless networks, in *Mobile Computing*, edited by Tomasz Imielinski and Hank Korth, Kluwer Academic Publishers, 1996, pp. 153–181.

[33] M. Jiang, J. Li, and Y. C. Tay, *Cluster Based Routing Protocol*, IEEE Internet Draft, http://www.ietf.org/proceedings/99mar/I-D/draft-ietf-manet-cbrp-spec-00.txt, 1999.

[34] Y.-B. Ko and N. H. Vaidya, Location-aided routing (LAR) in mobile *ad hoc* networks, in *Proc. ACM/IEEE Int. Conf. on Mobile Computing and Networking (MOBICOM '98)*, 1998.

[35] P. F. Tsuchiya, The landmark hierarchy: a new hierarchy for routing in very large networks, in *Proc. SIGCOMM '88*, 1988.

[36] J. Li, J. Jannotti, D. S. J. D. Couto, D. R. Karger, and R. Morris, A scalable location service for geographic *ad hoc* routing, in *Proc. Annual ACM/IEEE Int. Conf. on Mobile Computing and Networking (MOBICOM '00)*, 2000.

[37] A. Qayyum, L. Viennot, and A. Laouiti, *Multipoint Relaying: An Efficient Technique for Flooding in Mobile Wireless Networks*, Research Report RR-3898, INRIA, Rocquencourt, France, 2000.

[38] T. Camp, J. Boleng, B. Williams, L. Wilcox, and W. Navidi, Performance evaluation of two location-based routing protocols, in *Proc. INFOCOM '02*, 2002.

[39] Y.-C. Hu, A. Perrig, and D. Johnson, Ariadne: a secure on-demand routing protocols for *ad hoc* networks, in *Proc. ACM/IEEE*

Int. Conf. on Mobile Computing and Networking (*MOBICOM '02*), 2002.

[40] H. Xiao, W. Seah, A. Lo, and K. Chua, A flexible quality of service model for mobile *ad hoc* networks, in *Proc. IEEE Vehicular Technology Conference*, 2000.

6 Mobile Security and Privacy

The phenomenal growth of the Internet has given rise to a variety of network applications and services that are pervading our daily life at a staggering pace. This trend is being boosted by myriad mobile devices that essentially make it possible to access network resource anywhere, anytime. In parallel, security and privacy issues have surfaced in almost every aspect of the mobile computing paradigm, from wireless communication security to network denial of service (DoS) attacks, to secure network protocols, and to mobile privacy. Furthermore, the inherent characteristics of mobile computing have imposed greater challenges on mobile security and privacy solutions than on general wired network security approaches.

This chapter explores a wide range of mobile security and privacy issues, presents a big picture of this broad area, and offers some insight into the fundamental security problems surrounding the design of secured mobile wireless systems and applications. The chapter begins with a security primer summarizing a set of basic network security concepts and security schemes, followed by an in-depth coverage of security issues in cellular networks, wireless LAN, Bluetooth, and other emerging mobile wireless systems. When presenting each topic, we introduce technical aspects of each problem and discuss some proposed approaches for solving it. When possible, we then outline some real-world solutions to the underlying problems. Readers will be able to quickly obtain a solid understanding of key mobile security and the related privacy issues.

The security issues surrounding mobile wireless networks and applications can be categorized as follows:

- Message confidentiality
- Message integrity
- Message authentication
- Nonrepudiation
- Access control

When discussing differences between security and privacy, we consider this list to be comprised of security problems, whereas identity and location anonymity are topics relevant to mobile privacy.

6.1 Security Primer

Let us first consider a typical scenario in a mobile computing paradigm, where it is possible to use a mobile device (*e.g.*, cell phone, PDA, smart phone, laptop computer) to access a network service using a variety of wireless communication technologies, such as a wireless local area network (LAN) or cdma2000. This operation involves utilizing some type of hardware (*i.e.*, the mobile device being used), one or more wireless network devices, a back-end wired or wireless network infrastructure, and software, such as the application and supporting mobile operating system of the mobile device, operational and management software on wireless devices, and application software on destination servers. The scenario becomes much more complicated when group communication is being performed. Nevertheless, the fundamental question is how we can secure the entire communication environment. This problem can be approached from several different perspectives:

- *End user's perspective*—An end user may use the mobile device for many purposes, including online shopping, online banking, and personal communication with friends and colleagues, or the end user may utilize such services as online maps, weather forecasts, or online gaming. Because in many cases sensitive information is sent back and forth, the end user's major concerns are likely to include data confidentiality and integrity, as well as authenticity of the other party with which the user is connected.

- *Service provider's perspective*—A service provider has to provide a secure network infrastructure for various mobile applications and services that directly interface to end users. This implies secured communication over wireless networks and wired networks. The service provider and the end user have to authenticate each other, and the computing platform should guarantee that no information will be divulged during the communication between them. The service provider also has to protect the network infrastructure against attacks.
- *Employer's perspective*—Enterprise networks must be able to ensure the security of corporate assets. This is particularly crucial when the enterprise network provides both wired and wireless access. A well-defined, highly secured wired enterprise network may be completely open to attackers if a wireless access extension to the enterprise network is not secured. For example, a rogue access point in an enterprise network may essentially provide a means to bypass corporate firewalls and directly access network resources.

Many technical notions, terms, and technologies have been introduced to address security problems in common network environments. Table 6.1 provides a brief summary of this terminology.

Depending on the nature of security problems encountered in the mobile wireless world, they can be addressed in one or more layers of the network protocol stack. Radio modulation techniques such as FHSS (Frequency Hopping Spread Spectrum, see Chapter 3 for details) can be used to provide wireless signal transmission security at the physical layer. Link encryption is often used in wireless networks where an access point or master serves as the gateway for everyone. Internet protocol security (IPSec) is

Table 6.1 Security Terminology

Term	Description
Encryption	The transformation of some information (*cleartext* or *plaintext*) into a form (*ciphertext*) that is only readable by intended recipients who hold some decryption keys
Confidentiality	A security function that ensures that no one except the intended recipient who holds some key is able to obtain the message being transferred between the sender and the recipient
Integrity	A security function that allows the intended recipient to detect any modification to a message from a sender performed by a third party
Authentication	A security function that enables verification of the identity of a person, a data object, or a system
Nonrepudiation	A security function that ensures that a message sender cannot deny a message it sends previously
Cryptography	Mathematical foundations of security mechanisms facilitating the four security functions: confidentiality, integrity, authentication, and nonrepudiation
Secret key cryptography	A type of cryptographic mechanism that enables the sender and the intended recipient to use the same shared key for security functions
Public key/private key cryptography	Another type of cryptographic mechanisms in which two keys are used by an entity—a public key that is made available to anyone and a private key derived from the public key and known only to the owner and sometimes some trusted parties
Symmetric key encryption	An encryption mechanism that allows the sender and recipient to use the same secret shared key to encrypt and decrypt a message; also called *secret key encryption*
Asymmetric key encryption	An encryption mechanism in which the message sender uses the intended recipient's public key to encrypt a message and the recipient uses his or her private key to decrypt it
Cipher	The mathematical algorithm that is used to encrypt cleartext
Message digest	Fixed-size output of a one-way hash function applied to a message of arbitrary size
Message authentication code (MAC)	A code of a message that is computed based on the message and a secret key such that the intended recipient who holds the secret key can verify the integrity of the message
Hash MAC (HMAC)	A MAC that is computed using a one-way cryptographic hash function such as MD5 and SHA-1 and a key

continued.

Table 6.1 Cont'd

Term	Description
Digital signature	A code that is computed based on the message or a hash code of the message and the private key of the sender such that anyone can verify the integrity of the message using the sender's public key; the sender "signs" the message (digital signature is the public key equivalent of MAC)
Digital certificate	A form of electronic certificate document issued by a generally trusted certificate authority (CA) to certify someone's public key; a digital certificate, signed by the CA, contains the owner's identity, the owner's certified public key, the name of the issuer (the CA that issued the digital certificate), certificate expiration date, and some other information; a CA's public key is often distributed with software packages such as web browsers and e-mail software
Public key infrastructure (PKI)	A public-key-based architecture that uses digital certificate signed by a CA to create, manage, distribute, and verify public keys and their associated identity information
Pretty good privacy (PGP)	A technique developed by Phil Zimmermann that uses asymmetric key encryption for e-mail encryption and authentication between two entities
Authorization	The process of granting and denying specific services to an entity based on its identity and established policy.

an example of a network layer security mechanism. End-to-end security can be addressed at the transport layer. Applications usually have to deal with user authentication and access control. This chapter focuses on security solutions at the data link layer and above which invariably leverage cryptographic principles as building blocks.

A cryptographic system is the realization of a cryptographic scheme or mechanism that can be integrated into a general computer or network system to provide specific security services. The two types of cryptographic system are *symmetric key systems* and *asymmetric public key systems*. Symmetric key systems such as the Data Encryption Standard (DES) and Advanced Encryption Standard (AES) use the same *secret key* for encryption and decryption, thus requiring a secured way to distribute the key; for example, the Diffie–Hellman key exchange protocol (explained later in Section 6.1.4) specifies a method for symmetric key distribution. In contrast, public key systems use two different keys for encryption and decryption: a *public key*, which is known to public, and a corresponding *private key*, which is known only to the owner of the key pair. The public/private key pair generation algorithm ensures that it is mathematically impossible to deduce the private key based on a public key. An important characteristic of public key cryptographic systems is that the two keys are mathematically related in such a way that data encrypted by a public key can only be decrypted using the corresponding private key, and vice versa. Figure 6.1 depicts both symmetric key cryptography and asymmetric public key cryptography. Public key systems essentially provide a foundation for various security solutions to the problems listed earlier. The basic idea of these approaches is that a message from a sender can be encrypted using its private key and the recipient can verify that the message is, in fact, from

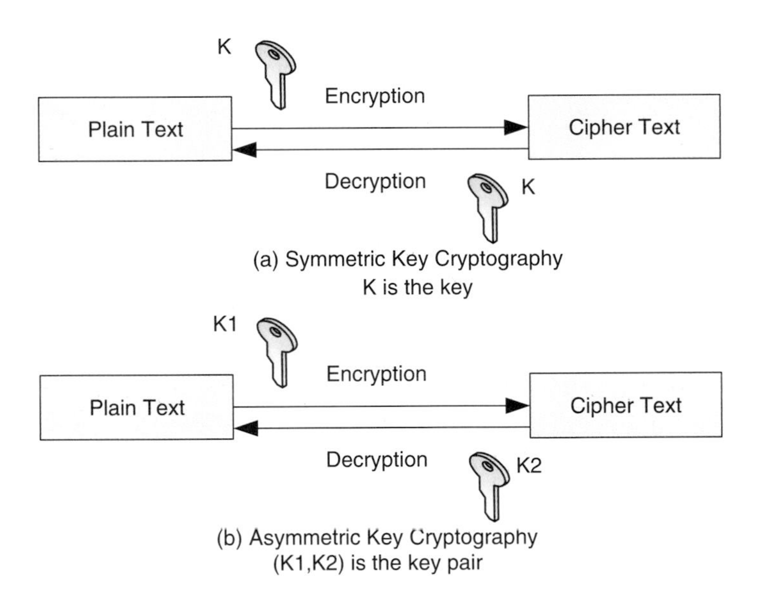

Figure 6.1 Symmetric cryptography and asymmetric cryptography.

the sender (sender authentication). Conversely, by using the recipient's public key to encrypt data, the sender can be assured that only the intended recipient is able to decrypt the scrambled data (recipient authentication). As discussed below, very often a public/private key pair is used in combination with other techniques to provide secure communication during a session. In order to ensure public key authenticity while it is being distributed in a network, the public key infrastructure (PKI) can be used (explained later in Section 6.1.3).

Public key cryptography was first proposed in 1976 by Whitfield Diffie and Martin Hellman as an encryption scheme. Public key cryptographic systems have been widely used to provide confidentiality and authentication between senders and recipients and to secure transmission of some negotiated secret such as a session key) between them. In the latter case, the cryptographic system is a hybrid system combining both asymmetric cryptography and symmetric cryptography. Popular public key cryptographic systems include RSA and elliptic curve cryptography (ECC).

6.1.1 Ciphers and Message Confidentiality

The first issue in message security is to encrypt the message such that no one except the intended recipient is able to recover the message content. In the context of symmetric key cryptography, this is often done by a block cipher using some secret key. A block cipher takes a fixed length of information (for example, a 128-bit block of cleartext) and uses a secret key to produce ciphertext, usually of the same length as the cleartext block. A block cipher also supplies a decryption function that takes the cipher text and the secret key and then produces the original cleartext. For messages that are larger than block size, a cipher may employ a particular mode to deal with the message. A mode defines the way a cipher is applied to cleartext. An important concept in data encryption is the well-known Kerckhoffs' principle, which states that an encryption scheme should be secure even if the algorithm used is known to the public. This means that an attacker is well aware of the algorithm and the ciphertext of a message but not the secret key.

Asymmetric encryption algorithms use public/private key pairs for encryption and decryption, thus they do not require the two parties involved to share the same secret key. A good cipher should make it computationally difficult for an attacker to decrypt a message without knowing the key (*i.e.*, the shared secret key or the private key being used for encryption). Popular symmetric block ciphers are DES/Triple-DES and AES, whereas well-known asymmetric ciphers include RSA and ECC. Generally, asymmetric ciphers are much slower than symmetric ones in terms of encryption speed. In addition to the common ciphers introduced below, a number of technology-specific ciphers such as the A5 algorithm are used in global system for mobile (GSM)/general packet radio service (GPRS) systems. Following is a brief introduction to these ciphers:

- Data Encryption Standard (DES) and Triple-DES—DES uses a 56-bit secret key to encrypt message blocks of 64 bits. There are 16 identical stages of processing, called *rounds*, and an initial and final permutation. The Feistel function determines how data are processed throughout those rounds using carefully generated subkeys for each round. DES has been a federal standard of data encryption for years but was finally superseded by AES in 2002, due to its weakness of using short 56-bit keys. In fact, as a result of the fast advancement in computing power, DES has been broken by brute force attacks in one to two days with the help of some powerful computers. Triple-DES is a relatively improved DES in that it uses three DES operations sequentially to compute the ciphertext. It performs a DES encryption, then a DES decryption, and then a DES encryption again. Triple-DES is generally considered a better cipher than DES. Its main drawback is computation overhead incurred by the three DES procedures.
- Advanced Encryption Standard (AES)—AES has a fixed block size of 128 bits and a key size of 128, 192 or 256 bits. A data block is organized into a 4×4 array, or *state*. AES may require 10 to

14 rounds of computation, depending on the key size. Because many operations in a single round can be performed in parallel, AES is comparatively easier to implement in both hardware and software and can be done much faster than DES. The real name of the cipher is Rijndael, a combination of the two designer's names: Joan Daemen and Vincent Rijmen. Rijndael was chosen by National Institute of Standards and Technology (NIST) to be the government standard. As of this writing, no attack has broken AES.

- Blowfish and Twofish—Blowfish is yet another block cipher developed by Bruce Schneier in 1993. It uses a key up to 448 bits over blocks of 64 bits. Blowfish has 16 rounds following the Feistel function. Blowfish is generally regarded as a compact and fast replacement of DES. Twofish specifies block size of 128 bits and uses a key size up to 256 bits. Twofish also made it to the final list of the AES contest but lost to Rijndael. There is no reported successful attack over Blowfish and Twofish.

Other well-known block ciphers are CAST-128, CAST-256, RC5, and RC6, among others. It is important to remember that, with regard to data encryption on mobile devices, computational overhead becomes a much more severe problem than on desktop computers; hence, while choosing a cipher to encrypt packets in a wireless network, those ciphers with low overhead such as RC5 will be advantageous.

In addition to block ciphers, another type of cipher is the stream cipher. Unlike block ciphers, a stream cipher encrypts one bit or one byte at a time. The two types of stream ciphers are synchronous and self-synchronizing ciphers. Synchronous stream ciphers require a key to produce a keystream, which in turn is used to compute the ciphertext. The computation is done by XORing (exclusive OR operation) the keystream with the cleartext. Decryption follows in the same manner. Self-synchronizing stream ciphers do not require a key. Instead, they use some bits of the previous ciphertext to produce the keystream. Stream ciphers are primarily used to secure network data transmission where the cleartext is a stream of bits rather than a static data block.

RC4 is the most widely used stream cipher, although it has been shown that RC4 is not always secure. RC4 was designed by Ron Rivest of RSA Security in 1987. RC4 (Rivest Cipher 4) is one of the four ciphers that Rivest developed. In RC4, a variable-length key is first used to perform a permutation of one byte according to a key scheduling algorithm. The result, along with two index pointers, is fed into a pseudo-random generation algorithm (PRGA) to produce the keystream, which will be XORed with the cleartext to obtain the cipher. RC4 has been found to have serious vulnerability in the key scheduling algorithm that in some special cases may enable an attacker to recover the encryption key [1]. This weakness has been leveraged by some researchers to break wireless equivalent privacy (WEP) encryption, the security mechanism of IEEE 802.11b wireless LAN which uses RC4 for data encryption. Details regarding this WEP vulnerability are provided in Section 6.3.

Most commercial security software supports a list of block or stream ciphers from which users can choose. A well-known open-source cipher implementation is the *libcrypto* library in the OpenSSL package (http://www.openssl.org/). Both Java and Microsoft .Net provide a package of these ciphers. In addition, they are also supported in the mobile platforms J2ME and .Net Compact Framework. Cryptographic schemes discussed in the rest of this section, such as hashing algorithms, digital signatures, and digital certificates, are generally supported by these libraries.

6.1.2 Cryptographic Hash Algorithms and Message Integrity

Aside from message confidentiality, another security problem is how to ensure message integrity—that is, how to protect data from being modified between the two parties. One-way hashing was introduced for this purpose. Simply put, a one-way hash algorithm, sometimes referred to as a *message digest algorithm*, makes sure that any modification to a message can be detected. A cryptographic hash algorithm or message digest algorithm in this regard must possess the following security properties:

- *Fixed-length output*—Given any size of message, it must produce a fixed size result, which is the hash code.
- *One-way*—Given a message m and a hash algorithm h, it is easy to compute $h(m)$; however, given a hash code x and hash algorithm h, it is computationally impossible to find m such that $h(m) = x$.
- *Collision resistance*—Because a hash algorithm is effectively a mapping between a large code space to a considerably smaller code space, collisions are bound to happen, meaning that brute force attacks are theoretically possible. The challenge is how to find collisions within a reasonable amount of time, given a state-of-the-art computing facility. The two types of collision resistance are strong collision resistance and weak collision resistance. Strong collision resistance means it is computationally impossible to find two different messages that can be hashed into the same code, whereas weak collision resistance means it is impossible to find a message that can be hashed into the same hash code of another given message.

Depending on how a hash algorithm operates, the two types of cryptographic hash algorithms are keyed and keyless. Keyed hash algorithms take a message and a key to compute the hash code, while keyless hash algorithms simply use the message to compute the hash code. Keyless hash algorithms are used to detect modifications to a message, assuming that the hash code of the original message is correctly transmitted to the recipient. Because of the collision resistance property, any change to the transmitted message can be detected immediately; however a problem arises when an attackers modifies the intercepted message, generates a hash code, and sends the tampered message and its hash code to the recipient. In this case, a hash code produced by a keyless hash algorithm fails to ensure message integrity. Message authentication code (MAC) algorithms solve this problem by including a key (either a symmetric secret key or the private key of the sender) in the computation of the hash code;

thus, attackers are unaware that the key cannot generate the correct hash code for a modified message. Hash algorithms can also be used in digital signatures (introduced in the next section). Following is a list of widely used cryptographic hash algorithms:

- *Message digests 4 and 5 (MD4 and MD5)*—MD5 splits a message into blocks of 512 bits and then performs four rounds of hashing to produce a 128-bit hash code. MD4 is a weaker hash algorithm that only performs three round of hashing. In August 2004, collisions for MD5 were announced by Wang *et al.* [2]. Their attack technique was reported to take only an hour; on a fairly powerful computer they were able to find an alternative message for a given message, yet both created the same hash code, proving that MD5 is vulnerable to a weak collision attack. Using the same technique, they also devised a method to manually attack MD4 and two other hash algorithms, HAVAL-128 and RIPEMD. MD5 is still widely used in existing systems, ranging from digital signature to file checksum; however, neither MD4 and MD5 should be considered for future systems due to the collision problem, especially for systems utilizing MD5 to generate digital signatures and digital certificates.

- *Secure hash algorithm 1 (SHA-1)*—SHA-0 was initially proposed in 1993 as a hashing standard by the National Security Agency (NSA) and was standardized by NIST. Later, in 1995, SHA-0 was replaced by SHA-1 after the NSA found a weakness in SHA-0. The weakness was also discovered by Chabaud and Joux. Based on MD4, SHA-1 works on blocks of 512 bits and produces a 160-bit hash code. SHA-1 adds an additional circular shift operation that appears to have been specifically intended to address the weaknesses found in SHA-0. The 160-bit hash code of SHA-1 may not be sufficiently strong against brute force attacks. It has been reported that the same team of Chinese researchers who broke MD5 has found a way to significantly reduce the computational complexity of discovering collisions in SHA-1. As it turns out, the problem of SHA-1 is the hash code size. NIST published three SHA hash algorithms that produce larger hash code: SHA-256, SHA-384, and SHA-512. These hash algorithms are able to generate hash codes of 256 bits, 384 bits, and 512 bits, respectively. Not surprisingly, they are significantly slower than SHA-1.

- *RACE integrity primitives evaluation message digest – 160 (RIPEMD)*—RIPEMD-160 was developed in 1996 by Dobbertin *et al.* . It is an improved version of the original RIPEMP, which was developed in the framework of the EU project RIPE (RACE Integrity Primitives Evaluation, 1988–1992). There are also variants of RIPEMD supporting hash code length of 128 bits, 160 bits, 256 bits, and 320 bits. RIPEMD collisions were reported in 2004 [2], and RIPEMD is not used as often as SHA-1.
- *Message digest and MAC (Message Authentication Code)*—Message digest ensures that if someone in the middle alters a message, the recipient will detect it. On the sender side, the sender will hash a message or a file (for checksum computation) to be downloaded using a one-way hashing algorithm (such as MD5 or SHA-1, described above), attach the result (the message digest) to the message, and send it out. Upon receiving the message, the recipient will apply the same hash algorithm to the received message body and compare the result with the received message digest. If they match, the message has been transmitted intact; otherwise, the message has been changed in some way on its way to the recipient, and the recipient may simply reject the message.

If an attacker forges a hash code of modified message, the hashing algorithm may utilize a cryptographic key as part of the input in addition to the message being transmitted. More generally, a MAC that is computed based on the message and a cryptographic key can be used to guarantee message integrity. If the computation is done using a hash algorithm, such a technique is referred to as HMAC, which essentially uses a keyless hash algorithm and a key to implement the algorithm of a keyed hash algorithm. Well-known HMAC algorithms include HMAC-MD5, HMAC-SHA1, and HMAC-RIPEMD. MAC can also be computed using symmetric block ciphers such as

DES; for example, a message can be encrypted using the DES CBC (Cipher Block Chaining) mode. The ciphertext can then be used as MAC. Furthermore, to prevent tampering of the message digest itself, the sender can encrypt the message digest using its own private key so the recipient, with the sender's public key at hand, can be assured that this message has come from the sender. This scheme is referred to as *digital signature* and will be discussed in the next subsection.

As a last note, an attacker may launch a message reply attack by simply resending a number of legitimate messages previously captured. The recipient may be fooled by such legitimate messages. To counteract these attacks, the sender can use a sequence number for each message that is contained in the integrity-protected part of the message. The sequence number keeps increasing so replayed messages will not be accepted.

6.1.3 Authentication

Common authentication mechanisms are digital signature, digital certificate, and PKI, which are described in the following text.

6.1.3.1 Digital Signature

Digital signature is designed to assure recipients that the senders of messages are really who they claim to be and the messages have not been modified along the way. Similar to a signature in the real world, the sender digitally signs a message, and the receipt is able to verify the authenticity of the message by looking at the digital signature. In other words, digital signature offers authentication of the sender and message integrity.

Digital signing and verification between two parties are conducted as shown in Figure 6.2. The sender:

- Prepares cleartext to send (*e.g.*, an e-mail or a packet).
- Hashes the data using a cryptographic hash algorithm to generate a message digest; hashing is not reversible.

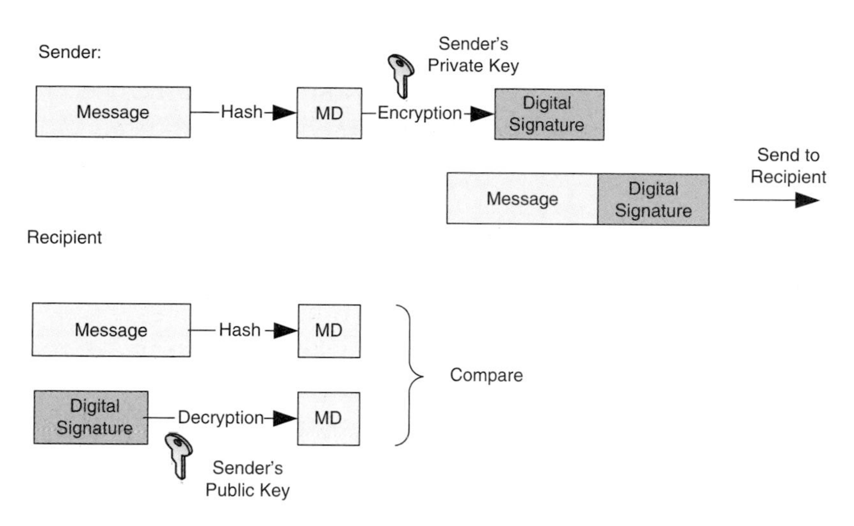

Figure 6.2 Digital signature.

- Encrypts the message digest with the sender's private key, which generates the digital signature that uniquely identifies the sender.
- Appends the digital signature to the original cleartext and sends it to the recipient. Of course, the cleartext can be encrypted using symmetric or asymmetric ciphers.

The recipient:

- Uses the sender's public key to decrypt the digital signature; the result is used in the next step.
- Hashes the received message body with the same algorithm used by the sender.
- Compares the decrypted message digest with the computation result from the previous step; if they are the same, the message must be originated from the sender, and the message has not been altered.

Now let's see if an attacker can impersonate the sender. Without the sender's private key, the attacker has no way to create a valid digital signature for the message because on the recipient side, after the message is hashed, the result will never be the same as the result after decryption of the digital signature. On the other hand, an attacker who chooses to tamper with the sender's message body will also fail, as the hash code of the received message will become inconsistent with that carried in the digital signature.

6.1.3.2 PKI and Digital Certificate

Asymmetric cryptographic systems (introduced above) assume that a party knows the other's public key. A problem with public authenticity is how someone holding the public key of someone else can be sure that the key does, indeed, belong to that person. What if the distribution of pubic keys is not at all secure? For example, an attacker could generate and publish bogus public keys of some victims.

The general architecture to address this issue is public key infrastructure (PKI). In a PKI system, the certificate authority (CA) has a public key but its private key is not known to everyone in the system. A single CA PKI is depicted in Figure 6.3a. To join the PKI system, a user must generate his or her own public/private key pair and ask the CA to certify the public key. The CA will then verify the identity and the associated public key. The CA then signs a digital document stating that the public key really does belong to the person in question. This digital document is a *digital certificate* and should be sent to a recipient whenever the person is about to communicate with some party with public key encryption or digital signing. Because everyone in the PKI system knows the public key of the CA, they can check the authenticity of the certificate and thus the public key of the sender. The certificate usually contains the owner's identity, a signature of the CA, and an expiration date. Table 6.2 shows common fields in a digital certificate. The X.509 standard defines the format of a digital certificate.

In reality, a PKI system is organized into multiple levels in a hierarchy to distribute certificate generation and verification among a number of CAs, as shown in Figure 6.3b. On the top of the tree is the root CA, who is trusted by every user and every other CA. In effect, a chain-of-trust relationship can be established regardless of which

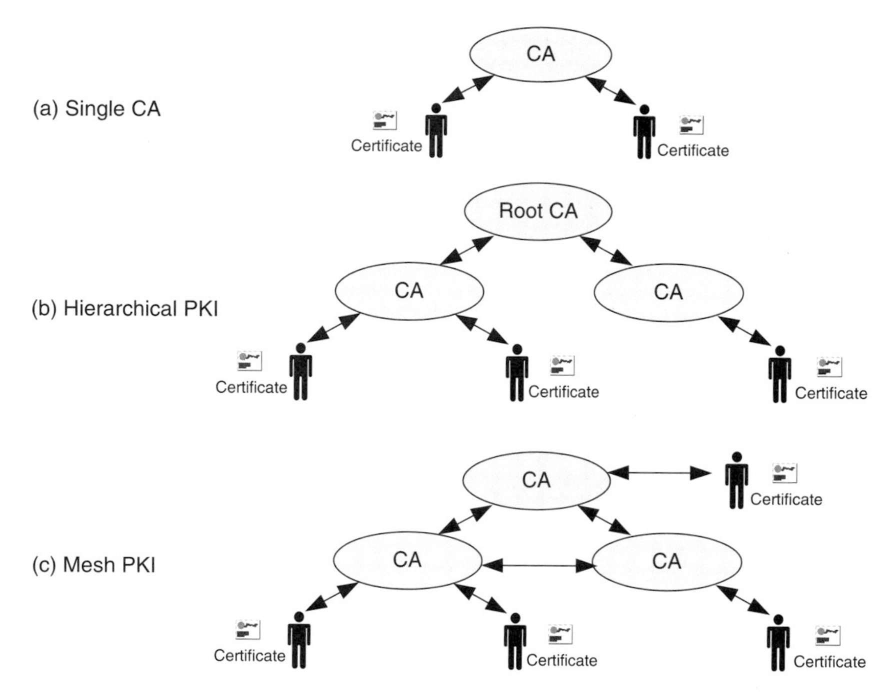

Figure 6.3 PKI architecture.

low-level CA a user selects, as those CAs can always find a common high-level CA within the hierarchy. Verification is done in the same way as DNS (Domain Name Server) resolution. For example, in a two-level CA system, the public key certificate of a user consists of two parts: (1) a message issued by a high-level CA to certify a low-level CA and (2) a message issued by the low-level CA who will eventually certify the public key of the user. This forms a trust chain of two CAs, and path validation can be conducted. Thus, any party who elects to receive a user certificate (as well as the certificate of the CA certifying the user certificate) must first compute the public key of the low-level CA serving that user and then obtain the user certificate. As the number of levels increases, certificate verification requires more computation. A variance of hierarchical PKI is a trust list architecture, in which some high-level CAs maintain a list of trusted CAs in another hierarchy. A trust chain is therefore established with the trust list instead of a root CA.

Table 6.2 Field in a Digital Certificate

Field	Description
Version	Version number
Serial number	Unique ID of the certificate
Certificate signature algorithm	Encryption and hashing algorithms used to create the signature in the certificate
Issuer	ID of the issuing CA
Validity	Duration for which the certificate is valid
Subject	Owner information
Subject public key info	Subject's public key algorithm (RSA, for example) and public key
Extensions	Additional information regarding the certificate
Certificate Signature Value	Signature of the CA

A third PKI architecture, mesh PKI, is shown in Figure 6.3c. There is no publicly trusted root CA in a mesh PKI. A CA in a mesh PKI may choose to trust a subset of other CAs. Users always trust the CA issuing the certificates. Path validation of a user certificate may involve a means to discover the path itself. A bridge CA can be used to link a hierarchical PKI to a mesh PKI. It is not a root CA trusted by everyone; rather, it serves as a common intermediate CA in a trust chain.

6.1.4 Key Management

Key management refers to the process of creating, distributing, and verifying cryptographic keys. It determines how an entity binds to a key. Here, we introduce the Diffie–Hellman (DH) key exchange protocol, RSA, and ECC.

6.1.4.1 Diffie–Hellman Key Exchange Protocol

The DH key exchange protocol provides a means for two parties to agree on the same secret key over an insecure communication channel. In its simplest form, each party send to the other a number that is computed with a chosen secret number respectively. The same secret key is thus determined based on the number received from the other party; however, if the two numbers are transmitted over an insecure channel, it is computationally difficult for any third party to recover the secret key. The DH key exchange protocol uses a pair of publicly available numbers (p and g) along with the user's random variables for the computation of a secret number. In this case, p is a large prime number and g is an integer less than p, where p and g satisfy the following property: For any number n between 1 and $p-1$ inclusive, there is a number m such that $n = g^m \mod p$. Each of the two parties engaging in the DH key exchange protocol will first generate a private random variable. Let's say the variables are a and b. Each party proceeds to compute $g^a \bmod p$ and $g^b \bmod p$ and they exchange results. Then, the shared secret key (k) can be obtained by computing $k = [(g^b) \mod p]^a \mod p$ and $[(g^a) \mod p]^b \mod p$ at each party. Note that $[(g^b) \mod p]^a \mod p = [(g^a) \mod p]^b \mod p = (g^{ab}) \mod p$. No one other than the two communicating parties will know a and b, so it is not computationally feasible to compute k using p, q, and the two public values $g^a \mod p$ and $g^b \mod p$.

Note that, although both sides are able to agree on a secret key, there is no way for each of them to be sure that the other side is indeed the person with whom they want to communicate, meaning that no authentication is being performed during the key exchange process. This opens up the protocol to a man-in-the-middle attack, in which an attacker is able to read and modify all messages between the two parties. Digital signature can be applied in this case to prevent man-in-the-middle attacks.

6.1.4.2 RSA

Designed by Ron Rivest, Adi Shamir, and Len Adleman [3], RSA is a public key algorithm that provides both digital signature and public key encryption. RSA is the public key algorithm used in pretty good privacy (PGP). Key generation in RSA is based on the fact that factoring very large numbers is computationally impossible. RSA keys are typically 1024 to 2048 bits long, much larger than the largest factored number ever. A message is encrypted using the public key of the recipient. To decrypt the ciphertext, one must know the private key corresponding to that public key. Given the public key and the cipher text, an attacker must factor a large number in the public key into two prime numbers so as to deduce the private key. In addition to message encryption, RSA also provides digital signature that allows senders to sign a message digest using their private keys. Thus, no one is able to forge a message from the sender unless he or she knows the private key. RSA was patented in the United States in 1983; the patent expired in 2000.

6.1.4.3 Elliptic Curve Cryptography

An alternative to RSA, elliptic curve cryptography (ECC) is another approach to public key cryptography. It was independently proposed by Victor Miller and Neal Koblitz in the mid-1980s. ECC is based on the property of elliptic curve in algebraic geometrics. An elliptic curve is defined by a set of points (x, y) over a two-dimensional space such that $y^2[+x \cdot y] = x^3 + a \cdot x^2 + b$, where the term in the square bracket can be optional. ECC allows one to choose a secret number as a private key, which is then used to choose a point on a non-secret elliptic curve. A nice property of an elliptic curve is that it enables both parties to compute a secret key solely based on its private key (the number chosen) and the other's public key. The secret key specific to these two parties is a product of those two private keys and a public base point. A third party cannot easily derive the secret key. NIST has published a recommendation of five different symmetric key sizes (80, 112, 128, 192, 256). ECC is generally used as an asymmetric scheme that allows for smaller key sizes than RSA. The drawback of ECC is the computation overhead associated with the elliptic curve.

Key management in symmetric cryptographic systems poses a different problem. Using stream ciphers, communication between two parties can be encrypted with a secret key only known to the two parties. Naturally it would be better to allow the two parties to frequently change the secret key to reduce the risk of message replay attacks and cipher breaks. For example, the two parties may agree on a new secret key for each new session between them. This secret key is referred to as a session key. In a network environment where many nodes have to communicate with others, a session key can be issued by a trusted third party every time two nodes are about to communicate. This simple scheme requires a node to have only one secret key shared with the trusted third party, relieving it from maintaining a secret key for every other node in the network. An example of such systems is Kerberos (http://web.mit.edu/kerberos/www/).

As a last note in the authentication section, GSP/GPRS systems employ a technology-specific authentication mechanism (the A3 algorithm) for authentication between a base station and a mobile station. The A3 algorithm, along with the A5 encryption algorithm and A8 key management algorithm, are introduced in Section 6.2.

6.1.5 Nonrepudiation

Nonrepudiation refers to a security function of a system that produces evidence to prove that an operation has been performed by an entity. For example, a message recipient should hold a piece of electronic documentation for the message such that the sender cannot deny message transmission. Conversely, the sender must be able to show that the recipient did indeed receive the message. Nonrepudiation of origin proves that the message was sent, and nonrepudiation of delivery proves that the message was received.

Nonrepudiation is generally considered a facet of the security function in electronic transaction settings, as neither sender nor recipient can repudiate a transaction after it is committed. A digital signature appended to a message sent by a sender or an acknowledgement generated by the recipient can be used to provide nonrepudiation. In this case, the digital signature serves as the evidence for nonrepudiation of origin and delivery. Because only the owner of the digital signature knows his or her private key, that person cannot deny transmission of any messages signed by his or her digital signature. One-time passwords are another scheme to realize the nonrepudiation function.

6.1.6 Network Security Protocols

We have discussed security schemes for message confidentiality, message integrity, and message authentication. Those schemes are generally used to secure a communication channel between two parties. Another level of authentication is concerned with user authentication (*i.e.*, verifying the identity of an entity to prevent unintended data access or impersonation). Recall that cryptographic keys are invariably used in those message-centric security mechanisms. Now, let us assume that point-to-point communication channels are secured and look at a network consisting of more than two nodes in which user authentication is associated with proper authorization with respect to data access. For example, in a typical setting, a user elects to log-in to a system (a group of machines) in order to read or write a file physically stored somewhere in the system. A user must be authenticated against some security tokens managed at a log-in server before the desired access is granted.

6.1.6.1 Password

Each account in a multiple-user system is assigned a password. Users can change their passwords but must obey some password creation guidelines to avoid the use of passwords that are too simple. For any network security protocols, cleartext passwords should never be sent over a network. This is the reason why the once popular Telnet protocol has been abandoned in today's networks. On a log-in server, users' passwords are usually hashed. A good password policy should force a user to change passwords once in a while. In addition, other means of human identity can be used to replace passwords. Recent developments in biometrics suggest that fingerprints, voices, faces, and irises can be utilized to identify humans with much better security. The term *biometrics* refers to systems and techniques that make use of features of a person's body for verification and identification. Features of a person include fingerprint, facial pattern, hand geometry, iris, retina, voice pattern, and signature. Biometrics systems are further discussed in more detail in Chapter 7 (Mobile Application Challenges).

6.1.6.2 Challenge and Response

For challenge and response schemes, the log-in server of a system sends a random message (the challenge) to a user who is willing to authenticate the system. The user applies a security function to the challenge and sends the result back to the log-in server, which performs the same security function and compares its results with that from the user. The challenge and response scheme can be applied in various settings. For example, it can be used to implement a one-time password, in which each password becomes invalid right after it is used. Additionally, the security function itself can be the secret. In effect, no password is transmitted over the network, and the messages subject to interception are different every time, thereby reducing the likelihood of success of an eavesdropping attack.

6.1.6.3 Kerberos

Kerberos (http://web.mit.edu/kerberos/www/) is a secret-key-based network authentication protocol (Figure 6.4). The name *Kerberos* comes from Greek mythology (Kerberos was the three-headed dog that guarded the entrance to Hades). Kerberos can be viewed as a distributed authentication service that allows a computer program

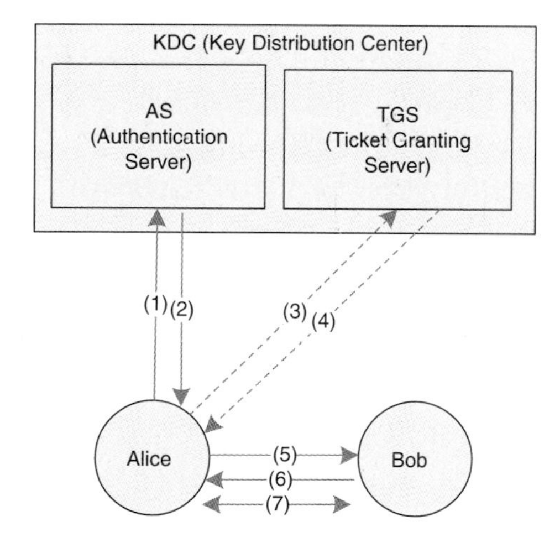

(1) Alice is authenticated by AS

(2) AS generates TGT (Ticket Granting Ticket)

(3) Alice sends TGT and her identity encrypted by the session key to TGS

(4) TGS verifies Alice's request and then generates a new session key for communication between Alice and Bob, and a service ticket for Alice to pass to Bob.

(5) Alice sends the service ticket to BOB and identity encrypted by the new session key.

(6) Bob verifies Alice's identity with the service ticket

(7) Alice and Bob start to communicate using the new session key

Figure 6.4 Kerberos (version 5).

(a client) running on behalf of a principal (a user) to prove its identity to a verifier (a server). In the heart of Kerberos is the key distribution center (KDC), which consists of two logically independently components: an authentication server (AS) and a ticket-granting server (TGS). A user (Alice) who wants to communicate with another user (Bob) must first be authenticated by the AS. To do this, Alice must use her secret key (*e.g.*, her password) to encrypt a challenge sent from the AS. The AS generates a ticket-granting ticket (TGT), which is comprised of (1) a session key encrypted by Alice's secret key (password) for the upcoming communication between Alice and the TGS and (2) a secured temporal credential used to identify Alice's request to the TGS encrypted by the TGS' secret key (which is unknown to Alice) and the session key. Alice then sends the TGT along with an authenticator (*i.e.*, Alice's identity encrypted by the session key of Alice and the TGS) to the TGS. It is the TGS that eventually generates a session key for the upcoming communication between Alice and Bob, after verifying the data in the ticket and the authenticator. At the same time, a service ticket (encrypted by Bob's secret key) for Alice to pass to Bob is also generated.

Finally, Alice sends the service ticket and a corresponding authenticator (her identity encrypted by their session key) to Bob, who verifies if the identity in the service ticket and Alice's authenticator match. If yes, Bob and Alice can begin to communicate with the session key. If Bob is a log-in server of a network system such as in a Windows domain, Kerberos is used to authenticate a user to access shared resources in the network. Kerberos relies on time-stamps and lifespan parameters to prevent message replay attacks. This requires clock synchronization among the participating machines.

6.1.6.4 Internet Protocol Security

Internet protocol security (IPSec) is a suite of protocols and mechanisms that collectively provide message confidentiality, message integrity, and message authentication at the IP layer. Depending on whether end systems support IPSec, an IP packet can be delivered in one of the two modes: transport mode or tunnel mode. In transport mode, the IP payload is secured in terms of message integrity

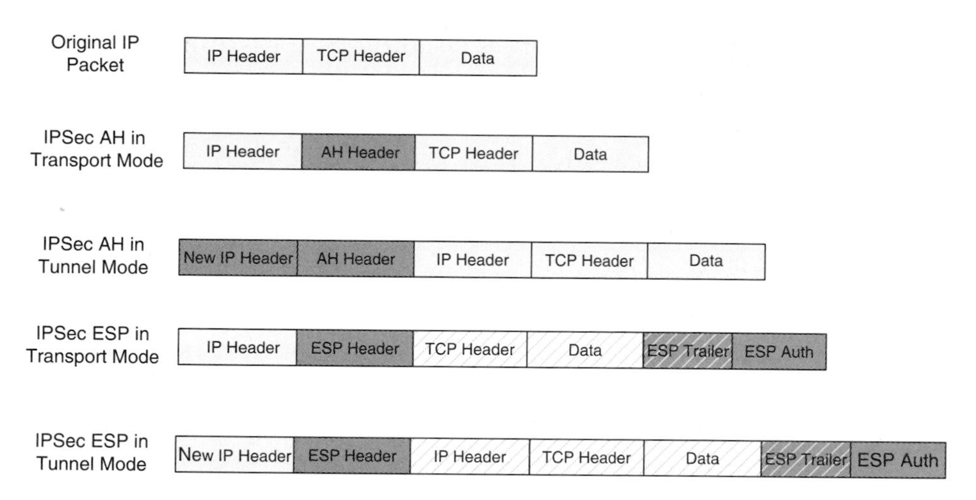

Figure 6.5 IPSec.

and authentication when the authentication header (AH) protocol is used or confidentiality when the encapsulating security payload (ESP) protocol is used; but the IP header is not protected. In tunnel mode, a new IP header is used, followed by the encrypted IP packet.

IPSec provides ESP and AH protocols for message security, as shown in Figure 6.5. The AH protocol determines how an IPSec system uses the AH header, which is a hash code of all immutable fields in the IP packet, for message integrity and origin authentication. In contrast, the ESP protocol provides message confidentiality in addition to message integrity and authentication. In transport mode, an ESP header is inserted after the original IP header, and the original IP payload is encrypted. In tunnel mode, a new IP header is used for tunneling, followed by an ESP header and by the encrypted IP packet. In both cases, message integrity and authentication are provided by an ESP authentication field appended to the end of the packet.

In addition, the Internet key exchange (IKE) protocol is supplied for symmetric key management. The IKE protocol is a hybrid of three key management protocols: Internet Security Association and Key Management Protocol (ISAKMP), Oakley, and SKEME (Versatile Secure Key Exchange Mechanism for Internet). These protocols work together to allow dynamic negotiation of cryptographic keys using the DH key exchange algorithm. IPSec is widely used to implement virtual private networks (VPNs), which enable secure access to a remote network via the public Internet.

6.1.6.5 Secure Socket Layer

Unlike IPSec, which works at the IP layer, secure socket layer (SSL) and its successor, transport layer security (TLS), are network security protocols at the transport layer. SSL and TLS support any connection-oriented application layer protocols such as HTTP (HyperText Transport Protocol), LDAP (Lightweight Directory Access Protocol), IMAP (Internet Message Access Protocol), and NNTP (Network News Transport Protocol). In reality, SSL and TLS are mainly used in conjunction with HTTPS protocol to secure communication between a web server and a web client, and TLS is being increasingly used with other application protocols such as POP3 (Post Office Protocol 3) and SMTP (Simple Mail Transfer Protocol). The default HTTPS port number on a SSL-enabled web server is 443. SSL requires a server certificate such that the server can be authenticated by a client or a browser according to an RSA public/private key encryption scheme. Subsequent web traffic is encrypted with a 128-bit or longer session key generated by a symmetric cipher such as DES, 3DES, RC2, or RC4. SSL can also be used to authenticate the client, in which case the client must obtain a public/private key pair and a digital certificate.

6.1.7 General Considerations of Mobile Security and Privacy

Mobile security and privacy essentially possess sets of unique characteristics that separate them from wired network security, such as open-air transmission of wireless signals, comparatively low computing power of mobile devices, high error rate of wireless signal transmission, security management for mobility, and location-sensitive security concerns. The need for security is much stronger than in wired networks, yet to build a secure mobile wireless system one must address a variety of constraints unique to the mobile wireless environment. Some security solutions such as those cryptographic ciphers and security network protocols may not be applicable to a mobile computing environment. For example, computationally intensive ciphers may not work on mobile devices, and in many cases the stable network connection required by many network authentication schemes is not always available.

Even if some security mechanisms can be ported to a mobile wireless system, they must be enhanced with sophisticated components so as to provide confidentiality, integrity, authentication, and non-repudiation in highly varying mobile wireless settings. In addition, many network protocols are designed without security in mind and must be augmented with security considerations. For example, routing protocols in *ad hoc* networks must offer some security to prevent eavesdropping and message tampering. The following is a list of threats in mobile wireless networks:

- *Loss and theft of mobile devices*—Every year hundreds of thousands of mobile devices are lost in airports, hotels, restaurants, etc. This is probably by far the most serious threat to enterprise data and individual privacy.
- *Channel eavesdropping*—An attacker may capture messages transmitted in a wireless channel without being detected.
- *Identity masquerading*—An attacker may impersonate a legitimate user or service provider.
- *Message replay*—An attacker may capture a series of messages between two parties and send them to someone later.
- *Man-in-the-middle attack*—An attacker may intercept and modify messages being sent between two parties or inject new messages without being detected.
- *Wireless signal jamming and interference*—An attacker may use powerful antennas to transmit noisy signals with appropriate modulation in order to disrupt the normal operation of radio receivers.

- *Denial of service*—An attacker may use rogue access points, mobile stations, or specific frequency jamming devices to generate a huge amount of network traffic toward a target computer.
- *War-driving and unauthorized access*—An attacker may use special radio equipment to pinpoint unsecured wireless access points in an area while driving around. Those unsecured wireless LANs, many of which are linked to corporate networks, are wide open to unauthorized users.
- *Virus and wireless spamming*—Small, malicious programs may propagate among mobile device users via short message service (SMS) messages or the wireless Internet. SMS spamming could be another big issue, as subscribers have to pay for that.

In the following sections, security issues in specific wireless networks are discussed in detail.

6.2 Cellular Network Security

As more mobile applications are being delivered to cell phone users, the security mechanism employed by underlying traditional cellular systems must be redesigned to adapt to various new network settings. Moreover, because of the extensive use of cell phones and smart phones, a security breach or a network attack may have an enormous impact on every aspect of the modern society, far beyond the scope of the Internet. Emerging cellular systems have provided the means to secure wireless data transmission and e-commerce transactions, in addition to providing a more general authentication, authorization, and accounting (AAA) solution.

6.2.1 Secure Wireless Transmission

Data transmission in a cellular network can be categorized as user traffic or signaling traffic. Four security issues with regard to cellular traffic are user traffic confidentiality, signaling traffic confidentiality, user identity authentication, and user identity anonymity. For user traffic between subscribers and the back-end system, encryption is necessary to ensure confidentiality. Aside from radio layer frequency hopping modulation and code-division multiple access (CDMA) coding schemes, GSM/GPRS, CDMA, universal mobile telecommunications system (UMTS), and cdma2000 all employ some encryption for user traffic to achieve end-to-end security. For user authentication, authentication schemes generally utilize some sort of identity module on the cell phone. We use GPRS and CDMA as examples to show how end-to-end security is implemented in cellular networks. GSM/GPRS (later Third Generation Partnership Project, or 3GPP) defined an encryption protocol based on an A5 algorithm (GEA3 for GPRS), an authentication protocol based on an A3 algorithm, and a cryptographic key management protocol based on an A8 algorithm. Table 6.3 provides a summary of these algorithms.

In a GSM/GPRS network, a subscriber is identified by a unique international mobile subscriber identity (IMSI) stored in the subscriber identity module (SIM) module along with the handset (the phone). The SIM module also has a secret key (K_i) associated with the IMSI. On the network side, IMSI and its K_i are stored in an authentication center (AuC). The subscriber authentication is carried out in a challenge-and-response fashion, whereby a random number as a challenge is generated and sent to the mobile station by a serving GPRS service node (SGSN). The mobile station uses its K_i to produce a code as the response according to the A3 algorithm. The encryption key (K_c) is derived from the same random number and K_i by the A8

Table 6.3 GSM/GPRS (3GPP) Security Algorithms

Type	Algorithm	Description
Key management	A8	Uses a 128-bit RAND and a 128-bit K_i to produce a 64-bit K_c.
Challenge and response authentication	A3	Uses 1280-bit RAND (the challenge) and a 128-bit authentication key K_i (allocated during user subscription) to produce 32-bit expected response SRES. Implemented on MS SIM and HLR or AuC.
Symmetric encryption	A5	Uses 22-bit COUNT (TDMA frame number) and 64-bit cipher key K_c to produce 140-bit cipher blocks on both BSS and MS for encryption and decryption, respectively.

algorithm. On the mobile station, this is performed by the SIM module. Data and voice traffic is encrypted using K_c by applying the A5 algorithm, a stream cipher. It is said that the K_c is 40 bits long, but no official document reveals its actual length. When the mobile station moves around, a temporary mobile subscriber identity (TMSI) is issued by the network to track the mobile station. Whenever a mobile station changes its associated mobile switching center (MSC), it will obtain a new TMSI that is only valid within the location area of the MSC in charge. A TMSI is encrypted with K_c as part of a TMSI reallocation request message and sent to the mobile station. After applying the A5 algorithm with K_c, the mobile station then confirms reception of the TMSI by replying with a TMSI reallocation confirmation message. Thus, the TMSI reallocation process is again a challenge-and-response scheme.

The universal mobile telecommunications system (UMTS)/wideband CDMA (WCDMA) improved GPRS mobile security by introducing large cipher keys of 128 bits and providing data integrity. Signaling messages and data messages are protected by a KASUMI block cipher protocol that uses the 128-bit cipher key. The same algorithm generates a 64-bit message authentication code to ensure data integrity. Unlike proprietary algorithms used in GSM/GPRS, KASUMI is publicly available for cryptanalytic review.

The Third Generation Partnership Project (3GPP) has formed a working group TSG SA (i.e., Technical Specification Group: Services and Systems Aspects) WG3 Security responsible for the investigation of security issues, and setting up security requirements and frameworks for overall 3GPP systems. The SA WG3 has published a number of technical specifications (TSs) and technical reports (TRs) of security issues ranging from 3G security threats to cryptographic algorithm requirement and specific algorithms to 3GPP and wireless LAN Internet security.

cdma2000 1x uses a 64-bit authentication key (A Key) and an electronic serial number (ESN) to derive two encryption keys for signaling messages and data messages, respectively. The encryption algorithm is AES. Cdma2000 1x EVDO uses a 128-bit A Key derived from a DH key exchange. The authentication protocol in cdma2000 networks is

Table 6.4 CDMA Security Algorithms

Type	Algorithm	Description
Key management	Cellular authentication and voice encryption (CAVE)	Uses a 64-bit reprogrammable authentication key (A Key, allocated with the handset); the electronic serial number (ESN) and a home location register (HLR)/authentication center (AC)-generated random number are used to derive a 128-bit subkey called the shared secret data (SSD), known to the mobile station and its MSC.
Challenge and Response Authentication	CAVE	Uses SSD- and MSC-generated random number (the challenge) to produce an 18-bit authentication signature and a key to replace a well-known value used for voice encoding.
Symmetric Encryption	Cellular message encryption algorithm (CMEA), ORYX, and AES	Uses 64-bit CMEA key derived from part of SSD for signaling message encryption. ORYX is a stream cipher for data messages.

cellular authentication and voice encryption (CAVE). Table 6.4 shows a summary of algorithms in CDMA networks. The 128-bit SSD generated by the CAVE algorithm has two equal-length parts: SSD_A and SSD_B. Using CAVE with SSD_A and a random number (the challenge) generated by the MSC, a mobile station is able to generate an 18-bit authentication signature (the response) and send it to the base station. A mobile station also uses SSD_B to generate a secret key that will be used to scramble the voice. In addition, using the CAVE algorithm, a mobile station can also generate a 64-bit CEMA key and a 32-bit data key. The CEMA key is used to encrypt signaling traffic, and the data key is used to encrypt and decrypt data traffic.

Authentication, authorization, and accounting (AAA) are an integral part of 3G cellular systems. In cdma2000, AAA functionalities are provided by home AAA servers and visited AAA servers along with other mobile IP components. The packet data service node (PDSN) (foreign agent) in a visited network forwards usage data of a mobile station to the home AAA, possibly through a broker AAA. In UMTS, the CN has a home agent and an AAA server. Using serving GPRS service nodes (SGSNs) and gateway GPRS support nodes (GGSNs) as gateways, a mobile station's visited AAA server can communicate with its home AAA server for usage updates, roughly the same procedure as for location updates with the exception that the data being transmitted are related to AAA functions.

6.2.2 Secure Wireless Transaction

Mobile applications are primarily deployed in a heterogeneous network environment in which wireless and wired networks, secured enterprise networks, and wide open home wireless networks coexist and interconnect. One cannot count solely on wireless communication security even though the underlying wireless network is highly secure. Higher layer (network layer or beyond) security mechanisms are invariably required when user traffic is exposed to the unsecured Internet or wireless networks fail to provide the desired security functions. In the following, the wireless transport layer security (WTLS) and WAP (Wireless Application Protocol) identification module (WIM) of WAP and IPSec or SSL VPNs are introduced as the most widely used security protocols on today's mobile devices. Note that they can also be used in other wireless networks, such as wireless LANs and Bluetooth.

6.2.2.1 Wireless Transport Layer Security

Wireless transport layer security (WTLS), as defined in WAP 2.0, provides message confidentiality, message integrity, and unidirectional or mutual authentication at the transport layer. It is logically identical to SSL/TLS but has been adapted to the wireless environment. Message encryption is performed using RC4, DES, and triple-DES or 3-DES. Message integrity is guaranteed using HMAC. Authentication is based on PKI, using RSA, ECC, or DH. A WAP server (also called WAP gateway) uses a WTLS certificate, a particular form of X.509 certificate. A WAP client may also use a digital certificate obtained from a CA for authentication, although it is uncommon. The following is a description of session establishment for the case when the WAP server must be authenticated (class 2 service of WTLS).

When a client and a server begin a handshake, they first exchange two random numbers in the "hello" messages. When the public key of the server has been verified with a certificate, the client sends a pre-master secret key encrypted by the server's public key. This pre-master secret key and the random numbers exchanged will be used on both sides to compute a 160-bit master secret key. For data encryption, an encryption key block is calculated based on the master secret key, a sequence number, random numbers exchanged, and a string indicating the party of the calculation. This key block will be eventually used to derive encryption keys for an algorithm such as RC4, DES, or triple-DES that has been negotiated during the "hello" message exchange.

WTLS specifies keyed hashing algorithms such as SHA-1 and MD5 for the computation of MAC over compressed data. For mobile devices with limited computing power, a light overhead SHA_XOR_40 algorithm is also provided in earlier version of WTLS. The key used during MAC computation, also known as the MAC secret, is also derived from the encryption key block. In order to make denial-of-service attacks more difficult to accomplish, the WTLS specification suggests that a WAP server should not allow an attacker to break up an existing connection or session by sending a single message in plaintext from a forged address.

Figure 6.6 depicts the WTLS architecture. At its heart is the record protocol, which interfaces with the wireless datagram protocol (WDP) and the wireless transport protocol (WTP) and is responsible for data encryption and integrity verification. The handshake protocol defines the negotiation of cryptographic parameter such as algorithms, authentication schemes, and compression methods. When the negotiation is complete, the change cipher spec protocol is performed, indicating that the party is ready to use the negotiated mechanism. After that application, data can be exchanged according to the application data protocol.

An earlier version of WTLS is not secure, as researchers have found some security problems [4]. For example, in WTLS predicable

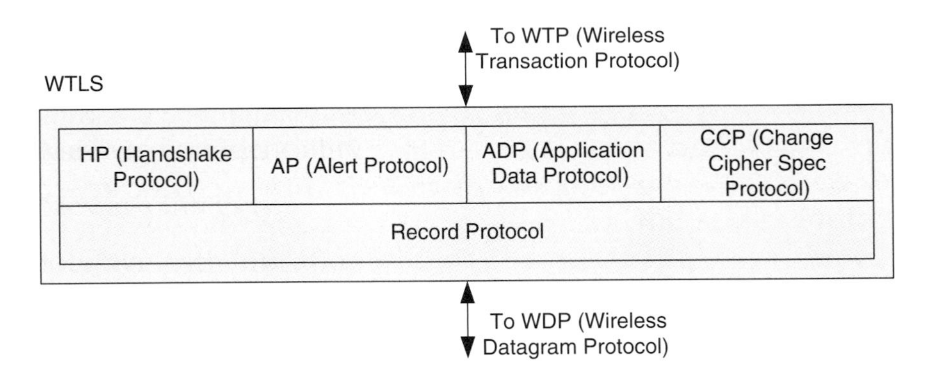

Figure 6.6 WTLS architecture.

initialization vectors (IVs) may lead to encryption key breach, and the SHA_XOR_40 algorithm does not provide message integrity if stream ciphers are used. In light of these problems, the latest version of WTLS (version 06-Apr-2001) has made significant changes; for example, the SHA_XOR_40 algorithm has been removed.

6.2.2.2 WAP Identification Module

In order to seamlessly integrate WAP into an e-commerce environment, a WAP client must be authenticated with respect to mobile device identity. A tamperproof WIM module can be embedded into a WAP client device for this purpose. It could be a component of the SIM card or an external smart card containing the following information: a public/private key pair of the device for signing and another pair for authentication, manufacturer's certificates, and user certificates or their URLs. A WIM module implements the WTLS class 3 service, allowing the WAP client associated with it to be authenticated. This class of service specifies that, in addition to server authentication during the handshake, the client must generate a digital signature using one of its public/private key pairs stored in the WIM module, enabling nonrepudiation of client messages.

As a similar wireless web platform, iMode also provides SSL-based server authentication, message encryption and integrity. Because iMode is a proprietary architecture, details of its security mechanisms are not publicly available. Other wireless web platforms have been developed by Japanese companies, such as EZWeb (KDDI) and J-Sky (J-Phone). Although internals of those systems are not revealed to the public, it is commonly believed that they offer the same set of security services based on SSL or TLS.

6.2.2.3 IPSec/SSL VPNs

IPSec/SSL VPNs are widely used in mobile wireless networks to allow for secure remote network access. These protocols are transparent to the underlying radio technologies used for wireless communication. As long as a network is IP based, theoretically IPSec will work without a problem, although in reality there are some problems with respect to the nature of wireless transmission and mobility. For example, a VPN tunnel may be interrupted during handoff. Unlike IPSec VPNs, which provide secure access to a network, SSL VPNs enables secure remote access to an application inside a network.

Mobile VPN is particularly useful when a mobile device is used by a salesperson, field engineer, or other type of mobile worker wishing to remotely access an enterprise network via the Internet. A mobile VPN, based on either IPSec or SSL, could solve the problem. Aside from a VPN client on the mobile device, a VPN gateway must be set up for client authentication and data encryption/decryption. A problem with using VPN is related to U.S. export control on cryptography, which basically imposes strict control over the export of cryptographic software and hardware for national security considerations. Strong cryptographic systems such as 128-bit key VPNs are not allowed to be exported unless certain licenses have been obtained. Worldwide corporate networks are at risk when VPN clients in overseas offices use 40-bit encryption.

Aside from these two protocols, smart phones running an advanced operating system such as Windows Smartphone allow for normal SSL to be used within a mobile web browser. It is expected that higher layer security protocols will be directly ported onto relatively powerful mobile devices such as smart phones.

6.3 Wireless LAN Security*

Because more cell phones and smart phones are being equipped with Wi-Fi interfaces, related security problems of IEEE 802.11 wireless LANs have become a hot topic, especially after numerous serious vulnerabilities of wired equivalent privacy (WEP), the security mechanism of 802.11, were discovered. Understandably, when the 802.11 wireless LAN standard was developed, security was apparently not a top priority. The "wired equivalence" design rationale essentially led to some earlier versions of wireless LAN security solutions that clearly failed to deliver security functions they were supposed to provide. Many Wi-Fi products in use are based on these flawed protocols and mechanisms. Fortunately, the IEEE 802.11 working group has offered several new standards with enhanced security. Wireless LAN products often incorporate enhanced security as an option in addition to wide-open configurations. For example, WAP has been required in all new Wi-Fi certified products since 2004, and WPA2 (for 802.11i) was required for Wi-Fi certification beginning in 2005.

Security risks in wireless LANs include eavesdropping, unauthorized access, masquerading, man-in-the-middle attacks, denial of service (DoS), and rogue access points:

- *Eavesdropping*—Eavesdropping is highly possible because the coverage of wireless signals is quite difficult to determine, and anyone within the range with an appropriate interface will be able to pick up the signal and intercept ongoing data transmissions at will. Weak encrypted signals can be cracked with modest effort. Powerful tools such as AirSnort and Kismet made wireless eavesdropping on unsecured wireless LANs much easier.
- *Unauthorized access*—Unauthorized access happens when a home or enterprise wireless LAN operates in default configuration mode, which permits anyone to use its Internet access as well as other resources shared in the network.

* Here, we use the most popular wireless LAN standard (IEEE 802.11) for discussion.

- *Masquerading*—Many wireless LANs use wireless adaptor's MAC address (physical address) as filters. Thus, attackers may masquerade themselves by spoofing MAC addresses. This can be done in conjunction with eavesdropping.
- *Man-in-the-middle attacks*—Wireless LANs are designed to allow an access point to authenticate a station but not the other way around; hence, a station cannot be sure that the access point in question is what it claims to be. Attackers may pretend to be an access point sitting between a station and a real access point to intercept, modify, and forge packets.
- *Denial of service (DoS)*—DoS is very common on the wired Internet. Many machines are organized to attack a single website, making it unable to service legitimate users. In wireless LANs, attackers may use rogue APs, their own stations, or other non-802.11 spectrum jammers to send a large amount of forged 802.11 management or control frames or broad-spectrum noise. The IEEE 802.11 MAC protocol also has been shown to be vulnerable to DoS attack [10].
- *Rogue access points*—Due to the ease of network setup and configuration, one may quickly build a small insecure wireless LAN and make it work instantly by connecting it to the wired back-end; hence, the entire wired network may become insecure because of the rogue wireless LAN.

6.3.1 Common 802.11 Security Myths

In practice, wireless LANs are often wide open without any access control or simply employ a MAC-based (here MAC refers to the adapter's physical address) access control list (ACL) to authenticate legitimate mobile stations. An ACL is essentially a list of MAC addresses that are permitted to access the network. Those data frames not originating from legitimate MAC addresses will be rejected by the access point without going through further authentication. As in a wired LAN, a MAC address in a frame header is always transmitted in cleartext regardless of encryption method in use, allowing anyone to gather a list of MAC addresses of stations associated with an access point. An attacker can forge data frames that use those authorized MAC addresses and gain access to the network; therefore, contrary to common belief, the MAC base access control solution does not solve the problem.

Another common security myth associated with 802.11 is the use of extended service set identifier (ESSID). Because an ESSID identifies an access point, many believe that by disabling beacon messages containing the ESSID of an access point an attacker will not be able to determine the ESSID and thus cannot associate to the access point. In fact, this does not prevent an attacker from getting the ESSID because it is still sent in probe messages when a client associates to an access point; also, many wireless LANs use default, well-known ESSIDs.

Given the fact that a large number of wireless LAN access points are being used and there is no effective way to prevent wireless LAN signals from traveling far, it is tempting to get free access to adjacent wireless LANs while walking, driving, or even flying by using appropriate wireless LAN equipment. In an effort to detect wireless LANs in a regional area, some people have been intensively engaged in activities known as war walking, war driving, and war flying [5]. In all cases, a PDA or a laptop computer with a wireless LAN interface and a global positioning system (GPS) receiver, a handy software tool such as Net Stumbler (http://www.netstumbler.com/) or Air Magnet (http://www.airmagnet.com/), and an optional high-gain antenna are all it takes to produce a so-called wireless access point (WAP) map of access points, either secured (using WEP/WPA/WPA2 or higher layer security measured) or unsecured. With a powerful antenna, a war driver could be many miles away from the physical location of a wireless LAN yet still manage to pick up its signals. Figure 6.7 is a Wi-Fi map of Seattle made by students at the University of Washington (http://depts.washington.edu/wifimap). The dots in the figure represent 802.11 access points (secured and unsecured) within reach of the war drivers. Unsecured wireless LANs detected by war driving not only offer war drivers a free ride on the Internet but also invite attackers to obtain remote access to a network without being filtered by firewalls or detected by intrusion-detection systems.

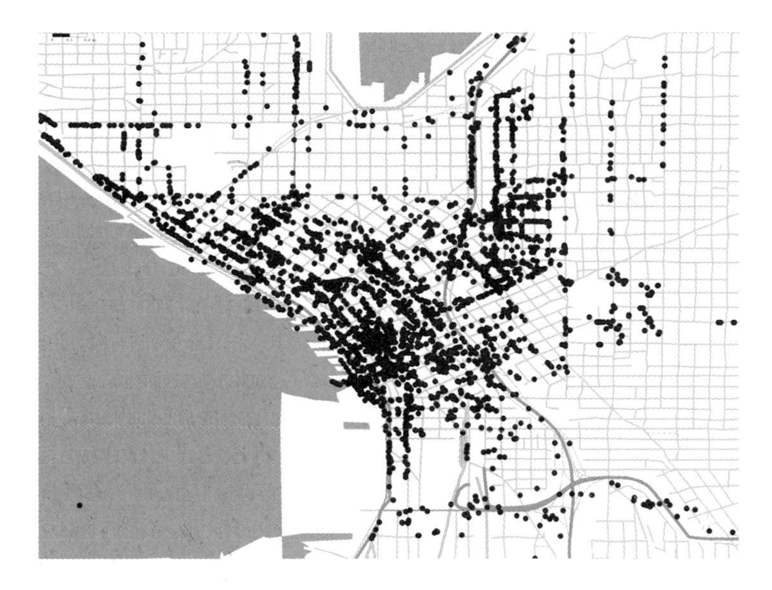

Figure 6.7 A Wi-Fi map of Seattle, WA (Courtesy of University of Washington.)

6.3.2 WEP Vulnerability

The service set identifier (SSID)-based access control indeed does not offer any security functions. Besides, it is common sense that wireless communication should be encrypted and properly authenticated. WEP is the first security mechanism for wireless LANs. A shared secret key of 40 or 104 bits is used by all participating stations within a BSS (Basic Service Set)bounded to the same access point. The encryption algorithm is RC4. For every packet sent between a station and the associated access point, a 32-bit integrity check value (ICV) is computed according to a CRC-32 algorithm. Then RC4 uses a 64-bit key to encrypt the data and the ICV. The encryption key is composed of a 24-bit randomly generated initialization vector (IV) and a 40-bit shared secret key, as shown in Figure 6.8a. Using the 64-bit encryption key, the pseudo-random generation algorithm (PRGA) of RC4 computes a keystream, which will be XORed with a plaintext message (see Figures 6.8b and c). To let the other party know the IV, it is added to the encrypted payload data as part of the packet (the ciphertext), as shown in Figure 6.8d.

Wired equivalent privacy (WEP) is known to have numerous security problems. The first problem is the lack of key management, such as the DH key exchange protocol. The secret key must be distributed by other means of communication and is subject to social engineering attacks, where attackers trick legitimate users of a system in order to obtain passwords, addresses, or other sensitive information. As the network grows, more stations must be informed of the same secret

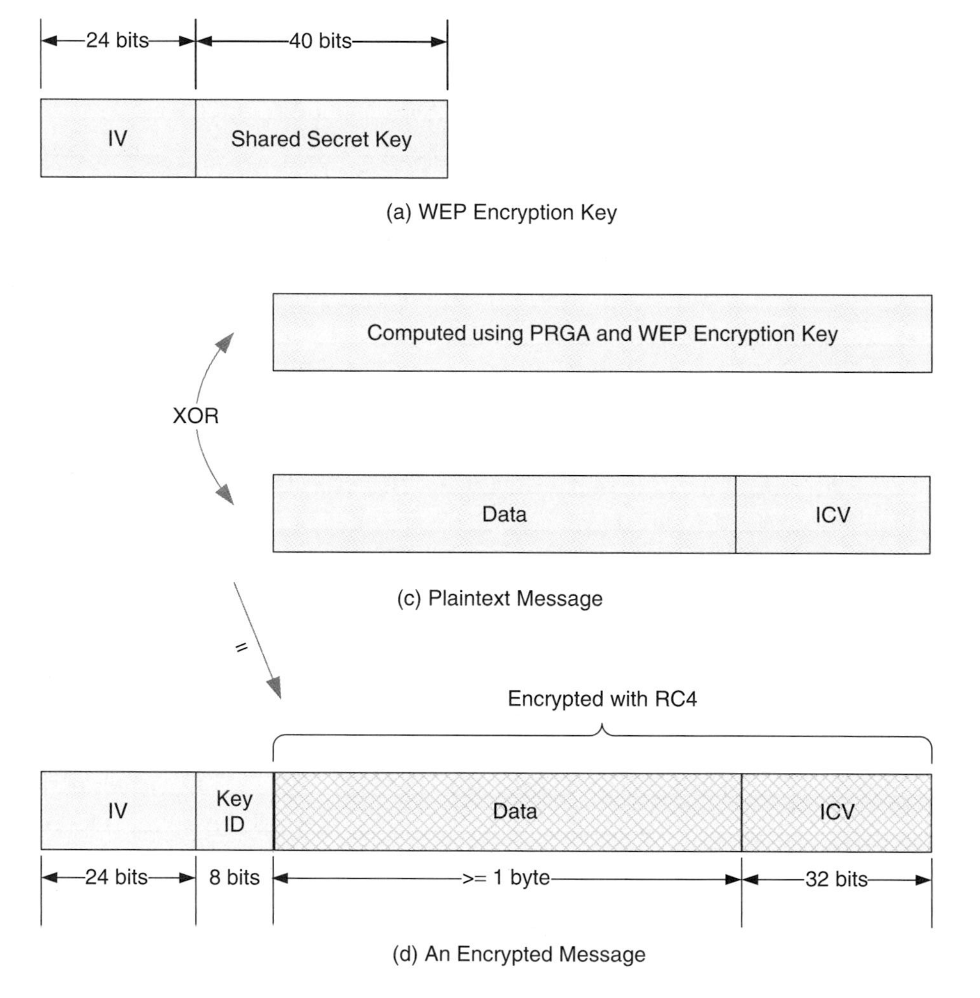

Figure 6.8 WEP encryption.

key, and it would be quite cumbersome to change the secret key for security reasons. Very often the secret key is not secret any more after some time.

The second problem with WEP is the 24-bit IV. Because each packet transmitted has an IV, it is possible that the same IV will be used again after some time. (The code space of IV will exhaust after 2^{24} packets have been sent.) On the other hand, RC4 has been found by Fluhrer *et al.* [1] to have a severe weakness in its key scheduling algorithm; when an encryption key is constructed by the above-mentioned method, an attacker will be able to derive the 40-bit secret part of the encryption key by analyzing those packets that share the same encryption key (secret key + IV) [1]. This attack is referred to as the FMS attack. It has been shown that a WEP key can be cracked in a matter of several hours.

The third problem with WEP is the CRC-32 algorithm used to calculate the ICV [7]. CRC in itself is a simple mechanism for detecting random errors; it was not designed to detect deliberate data falsification. In fact, it has been shown that it is possible to modify the encrypted payload of an 802.11b message without disrupting the checksum (ICV). Furthermore, the CRC-32 algorithm does not involve any keying function, such as HMAC. Thus, an attacker who knows a keystream that corresponds to an IV can safely inject forged packets into the BSS.

6.3.3 802.11 Authentication Vulnerabilities

The IEEE 802.11 wireless LAN specification defines two authentication modes: open and shared key authentication. The default open authentication imposes no authentication on a station that wants to communicate with the access point. In the shared key mode, a challenge-and-response scheme is used. Upon receiving an authentication request from a station identified by its MAC address, the access point responds with a 128-byte randomly generated challenge text in cleartext. The station then encrypts the challenge text with a shared key using RC4 and sends the result back to the access point. The access point uses the same shared key to decrypt the response. If the decrypted value matches the challenge text, the station is authenticated and can proceed to send and receive messages in the BSS; otherwise, the station is rejected.

As mentioned earlier, the problem of this authentication mechanism stems from RC4, stated in a paper by Fluhrer *et al.* [6]. An attacker who obtains a large number of challenge-and-response authentication sequences corresponding to WEP encryption keys (the same IV) can easily deduce the keystreams produced by RC4 by leveraging those weaknesses described in the previous section. From that point, the attacker can authenticate himself to the access point by correctly responding to any challenge texts using the keystream without knowing the shared secret key. Even worse, with the keystream, the cleartext of those messages being analyzed can be revealed by simply XORing the ciphertext against the keystream, exactly the same operation that the associated access point should perform. Using tools such as WEPCrack (http://wepcrack.sourceforge.net/) or AirSnort (http://airsnort.shmoo.com/), it would not take long to crack a WEP key.

6.3.4 802.1X, WPA, and 802.11i

To address the security issues of WEP, one method suggested is to build security overlay on top of the insecure wireless LAN. VPN is often used in practice. 802.11i, which was complete in 2004, was designed to address wireless LAN security issues. 802.1X is a security standard for a more general LAN environment. The Wi-Fi protected access (WPA) protocol has been developed by the Wi-Fi Alliance as an interim solution for 820.11i; hence, 802.11i includes WPA features and some new features, such as AES, CCMP (see discussion below), preauthentication, and key caching for fast handoff.

The IEEE 802.1X standard enables port-based mutual authentication and flexible key management in an IEEE 802 local area network. It DOES not specify a single authentication method but uses the extensible authentication protocol (EAP) as the underlying authentication framework to support various authentication methods such as smart cards, one-time passwords, and certificates. When an unauthenticated supplicant (a client) attempts to connect to an authenticator (a wireless access point), the authenticator opens a port for the supplicant to pass only EAP authentication messages to the back-end authenticator server, which could be, for example, a remote dial-in user service (RADIUS) server. Initially designed for authentication and authorization of dial-in modem access, RADIUS as a protocol (standardized in RFC 2058) has been augmented to facilitate any form of secure remote access with respect to authentication, authorization, and accounting. The supplicant submits its identity to the authentication server, which makes the decision as to whether or not the supplicant should be granted access to the LAN. The authentication server will send either "accepted" or "rejected" to the authenticator. If the result is "accepted," the authenticator will change the client's port to an authorized state, meaning that the port can be used to pass any other additional traffic. As shown in Figure 6.9, 802.1X can be integrated with an existing AAA infrastructure such as RADIUS to provide user-based centralized authentication.

Wireless protected access (WPA) is an interim solution to wireless LAN security that is required by the Wi-Fi Alliance. WPA is backward compatible with WEP in place on widely deployed wireless LAN devices; WPA only requires software or firmware upgrades to existing systems. Each station using WPA will use a different 128-bit encryption key for RC4 data encryption, which can be "refreshed" frequently. The protocol enabling these features is

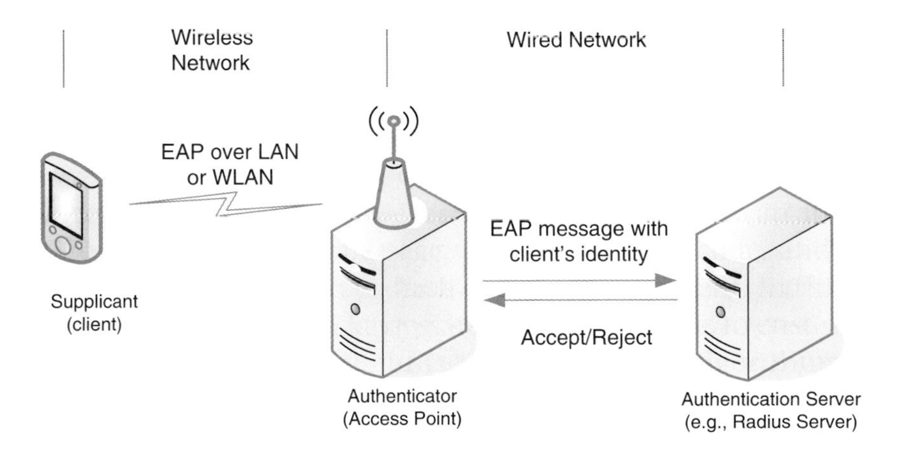

Figure 6.9 802.1X in a wireless LAN setting.

temporal key integrity protocol (TKIP). Key elements of TKIP are listed as follows [8]:

- *Michael*—Michael is a message integrity code (MIC) algorithm that uses a 64-bit key, called the MIC key, to produce a 64-bit tag (a MAC) for a packet, in addition to ICV. Michael is designed to impose dramatically lowered computational overhead on a mobile station than are other MAC algorithms.
- *Per-packet key mixing*—TKIP employs a key mixing function that takes the base WEP key, source MAC address, and packet sequence number as inputs and produces a new 128-bit WEP key for each individual packet. The mixing function is carried out in two phases to reduce computational overhead.
- *Packet sequencing*—Each packet has a 48-bit sequence number, which will be further used to compute the encryption key. This feature defeats message replay attacks.

WPA adopted IEEE 802.1X to provide both authentication and key management. For enterprise networks where a separate AAA server such as a RADIUS is in place, WPA can be integrated with the AAA server for authentication and key distribution. In WPA and WPA2-Enterprise (WPA2 is the product certification available through the Wi-Fi Alliance for 802.11i compatible products. Both WPA and WPA2 have two authentication modes: Enterprise and Personal), the AAA server authenticates individual users and then delivers per-session pairwise master keys (PMKs). In WPA and WPA2-Personal, all stations and the AP have the same pre-shared secret key (PSK) used for both group authentication and PMK. In both cases, the PMK is not used for encryption; it is mixed with the station's MAC and an IV to derive a pairwise temporal key (PTK), which in turn will be used to deduce the AES encryption key.

The encryption key and MIC key used by TKIP are derived from a master key generated by 802.1X. Frequent key changes enabled by 802.1X allow the encryption key and MIC key used by the TKIP to be refreshed every once in a while, thus reducing the risk of key breach due to eavesdropping. Created by the Wi-Fi Alliance, WPA is supported by a large number of devices vendors.

Because WPA serves as a quick patch to WEP, it effectively makes it more difficult to compromise a wireless LAN. The downside of WPA is that it is rather complicated to implement, which could give rise to more security risks. It is also not efficient to introduce an additional MIC key other than the encryption key and use both ICV and MIC for message integrity. Unlike WPA, 802.11i is designed to provide a long-term solution to 802.11 security. Like WPA, it employs 802.1X as the underlying authentication mechanism. Other key features of 802.11i are:

- Countermode–CBC–MAC protocol (CCMP)—Like TKIP, CCMP provides message confidentiality and integrity but uses AES as the cipher instead of RC4. The cipher block chaining message authentication code (CBC–MAC) protects both header and data integrity. The 128-bit encryption key is also used for computation of the 64-bit MAC. The IV is still 48 bits.
- *Pairwise key hierarchy*—802.11i does not compute an encryption key for each packet; instead, the same PMK generated by the 802.1X authentication procedure is used for all packets during an association. PMK is first used to derive a PTK by the access point and the station after proper handshakes between them. The AES encryption key is further deduced from PTK.
- *Key caching and preauthentication*—A user's credentials are kept on the authentication server; thus, when the user leaves and returns shortly, it is not necessary to prompt the user for log-in information; the reauthentication is done transparently. Preauthentication enables a station to be authenticated to an AP before moving to it. Both schemes are designed to speed up authentication in supporting fast handoff.

It should be noted TKIP is also part of 802.11i, but it should only be considered as a short-term solution. Table 6.5 presents a comparison of the three security protocols for 802.11 wireless LAN.

Table 6.5 802.11 Security Protocols Comparison

	WEP	WPA	802.11i
Stage	Initial security mechanism; insecure	Intermediate solution (a snapshot of 802.11i taken in 2002)	Long-term solution (completed in 2004) (WPA2 certifies 802.11i products)
Encryption algorithm	RC4	Enhanced RC4	AES
Key length	40 bits	128 bits refreshable	128 bits
Key management	None	802.1x EAP	802.1x EAP
Message integrity	CRC-32	Michael (including header)	CCMP (including header)
Logical equivalence	None	802.1X + TKIP + RC4	802.1X + CCMP + AES

6.4 Bluetooth Security

As a simple personal area network (PAN) solution, Bluetooth has become the *de facto* standard interface on cell phones and PDAs. People use Bluetooth to transmit files between a mobile device and a desktop computer or between two Bluetooth-enabled devices. Bluetooth earphones enable voice over Bluetooth channels within a short range. Even though the Bluetooth signal can travel only a very limited distance (usually less than 10 m), there are still security issues with respect to data confidentiality and authentication. The Bluetooth SIG (Special Interest Group) has incorporated a security architecture into the official Bluetooth specification.

6.4.1 Bluetooth Security Architecture

Recall that the Bluetooth specification defines a number of "profiles" for different types of typical usages, such as dial-up networking profile, fax profile, headset profile, LAN access profile, file transfer profile, and synchronization profile. Each profile has been specified with a set of protocols suitable for those applications falling into the profile. In providing security for various applications, the Bluetooth SIG has defined a number of profile security policies, each of which specifies recommended baseband security options and protocols for different usage models and profiles. Aside from frequency hopping, the basic Bluetooth baseband security mechanisms are list below:

- *Challenge-and-response authentication*—If device A wishes to be authenticated by device B, device B will send a 128-bit random number (RAND) to device A upon being requested to do so by device A, which uses a 128-bit secret authentication key (link key), RAND, and its 48-bit device address (BD_ADDR) to compute a response according to an algorithm called E_1. When the response is received at device B, device B performs the same computation and compares the result with the response. If they match, then device B is authenticated. Bluetooth devices in a piconet of multiple devices use a shared link key for mutual authentication between two devices. The same link key is also used to derive the encryption key.
- *Per-packet encryption using E_0*—Bluetooth devices may use an encryption key of length 4 to 128 bits, subject to an individual country's regulations. The encryption key is generated by an E_3 algorithm each time the device enters encryption mode. Because communication is always between a slave and a master, the master should initiate the encryption sequence by sending a RAND to the slave. On the slave side, it performs the E_0 algorithm that takes the encryption key, the device address of the master, current clock value, and RAND to compute a keystream. The keystream is then XORed with the packet payload to produce to ciphertext.

In a Bluetooth piconet, a session refers to the period of time a devices stays in a piconet. A link key can be either a semipermanent key or a temporary key, depending on the application.

A semipermanent link key allows a device to use the same link key to connect to other devices in a piconet after a session is over. This is useful when some devices must communicate frequently once in a while. A temporary link key is valid only within a session and will be discarded when the session is over. For different scenarios, four different types of link keys are defined. Below is a summary of these keys and when they should be used:

- *Combination link* key is used for each new pair of Bluetooth devices if they decided to use this type of link key. The procedure to establish a combination link key between two devices is called pairing, in which both devices generate a random number and use it to produce a key. They then exchange those random numbers and compute the combination key. It is used for multiple connections from a single device.
- *Unit link* key is specific to a single device and is stored in non-volatile memory. It is used in installation or when the device is first activated and is never changed afterwards. A device can use another device's unit key as a link key. Which link key should be used is determined during initialization. It is used for communication between two trusted devices.
- *Master link* key is a temporary link key generated by a master device to replace the current link key. It is used for point-to-multipoint communication such as a master broadcasting to its slaves.
- *Initialization link* key is generated using a shared PIN code and device address. The PIN code must be entered to both devices. It is used only to protect initialization parameters transmission when no other keys are available during Bluetooth pairing.

Bluetooth security profile policies have provided general recommendations as to what protocols and algorithms as well as keys should be used in different settings. For specific applications, however, care must be taken to ensure that desired security functions or countermeasures to possible attacks are implemented.

6.4.2 Bluetooth Weakness and Attacks

The use of a PIN code during pairing presents some security risks [9]. The length of a PIN can be between 8 and 128 bits. It could come with the device or can be selected by the user. Prior to link key exchange, an initialization key will first be computed, which in turn uses the PIN code. An attacker may make an exhaustive search over all possible PINs up to a specific length. To verify its guess, the attacker only needs to eavesdrop on the communication channel between two victims to capture random numbers in cleartext and perform the initialization key algorithm. When the PIN code is obtained, the attacker can compute the initialization key and the link key. Eventually, the encryption key can also be obtained, and the communication between those two devices is completely compromised. For this reason, longer PIN codes are strongly suggested by the Bluetooth SIG. An even better countermeasure to PIN attacks is to conduct initialization of two devices in a private and closely secured environment where no wireless communication can be eavesdropped.

The nature of the Bluetooth technology allows mobile device manufacturers to choose a set of configurations optimized for a specific application model. Although this does offer some flexibility to mobile device manufacturers and effectively promote the technology, it also results in security risks to some extent because in some cases security mechanisms are not well implemented or not taken into account, even if the security building blocks are clearly specified in Bluetooth specification. The five types of attacks targeting Bluetooth implementation problems are:

- Bluesnarfing
- Bluebugging
- Bluejacking
- Back-door attack
- Virus and battery draining attack

In a Bluesnarfing attack, an attacker uses modified Bluetooth equipment and directional antennae to capture data from some Bluetooth devices that could be a mile away. The weakness being leveraged in this case is a default insecure mode enabled by some mobile device manufacturers (see below for details). After successful Bluesnarfing, everything on the device is exposed to the attacker.

In a Bluebugging attack, an attacker may remotely control a Bluetooth device, intercepting or rerouting communication without a trace. Bluesnarfing and Bluebugging attacks are mainly targeting cell phones with a Bluetooth interface. They usually require the victim devices to be in "discoverable" mode; that is, the device will respond to discovery queries sent from other Bluetooth devices. It turns out that many cell phones are in this mode by default, which makes them susceptible to these attacks. Worse, a brute force MAC address scan could possibly discover those devices that are not in "discoverable" mode, aided by tools such as RedFang and Bluesniff (http://bluesniff.shmoo.com/); thus, Bluetooth war walking or war driving (*i.e.*, the activity of discovering Bluetooth devices in the proximity) are also possible using these tools.

Bluejacking involves sending unsolicited messages to a Bluetooth cell phone utilizing a security vulnerability in the Bluetooth handshake protocol when two devices are pairing for mutual authentication. During the handshake, the other party's device name will be displayed. Thus, by manipulating the device name, an attacker can send anonymous messages or broadcast messages (proximity spamming) among visible devices. Contrary to public perception, Bluejacking does not imply hijacking of a Bluetooth device. Personal data on a device remain secure and the device is still under the total control of the user, but it does make the victim worry about the security of the device because unwanted messages from someone are being displayed on the device.

A back-door attack allows an attacker to take advantage of a secretly established trusted "pairing" relationship such that the target Bluetooth device can be remotely monitored and controlled without the user's notice. Not only can personal data such as phone books, business cards, calendar, pictures, and e-mail be downloaded from the target device, but also all services available on the device, such as the cellular network connection, built-in camera, audio recorder, or music player, may be accessed and surreptitiously controlled.

The insecure "discoverable" mode of Bluetooth provides a vehicle for mobile virus and worm propagation. Although today's mobile operating systems have imposed strict security mechanisms whereby users are prompted when any installation of programs is about to occur, most people do not even bother to read the warning message and simply click "OK." Worms such as the Cabir worm have certainly demonstrated that cell phones can easily be infected by a mobile virus. Several variants of the Cabir worms have spread among smart phones running Symbian OS with Bluetooth configured in the "discoverable" mode. As a worm, the program tries to propagate by scanning for vulnerable cell phones using Bluetooth and then sends itself to those victims. A side effect of this worm is that the device's battery drains quickly while the worm is constantly scanning for other devices. Other forms of battery draining attacks use some properly powered attacking Bluetooth device to query a victim repetitively, effectively disabling the device after some time. Code signing is a defending technique against these threats. Only those programs developed by trust vendors will be registered and digitally signed, so users have the chance to reject any unsigned downloaded code.

It is clear that, in fighting with mobile viruses, users have to bear the responsibility to be alert to any suspicious programming. Mobile antivirus software may also help users detect any possible infections.

6.5 *Ad Hoc* Network Security

We introduced secured routing protocols for mobile *ad hoc* networks (MANETs) in the last chapter. In fact, security issues in mobile *ad hoc* networks encompass a much broader range of challenges, in addition to secured routing at network layer. As communication in a MANET involves one-hop link layer protocols between two directly connected nodes and multihop packet routing protocols across a set of nodes, security mechanism in MANET should also take into account both the link layer and network layer accordingly, assuming the wireless physical layer is properly secured.

6.5.1 Link Layer *Ad Hoc* Security

For a MANET application, end-to-end security service can be provided by authentication and encryption, which in turn rely on lower layer security protocols to function. IEEE 802.11 WEP is an example of a link layer security mechanism that unfortunately fails to protect one-hop communication between a mobile station and an access point. As discussed earlier, 802.11i has been designed to address the problem of WEP. Specifically, in the distribution coordination function (DCF) mode, when a node senses the channel and finds out it is used by other transmission, it will initiate a binary exponential back-off procedure waiting until the next try. This scheme does not guarantee any fairness over channel access. In fact, it favors the last node among contending nodes. Therefore, one heavily loaded node may keep occupying the channel whereas a lightly loaded node may have to back off many times. Modifications to the back-off scheme have been proposed, mainly to penalize those misbehavior nodes with a large back-off value.

Secured *ad hoc* routing protocols were discussed extensively in the previous chapter. The principle idea of those protocols is to add security extensions to traditional *ad hoc* routing protocols. Note that secured *ad hoc* routing protocols can be categorized as "proactive" security services that are based on node authentication and message confidentiality [11] and the assumption that a node will forward messages according to its routing table or routing mechanism. When a node is compromised and does not forward messages as expected, "reactive" schemes such as ACK (Acknowledgement)-based malicious node detection and coordinated rating are needed.

6.5.2 Key Management

Node authentication in MANET is much more complicated than in a fixed network because of the nature of transient network organization and dynamically changing network topology. Indeed, there is hardly a centralized trusted authority in MANET. And, even when there is, it may not constantly be accessible to every node in the network. Thus, a PKI-based authentication scheme is not directly applicable to MANET. To provide authentication among mobile nodes in such a distributed environment, threshold cryptography can be used.

Threshold cryptography essentially distributes cryptographic functions of an individual node to each node in a group, thus eliminating central authority. It is based on the idea that, even if some individual nodes may be compromised, the majority of a group can be always trusted. In its simplest form, in the context of CA, each node in a group of n nodes holds a distinct piece of the group's private key, and any t nodes can work together to perform the security function as a whole for the group, but any $t - 1$ nodes cannot. This scheme can be used to distribute the security function (*i.e.*, providing certificate for a node's public key) of a single CA over a number of servers [12]. Each server (a fairly stable node in an *ad hoc* network) holds a share of the private key (k). It computes a public key corresponding to its private key share. The public key (K) corresponding to the private key (k) is known to each server. To sign a digital certificate, each server generates a partial digital signature using its private key share. A combiner (a server that directly interfacing service requester) needs to gather t such partial signatures in order to produce a signed digital certificate. Hence, compromised servers will not affect the digital signature service provided by these servers as a whole because they can only generate at most $t - 1$ partial digital signatures. A combiner also verifies the combination using its public key. Tampered partial signatures from compromised servers will be detected by the combiner.

Constructing partial signatures for CA certification is highly computationally intensive and cannot be performed on mobile devices with inherent resource constraints. To adapt threshold cryptography to MANET, a scheme that combines ID-based cryptography and threshold cryptography has been introduced [13]. An ID-based cryptosystem provides public/private key encryption using node ID to derive the effective public key of each node. An ID-based encryption scheme consists of four algorithms as follows:

- *Setup* takes an input security parameter and returns a master public/private key pair for the system. Every node in the system knows the master public key but not the private key.
- *Encrypt* takes the master public key, the identity of the recipient, and a plaintext message and returns a ciphertext. Note that in normal encryption, the recipient's public key and the plaintext are fed into a cipher.
- *Extract* takes the master private key and an ID (an identity string,such as a MAC address) and produces a personal private key to the identity. Every node must obtain its private key from a private key generation (PKG) service.
- *Decrypt* takes the master public key, a cipher text, and a personal private key and returns the plaintext.

It is obvious that an attacker cannot decrypt an intercepted message without knowing the master private key or personal private key of the node to which the message is headed. The combined key management approach aims at leveraging ID-based public/private key pair generation to reduce computational overhead. It works as follows. First, the initial participating nodes decide on a set of security parameters, such as threshold t, and their identities. Then a threshold PKG is performed by these initial nodes to compute a master public/private key pair in a distributed fashion. The master public key is known to everyone. The personal private key of each node is generated based on the node's identity conforming to t-out-of-n threshold cryptography such that fewer than t nodes cannot recover the master private key. Nodes joining the system later must communicate with at least t nodes serving the PKG to obtain t shares of the personal private keys (not the private keys) and compute the personal private key. Because node ID is commonly available in a message header, this approach does not require any specific public key propagation mechanism, as is the case in the CA approach. Another way to introduce low-overhead asymmetric cryptography to mobile devices is using ECC. To this end, some ECC-based distributed key generation schemes have been proposed, such as that of Boneh–Franklin [14].

6.5.3 Wireless Sensor Network Security

Wireless sensor networks have been used in a number of application scenarios, including wild habitat monitoring, lighting and temperature control of a building, and glacial monitoring. More wireless sensor applications that closely relate to our daily life are on the way. As a consequence, security problems of wireless sensor networks have surfaced in response to concerns that potential data interception or tampering could result in serious damage to a system.

The principle challenge of security in a wireless sensor network is the seriously constrained sensor hardware that cannot facilitate generally used security mechanisms on regular desktop computers. Below is a summary of the hardware capability of a SmartDust node developed at the University of California, Berkeley (see Table 6.6) [15]. It is worth noting that new wireless sensor modules tend to have significantly improved hardware components as a result of the rapid advancements in wireless sensor technology but still lag behind regular desktop computers and even PDAs.

Table 6.6 Characteristics of Prototype SmartDust Nodes

CPU	8-bit, 4 MHz
Storage	8-KB instruction flash 512-bytes RAM 512-bytes EEPROM
Communication	916-MHz radio
Bandwidth	10 Kbps
Operating system	TinyOS
Operating system code space	3500 bytes
Available code space	4500 bytes

A wireless sensor is generally expected to operate for years without battery replacement, thus reducing power consumption is always a key design objective. Even if it is possible to incorporate powerful processors and communication capabilities into sensor nodes, their power consumption may exceed what a small battery can support. Consequently, given such a hardware configuration, it would be impractical to use traditional security mechanisms in a wireless sensor network, as they usually require a large amount of memory during operation and impose significant communication and computing overhead on the sensor nodes; for example, asymmetric digital signatures for authentication are too expensive for sensor nodes because they may drain a battery too quickly. One way to provide authentication is to employ symmetric key cryptographic systems between sensor nodes, each sharing a secret key with the central trusted base station. To establish a new key, two nodes use the base station as a trusted third party to set up a secured communication channel.

Another security problem in this domain is secured routing in both static wireless sensor networks and future mobile wireless sensor network. *Ad hoc* routing protocols such as DSR (Dynamic Source Routing) or AODV (Ad hoc On-Demand Distance Vector) (described in Chapter 5) are again unsuitable for wireless sensor networks because of the communication overhead and requirement for state maintenance at each node. In addition, message routing in a wireless sensor network often follows a pattern of many-to-one, meaning that many sensor nodes communicate back to a base station, and, as opposed to routing in *ad hoc* networks of mobile devices, in-networking processing (intermediate nodes processing messages being forwarded) for data aggregation makes secured routing in a wireless sensor network more challenging. Commonly used end-to-end security mechanisms cannot be applied in this case because the contents of messages are subject to modification. Karlof and Wagner [16] compiled a list of attacks on sensor network routing:

- Spoofed, altered, or replayed routing information
- Selective forwarding (a sensor node does not forward messages faithfully)

- Sinkhole attacks (a compromised node spoofs messages to attract traffic from adjacent nodes according to the routing algorithm)
- Sybil attacks (a compromised node pretends to be multiple nodes, thereby confusing routing algorithms and resulting in potential identity theft)
- Wormholes (multiple compromised nodes can collude to establish out-of-band channels, effectively disrupting network topology)
- "Hello" flood attacks (a node simply broadcasts bogus "hello" messages or overheard messages in the network hoping to manipulate topology)
- Acknowledgment spoofing (a node sends spoofed link layer acknowledges to senders of overheard messages)

Among these types of attacks, bogus routing information, Sybil attacks, "hello" floods, and acknowledgment spoofing can be defeated by employing link layer encryption and authentication along with identity verification and authenticated broadcast. Multipath routing can be used to defeat selected forwarding attacks. In order to provide symmetric key cryptography, the network must ensure that no other nodes can impersonate the trusted base station, as each node will obtain a symmetric key from the trusted base station, which also initiates an authenticated broadcast to perform a query. Asymmetric authentication is needed to make sure compromised nodes cannot perform authenticated broadcasts. One way to achieve this is to use delayed disclosure of a series of keys derived from a one-way symmetric key chain [15], which requires the base station and nodes to be loosely synchronized. The base station uses a secret key (*K*n) as the last key in the key chain and computes K_{n-1} using a one-way function *F*: $K_{n-1} = F(K_n)$. Then, it uses K_1, K_2, ... during a specific time period subsequent to computing the MAC (Message Authentication Code) of packets sent within that time slot. Nodes receiving those packets can verify the integrity as well as authenticity of those packets later when the base station discloses keys in the same order as they were used to compute the MAC.

6.6 Mobile Privacy

As in wired networks, security issues in mobile computing environment are closely related to privacy issues. Generally, the notion of privacy encompasses two types of problems. One is data privacy: the protection of sensitive user information that by all means should be secured during transmission or in storage, such as a credit card number being transmitted over a secure socket layer (SSL) connection, or Social Security number stored in a database on disks and tapes. These problems also fall into the mobile security domain, and various security mechanisms to ensure data privacy have already been discussed in this chapter. The second type of privacy issue—namely, privacy services—is primarily concerned with adjustable privacy exposure and enabling mechanisms. The key challenge to this type of security issues is the conflict between more pervasive mobile applications utilizing the sensitive information of users and the need for privacy protection in a computing environment of many such applications. Three approaches have been proposed to offer general privacy-related services in a pervasive mobile computing environment [17]:

- *Increasing awareness of potential privacy breach*—Let the system notify the user whenever sensitive information is being revealed to an external service or system that the user cannot control. For example, a smart phone user should be notified when the user's location and identity are being tracked by a location-based service.
- *Maintaining an audit trail*—The system keeps an audit log of all privacy-related information exposure, interactions, and data exchanges. This does not prevent privacy violation but at least provides some record of what information has been exposed, how, and when.
- *Intelligent alert*—In some cases a user's privacy is exposed not by the system but by an adversary; for example, a smart phone user engaging in a Bluetooth data transmission may be detected by someone nearby and the identity of the user may be revealed because of the data transmission. In this case, ideally the system

should be able to detect such a privacy breach even if it is directly involved.

Mobile privacy is more complicated than mobile security because you cannot just draw a line between what information can be used or shared and what cannot, whereas in security we know that a set of security functions should be implemented in a system. Moreover, legislation is very often involved, because privacy is indeed surrounded by sensitive legal issues. Because a system may surreptitiously detect, use, expose, distribute, or abuse people's privacy-related information, it seems quite reasonable to regulate the use of personal data by credit card companies, telecoms, banks, etc. For example, laws pertaining to privacy include the Privacy Act of 1974, Electronic Communications Privacy Act (1986), and No Electronic Theft (NET) Act (1997). Also, in response to security challenges on the web, W3C has been working on a project called P3P (Platform for Privacy Preference Project, http://www.w3.org/P3P), aimed at developing a framework of protocol for an international privacy policy.

Technologically, the fundamental challenge in this domain is to provide more new services to improve productivity and the user's experience while still guaranteeing a minimal, satisfactory level of privacy exposure. Below we introduce two major mechanisms in this field: identity privacy and location privacy.

6.6.1 Identity and Anonymity

Anonymity in the context of computing refers to a service that prevents the disclosure of the identity of someone who is engaged in network communication or interaction to a system. Anonymity is not always a priority; in many cases, we are not particularly concerned that when we surf the web our travels on the Internet are being logged by nearly all web servers. In some cases, however, anonymity is required, such as:

- Users do not want to be censored when accessing some websites. Information censorship is largely due to political reasons.

- Users do not want to reveal information about operations such as file sharing being performed with a computer or a cell phone.
- Users do not want to expose personal information to an untrustworthy online community or they simply do not want to be traced in a network.
- Users want to remain anonymous to prevent identity theft.

Many people think that the Internet offers anonymity. This was reflected by a now-famous cartoon in an issue of *The New Yorker* magazine published in 1993. It showed a dog sitting at a computer, talking to another one: "On the Internet, no one knows you are a dog." Unfortunately, without using a specially designed privacy-enhancing system, nearly every action of a user, as well as the user's identity, is traceable as long as interested parties such as law enforcement agencies, Internet service providers (ISPs), and network administration authorities consider it worth the time and money to do so. In the mobile wireless world, we are well aware that every phone call is logged and can be tapped, and technically every bit can be traced back to the sender. Thus, the challenge of mobile privacy lies in the fact that a mobile system must provide both privacy and accountability.

When there is a direct logical mapping between the user's identity and network location such as IP address, cell phone number, or processor identifier, the goal of anonymity is thus reduced to protecting the network location from being exposed to unintended parties. A simple solution is to introduce a proxy between the user and the place of interest (a website, for example) to hide the network location of the user. An example of such a proxy-based anonymity system is Anonymizer (http://www.anonymizer.com). A user always goes through the proxy (the anonymizer) in order to reach the destination using a URL such as http://www.anonymizer.com:8080/www.yahoo.com. The proxy acts on behalf of the user when visiting a website. Although a user's identity is hidden from the visited server, this approach does not protect the anonymity of the server. To this end, a proxy for the server could be a solution, which hides the real URL of the server and simply exposes a cryptic URL to users. Rewebber (http://www.rewebber.de/) is an example of such a system. In both user–proxy and server–proxy setups, a user has to trust the proxy, and communication between a user and the proxy is not protected in terms of privacy. For this reason, cryptographic mechanisms are introduced in some systems such as Mix-Net and Onion Ring.

A mix network is a set of router nodes (mixes) that allow for anonymous message transfer using a layered public-key encryption. They have been used to maintain privacy during e-mailing, web surfing, electronic voting, and electronic payment. The idea of mix networks as a solution to e-mail privacy was first proposed by David Chaum in 1981. In its initial design, a computer (called a *mix*) processes each e-mail (or any type of data item) before it is delivered. A sender may choose intermediate mix nodes to form a path across a mix network, or the mix network can enforce a path for every message. The latter approach is referred to as a *cascade*. Figure 6.10 depicts the logical architecture of a cascade. Depending on the number of mix nodes a message will traverse, a message (M) is encrypted first using the public key of the recipient (K_a) and a random number (R_1). The result is then appended with the address of the recipient and further encrypted using the public key of the last mix node along the path (K_n). Each intermediate mix node decrypts the message using its private key and forwards the result to the next mix

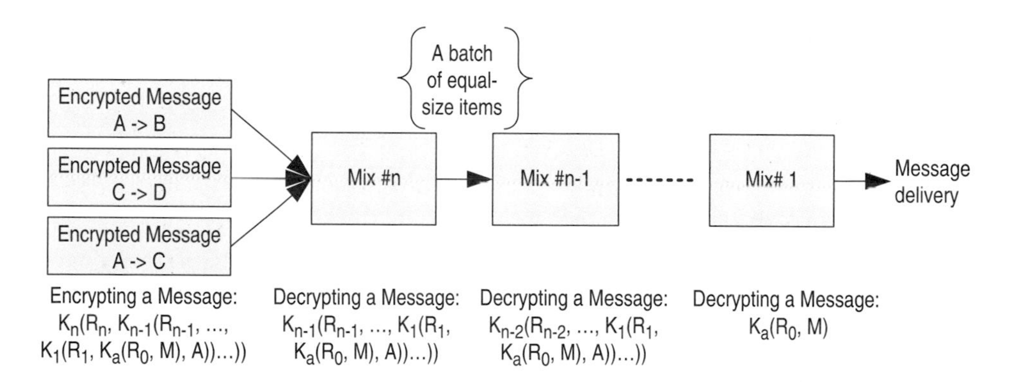

Figure 6.10 A cascade mix network.

node. Output messages at a mix node are also permutated to disguise the order of arrival. In effect, no single mix node knows both the sender address and recipient address. Input messages to a mix node are reordered; therefore, correspondence between items in its input and those in its output at a mix node is protected. The downside of a mix network is that it requires mix nodes to trust each other, meaning that everyone will perform normally. Later improvements to the approach have employed credit-based mix node selection and threshold cryptography to relax this requirement while still ensuring message anonymity.

A similar idea—namely, Onion Routing (http://www.onion-router.net)—is designed to use a collection of widely distributed routers (Tor nodes) to create random paths for the sender such that no individual server knows the complete path. Before sending data over an anonymous path, the first Tor router adds a layer of encryption for each subsequent one in the path. As the message traverses the network, each Tor router removes one layer of encryption.

Unlike Mix-Net and Onion Ring operating at the network layer, Crowds [18] is an application layer protocol designed for web traffic anonymity, utilizing a crowd of proxies to hide the network location of a message. The basic idea is to blend a user's traffic with that of many others such that it is not possible to trace a single web request or reply back to the sender. Any user willing to participate in the crowd could be a proxy in the crowd. The user's traffic will first be forwarded to the crowd along a probabilistic virtual path before going to the public Internet.

Freenet is an example of peer-to-peer-based anonymity network (http://freenet.sourceforge.net/). It allows anybody to publish and read information with complete anonymity. Freenet achieves this by pooling the nodes' storage for data replication services while the true origin or destination of the data remains completely anonymous. In Freenet, shared files are mapped into a key space. Aside from locally stored files, a node maintains a local key routing table allowing the node to forward a query message from one neighbor (the predecessor) to another appropriate neighboring node on behalf of that predecessor, in case the key in question is not locally served. Note that, unlike

IP routing, where the source IP address is always forwarded hop-by-hop as part of the IP header, query routing messages in Freenet do not carry the request's identity along the path. Therefore, requestor (a node querying a file) anonymity is preserved because a node forwarding or replying to a query does not know the requester's identity (a node ID in Freenet). In order to maintain the insertor's anonymity (or, more precisely, key anonymity, as a file is identified by a routable key), Freenet employs a variation of the mix network approach for inserter (a node that shares a file in the network) anonymity: messages between a sender and a recipient must go through a chain of prerouting nodes, each acting as a mix to impose public-key-based encryption over links along the chain. After going through the mix network, a message is disguised as if it is originated from the last mix. Then the message is sent to Freenet for normal routing.

In the context of mobile wireless services, identity anonymity is sometimes necessary in mobile payment, mobile trading, and information sharing. Considering the amount of web traffic in current mobile Internet and wireless network applications, an application layer anonymity system is preferable to network layer solutions. For example, we can use a set of WAP proxies acting like a crowd to mix WAP traffic from many mobile users. Alternatively, depending on the requirements of a specific mobile service, a service anonymizer can be introduced as part of the back-end system by a mobile service provider. The predominant task of a service anonymizer in an m-commerce environment is hiding one user's real identity from the other during a transaction. In addition, a service anonymizer can be integrated with an authentication system. An example of such systems used for mobile micropayment is described in Hu *et al.* [19]. We will discuss micropayment in more detail in the next chapter.

6.6.2 Location Privacy

A particularly significant class of privacy issues is location privacy in a mobile wireless environment. Location privacy refers to the capability of a mobile application or service to prevent unintended parties from obtaining a person's current or past location. The fact that more location-based services, including GPS, Wi-Fi, radiofrequency identification (RFID), and wireless sensor network technologies, will have the capability to monitor a user's location has led to increasing concerns as to how to protect the location information from unintended access. Technical details of location-based services are discussed in the next chapter; here, we focus on a subsystem in a location-aware system that enables location privacy.

There are three categories of problems surrounding location privacy for a mobile system, each solving the problem from a different viewpoint:

- *Category I*—Location information security (secure location data gathering and transmission with respect to privacy requirement)
- *Category II*—Identity pseudonym (applying identity anonymity schemes to location service)
- *Category III*—Location information policy (building interactive social and legal privacy aware framework)

The first category, location information security, is mainly concerned with the formatting and secure transmission of location information in order to protect user privacy. The IEEE Work Group Geographic Location/Privacy (Geopriv) [20] has provided a location privacy framework that is independent of the underlying location determination mechanism. The framework defines a location object that conveys location information and possibly privacy rules to which Geopriv security mechanisms and privacy rules are to be applied. Geopriv recommends the use of security mechanisms of the location object itself, such as MAC (Message Authentication Code) and encryption as part of the location object. In addition, secure transport of location objects should be used whenever possible in protocols carrying location objects to ensure appropriate distribution, protection, usage, retention, and storage of location objects based on the rules that apply to those location objects. One example of such a privacy-preserving communication protocol is Mist [21]. This approach is based on an overlay network in the form of a hierarchy of

Mist routers that perform limited PKI-secured handle-based routing to hide the location of a connection (here, location is the addresses of the source and the destination). A handle is an ID that uniquely identifies an upward Mist router in the hierarchy. Intermediate Mist routers are unaware of the endpoints of a connection (source and destination addresses). The protocol effectively prevents insiders, system administrators, and the system itself from tracking a user's location without affecting normal secured communication.

The second category, identity pseudonym, hides the user's identity by making network traffic anonymous in a location-based application. A broad range of anonymity techniques used in wired network applications could be adopted to location-based applications. For example, Beresford and Stajano [22] have designed a privacy-protecting framework based on frequently changing pseudonyms, thereby effectively mapping the problem of location privacy onto that of anonymous communication. An anonymizing proxy is introduced to leverages the idea of mix networks in the general anonymity service domain to delay and reorder messages when users exit mix zones.

The last category of solutions, location information policy, focuses on building a framework of privacy policies and mechanisms that allow users to interact with location-based applications to control location information release with respect to corresponding privacy policies. Privacy solutions in this category in essence rely on respect and social and legal norms to enforce privacy. The most notable effort in this direction includes the Privacy Preference Project (P3P) [23] and pawS [24], which provides an industry standard of privacy policies that websites can use to announce their specific privacy practices. The goals of P3P include simplifying the process of reading privacy policies, minimizing latency delays, and making polices conforming to the law. The P3P architecture consists of user agents, privacy reference files, and privacy policies. User agents can be part of a web browser or a browser plug-in. A user agent automatically fetches the P3P policies of a website when the user visits the site and checks these policies against the user's predetermined preferences. A policy reference file is used to collect the P3P policies of certain regions of a website (such as a web page), portions of a website, or the entire website. P3P employs an XML encoding scheme for P3P policies. pawS [24] is a similar approach. Both P3P and pawS are specifically designed to address privacy issues on the Web. A more general approach utilizing the same basic idea has been proposed to protect privacy when arbitrary location-based applications request a user's location [25].

6.7 Summary

Mobile security and privacy are by all means interrelated issues that must be addressed as a whole. Because of the potentially pervasive nature of future mobile computing applications, people are far more concerned with these issues than common security risks in a wired network environment. A mobile wireless system must take security and privacy into account at the very beginning of the design phase and utilize appropriate security service building blocks to provide data confidentiality, integrity, authentication, and nonrepudiation, as well as efficient access control. Different mobile wireless systems and applications may employ a set of security mechanism at different layers, due to the intrinsic restrictions of the underlying network and mobile devices. In this chapter, we have explored security issues in some widely deployed mobile wireless systems, such as cellular networks, wireless LAN, and Bluetooth. We also introduced interesting yet challenging security issues in emerging mobile *ad hoc* networks. 3G cellular networks by design provide strong low-layer security for mobile applications and services. On the other hand, wireless LAN is an excellent example of bad design strategy to demonstrate that security has to be considered a high priority when it comes to designing a mobile wireless system. The well-known WEP vulnerabilities have largely hindered the widespread implementation of 802.11 wireless LAN in business organizations and government agencies. The IEEE 802.11 working group has designed a new standard, called 802.11i, to address these weaknesses. Bluetooth security concerns grew significantly after researchers demonstrated that they could use Bluetooth equipment to hack into a Bluetooth cell phone up to a mile away. Although this particular security problem is merely an implementation issue rather than a serious protocol design issue, some researchers have pointed out several weaknesses in the official Bluetooth specifications that may lead to personal information breaches and device compromise. The Bluetooth specification was largely based on the assumption that within its limited signal range of <10 m security was not a significant problem. This turned out to be a false assumption. Security services in *ad hoc* networks lead to new challenges due to the absence of a fixed network infrastructure in MANET.

Problems such as secured routing, link layer security, and key management were examined. We introduced two problems in the domain of mobile privacy: anonymity and location privacy. Anonymity is a critical problem because people are seeking technological ways to ensure freedom over the Internet. Location privacy is particularly important to mobile users who wish to take advantage of emerging location-based services but do not want to be traced for whatever reasons. Technical, social, and legal solutions have been proposed to address this problem to some extent.

Aside from wireless network security mechanisms, many of the security and privacy problems discussed in this chapter are closely related to requirements of the underlying mobile applications and services such as location-based services, mobile commerce, and instant messaging. The next chapter looks at the challenges and solutions surrounding the design and implementation of a wide range of mobile applications and services.

Further Reading

3GPP SA3 Security Working Group, http://www.3gpp.org/TB/SA/SA3/SA3.htm (3GPP technical specifications).

3GPP2's Security Working Group (3GPP2 TSG-S Working Group 4), http://www.3gpp2.org/Public_html/specs/tsgs.cfm

AES/Rijndael, csrc.nist.gov/CryptoToolkit/aes/rijndael/

EEF DES Cracker, http://www.eff.org/Privacy/Crypto/Crypto_misc/DESCracker/.

For a list of practical ways to protect a Wi-Fi network, see http://www.wi-fi.org/OpenSection/secure.asp?TID=2 (the site also introduces WPA2, a Wi-Fi certified security solution based on 802.11i).

IEEE AAA Working Group, http://www.ietf.org/html.charters/aaa-charter.html.

IETF Geographic Location/Privacy Working Group, http://www.ietf.org/html.charters/geopriv-charter.html; Geopriv Requirement, http://www.ietf.org/rfc/rfc3693.txt.

IETF Internet Key Exchange, http://www.ietf.org/rfc/rfc2409.txt.

IETF IPSec Working Group, http://www.ietf.org/html.charters/ipsec-charter.html.

IETF PKI (X.509) Working Group, http://www.ietf.org/html.charters/pkix-charter.html.

References

[1] S. R. Fluhrer, I. Mantin, and A. Shamir, Weaknesses in the key scheduling algorithm of RC4, in *Proc. of the Eighth Annual Workshop on Selected Areas in Cryptography*, August 2001.

[2] X. Wang, D. Feng, X. Lai, and H. Yu, "Collisions for Hash Functions MD4, MD5, HAVAL-128, and RIPEMD," in *Proc. the 24th Annual International Cryptology Conference (Crypto'04)*, Santa Barbara, CA, 2004.

[3] R. Rivest, A. Shamir, and L. Adleman, A method for obtaining digital signatures and public-key cryptosystems, *Commun. ACM*, 21(2): 1978, pp. 120–126.

[4] M.-J. Saarinen, "Attacks Against the WAP WTLS Protocol," in *Proc. Proceedings of the IFIP TC6/TC11 Joint Working Conference*

on Secure Information Networks: Communications and Multimedia Security, Leuven, Belgium, 1999.

[5] H. Berghel, Wireless infidelity I: war driving, *Commun. ACM*, 47(9):21–26, 2004.

[6] S. Fluhrer, I. Mantin, and A. Shamir, Weaknesses in the key schedule algorithm of RC4, in *Proc. the 4th Annual Workshop on Selected Areas of Cryptography*, August 2001.

[7] N. Borisov, I. Goldberg, and D. Wagner, "Intercepting Mobile Communications: The Insecurity of 802.11," in *Proc. the 7th Annual International Conference on Mobile computing and networking (MOBICOM'01)*, Rome, Italy, 2001.

[8] N. Cam-Winget, R. Housley, D. Wagner, and J. Walker, Security flaws in 802.11 data link protocols, *Commun. ACM*, 46(5):35–39, 2003.

[9] M. Jakobsson and S. Wetzel, Security weakness in Bluetooth, in *Proc. RSA Conference 2001*, San Francisco, CA, 2001.

[10] V. Gupta, S. Krishnamurthy, and M. Faloutsos, "Denial of Service Attacks at the MAC Layer in Wireless Ad Hoc Networks," in *Proc. IEEE Military Communications Conference (MILCOM)*, Anaheim, CA, 2002.

[11] H. Yang, H. Luo, F. Ye, S. Lu, and L. Zhang, Security in mobile *ad hoc* networks: challenges and solutions, *IEEE Wireless Commun.*, 11(1):38–47, 2004.

[12] L. Zhou and Z. J. Haas, "Securing Ad Hoc Networks," *IEEE Network Mag.*, 13(6):24–30, 1999.

[13] A. Khalili, J. Katz, and W. A. Arbaugh, Toward secure key distribution in truly *ad hoc* networks, in *Proc. 2003 Symposium on Applications and the Internet Workshops (SAINT'03 Workshops)*, Orlando, FL, January 27–31, 2003.

[14] D. Boheh and M. Franklin, Identity-based encryption from the Weil pairing, *SIAM J. Computing*, 32(3):586–615, 2003.

[15] A. Perrig, R. Szewczyk, V. Wen, D. Culler, and J. D. Tygar, "SPINS: Security Protocols for Sensor Networks," in *Proc. the 7th Annual International Conference on Mobile Computing and Networking (MOBICOM'01)*, Rome, Italy, 2001.

[16] C. Karlof and D. Wagner, Secure routing in wireless sensor networks: attacks and countermeasures, in *Proc. of the First IEEE International Workshop on Sensor Network Protocols and Applications*, Anchorage, AK, May 2003.

[17] M. Satyanarayanan, "Privacy: The Achilles Heels of Pervasive Computing?," *IEEE Pervasive Comput.*, 2(1):2–3, 2003.

[18] M. K. Reiter and A. D. Rubin, Anonymous web transactions with crowds, *Commun. ACM*, 42(2):32–48, 1999.

[19] Z.-Y. Hu, Y.-W. Liu, X. Hu, and J.-H. Li, "Anonymous Micropayments Authentication (AMA) in Mobile Data Networks," in *Proc. the 23rd Annual Joint Conference of the IEEE Computer and Communications Societies (INFOCOM'04)*, Hong Kong, 2004.

[20] IEEE Geographic Location/Privacy, http://www.ietf.org/html.charters/geopriv-charter.html, 2004

[21] J. Al-Muhtadi, R. Campbell, A. Kapadia, M. D. Mickunas, and S. Yi, Routing through the mist: privacy preserving communication in ubiquitous computing environment, in *Proc. of the 22nd International Conference on Distributed Computing Systems (ICDCS'02)*, Vienna, Austria, July 2002.

[22] A. R. Beresford and F. Stajano, Location privacy in pervasive computing, *IEEE Pervasive Comput.*, 2(1):46–55, 2003.

[23] Platform for Privacy Preference 1.0 (P3P1.0) specification, World Wide Web Consortium, http://www.w3.org/TR/2002/REC-P3P-20020416/, 2002.

[24] M. Langheinrich, Privacy by Design: Principles of privacy-aware ubiquitous systems, in G. D. Abowd, B. Brumitt, and

S. A. Shafer (Eds.), *Proceedings of the Third International Conference on Ubiquitous Computing*, Springer–Verlag, Atlanta, GA, 2001.

[25] G. Myles, A. Friday, and N. Davies, Preserving privacy in environments with location-based applications, *IEEE Pervasive Comput.*, 2(1):56–64, 2003.

7

Mobile Application Challenges

In parallel with the amazing evolution of mobile wireless technologies, a full spectrum of next-generation mobile applications and services have begun to surface in countries where the penetration of cell phones and networked personal digital assistants (PDAs) is high. New mobile applications are being driven by mobile service providers seeking new value-added applications and services to take advantage of 3G systems and other wireless data infrastructures. Traditional mobile applications, such as voice, and simple data services, such as Internet surfing, are not enough today; consumers are looking forward to better leveraging the mobility supplied by low-power, wireless-capable mobile devices to enjoy rich-content entertainment, ubiquitous information access, and agile business operations.

This chapter introduces several mobile applications that will expand greatly over the next several years. To simplify the discussion and provide an overview of the mobile applications landscape, the chapter categorizes these applications as location-based services and location-aware mobile computing, mobile messaging, mobile multimedia, mobile commerce, and telematics, each category encompassing a wide range of mobile applications. For each category of mobile applications, the discussion focuses on the technological and economic challenges encountered in building systems for specialized applications. We also want to link the needs of consumers and enterprise users for mobile computing to enabling wireless technologies in an effort to shed some light on potential mobile applications that so far do not exist.

7.1 Location-Aware Mobile Computing

Mobility support of mobile wireless systems naturally leads to location-aware computing that encompasses a number of techniques, protocols, and enabling mobile wireless technologies and mobile devices utilizing an object's location information to provide augmented consumer- or business-oriented applications and services—namely, location-based services (LBSs). To mobile service providers, LBSs seem to be good candidates for value-added services that have the potential to grow significantly as more location-aware mobile devices and network capabilities are being used. An object's location is indeed a key component of context, meaning that location-aware computing is actually concerned with making a mobile wireless system context aware in terms of location. Location-based services and applications are therefore the first step toward context awareness in a pervasive mobile computing environment. The role of smart phones in the context of location-aware computing is crucial, as a smart phone is very likely to have multiple wireless interfaces; thus, with some supporting software, a smart phone could be used as a location-sensing device for object localization.

We begin our discussion with two fundamental issues of location-aware computing: location representation and localization. The first issue is concerned with choosing the best location data model for a location-aware computing system, whereas localization is generally defined as involving location determination schemes and algorithms. Enabling technologies for localization in the wireless world are referred to as location sensing technologies, including global positioning systems (GPSs), cellular base stations, wireless local area networks (LANs), ultrasound, radio frequency identification (RFID) tags, and wireless sensor networks. An assortment of location sensing systems using one or more of these technologies has been developed. In particular, indoor location sensing is in great demand by many industry sectors, such as healthcare facilities and warehouses, yet numerous challenging issues remain to be addressed. Localization in wireless sensor networks is another difficult but interesting issue, and location awareness is becoming a building block for many mobile applications and services. To foster future location-aware computing systems and location-based services, we explore location-related schemes and techniques used in emerging location-based services and applications.

7.1.1 Location Representation

In location-aware computing, the terms *position* and *location* are used frequently. In most cases they are interchangeable, but they do not always refer to the same thing. Position is defined as the geospatial place of an object, often represented by coordinates. Location has a more general, semantic meaning with respect to the position of an object; for example, the location of a person could be "room 113 of the engineering building." The process of determining an object's location is called *localization*, or positioning. Based on location information collected at different granularities over time and on the context of movement, object movement tracking and path projection can also be performed.

A core element of location-based services and location sensing systems is location representation. Localization algorithms generate data in the form of a location representation model, and applications and services must make location queries conforming to the selected location representation model. Depending on the circumstance of the application, different models can be used, such as the geometric model, symbolic model, and location graph model. A *geometric model* uses location properties of an object to pinpoint it in two-dimensional or three-dimensional space. The properties can be distance, angle, time, etc. The coordinates in a geometric model can be either absolute coordinates, such as latitude, longitude, and altitude used by GPS, or relative coordinates, such as an application-specific grid coordinates. Geometric models are fairly simple to process mathematically in Euclidian space, but in some cases they may be overkill and may not provide truly useful information to an application, as raw coordinates are in effect meaningless without referencing objects such as rooms, buildings, and streets. *Symbolic or semantic models* seem to be a better solution to relative localization, which simply identifies an object as being within or close to some known objects. For example, in an indoor location sensing system, a symbolic location representation could be "John Doe is now in room 3353 on the third floor of the engineering building." This type of location information is more meaningful than the person's geospatial coordinates. The symbolic model does not require fine-grained position information of the object. A third type of model, a *location map*, is based on a map of symbolic representations rather than geometric ones. It is essentially a conceptual combination of the other two models. Consequently, it is more complicated than the other two models because of the integration and mapping between the geometric location information and symbolic location information. Choosing a location representation model is largely dependent on the requirements of the location sensing system in terms of overall accuracy, location update frequency, and infrastructure cost.

7.1.2 Localization Techniques

Location sensing systems invariably employ some localization techniques or algorithms that generally require some sort of wireless signal measurement to compute an object's position. Localization techniques can be divided into three categories [3]: triangulation, scene analysis, and proximity.

7.1.2.1 Triangulation

Triangulation utilizes geometric properties of a triangle to compute an object's position based on the positions of two or three vertices of a triangle. The two types of triangulation are lateration and angulation. Lateration relies on distance measurements from the object being localized to three reference points and distances between reference points, whereas angulation achieves the same task using angular or bearing measurements and a known distance between two reference points. Figure 7.1 shows these two basic techniques on a two-dimensional plane. In Figure 7.1a, three noncollinear reference points are required to obtain a total number of six

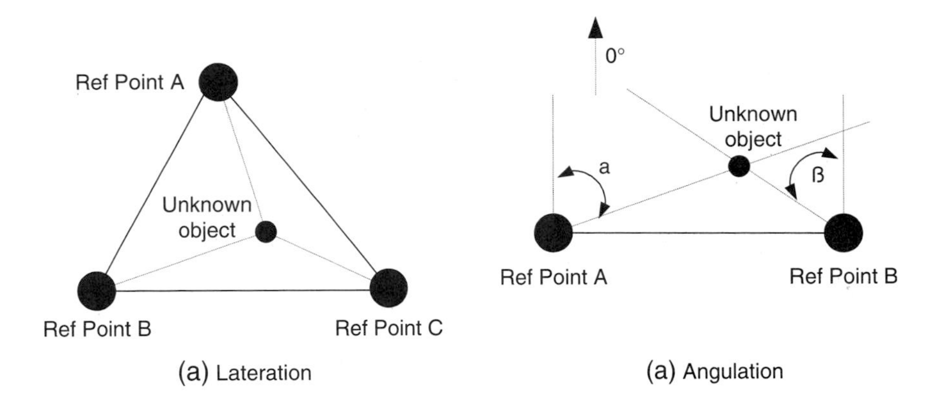

Figure 7.1 Triangulation techniques.

distance measurements for lateration. For three-dimensional localization, four noncoplanar reference points are needed. In Figure 7.1b, only two reference points, one distance measurement between the two reference points, and two angular measurements are necessary for angulation. In three dimensions, an additional azimuth measurement (horizontal angle between a reference point and the unknown object) is necessary. Note that in angulation the magnetic north is often chosen as the base of angle measurement.

Distance measurements for lateration in wireless communication are often conducted by measuring the time of flight (TOF), time of arrival (ToA), or angle of arrival (AOA) of wireless signals. Given the speed of a radio signal or a sound wave traveling in space, the distance between a transmitter and a receiver can be computed by multiplying the speed and the time it takes to reach the receiver (TOF). Because the speed of light and radiowaves is extremely high (approximately 300,000 km/sec), TOF measurement must have a very high resolution to achieve reasonable accuracy. For example, in order to achieve a location accuracy of 5 to 10 m, the clock resolution of a distance measurement must be at the level of 15 to 30 nsec. If a radio signal is unidirectional, the transmitter and the receiver must have their clocks synchronized to measure TOF. This is often difficult to realize in a distributed network environment. A solution to this problem is to enable precise clock synchronization among reference points but not the unknown object. Thus, the clock drift of the unknown object is an additional variable in lateration or angulation computations and requires an additional reference point.

Distance measurement for lateration can also leverage a property of radio signal propagation: attenuation. According to the path loss model introduced in Chapter 3, in a free open space, the radio signal strength (RSS) attenuates along the propagation path by a factor proportional to $1/r^2$, where r is the distance between the transmitter and a point on the path. Therefore, if the received signal strength at point P can be measured, one can estimate the distance between P and the transmitter in accordance with the path loss model. In reality, environmental factors such as rain, lightning, atmospheric conditions, and static or moving obstacles play a crucial role in signal propagation. Multipath propagation and signal fading further complicate the problem of indoor location sensing. Consequently, in practice, it is almost impossible to accurately and uniformly model the relationship between received signal strength and the distance. In some fully distributed mobile environment, however, actual geospatial distance measurement is not required by a location-based application. Instead, what really concerns the application is whether or not an object is within the transmission range of another or the number of hops (intermediate objects forwarding messages) by which they are separated. In such cases, distance measurement can be reduced to simply determining the direct reachability—the capability of a wireless device to directly reach another one. The wireless transmitter of an object may simply use its maximum signal strength to probe other objects for reachability measurements.

Angulation utilizes angle of arrival (AOA) measurement provided by specialized antennae (antenna receiving signals from the unknown object) installed on reference points (for example, magnetic north). An AOA is the angle between the arrival signal and a fixed reference direction, such as α and β in Figure 7.1b.

Both the unknown object and a location sensing infrastructure can initiate the localization process. For a resource-constraint mobile device to be localized, computation overhead incurred by periodic triangular processing must be considered. On the other hand, mobile privacy is guaranteed if localization is done at the object itself.

Triangulation can be performed in an iterative fashion to allow distributed localization. When an object's position has been determined, it can be used as a reference point for subsequent triangular computation. This allows a location sensing system to have only a small number of known objects while bootstrapping. As localization continues, more and more objects will become reference points to further localize other adjacent objects. Such a scheme has been used in localization for *ad hoc* networks, where it is difficult to obtain a large number of known objects in early stages of deployment. The problem with this scheme is that, if reference objects move around, significant computation and communication overhead due to position updates may inevitably increase dramatically and overwhelm the normal communication operation of those wireless devices.

7.1.2.2 Scene Analysis

Scene analysis for location sensing circumvents geometric computation but requires a map of the scene where known objects are located. The map is a dataset of features, such as signal strength across a space, and visual scenery taken from a vantage point. *Static scene analysis*, sometimes referred to as *calibration*, compares the observed features at a location to an existing feature map to identify the position in question. In *differential scene analysis*, changes between subsequent scenes are used to track the observer's position. In both cases, some imaging processing and information retrieval (IR) techniques are likely to be used. In addition, because of the inherent uncertainty in scene analysis, sophisticated statistical models such as Bayes filters have been introduced mostly in indoor location sensing system that use electronic features for scene analysis.

Aside from asset or person localization, scene analysis is used in wireless networks, mostly 802.11 wireless LANs, for site surveys to optimize access point channels and locations in a given location, utilizing onsite measurement along with floor plans. The technique can be also used in wireless intrusion detection systems (IDSs) to detect and estimate the location of rogue access points and possible wireless attacks. Generally, wireless IDSs are able to monitor and analyze wireless traffic in the network and identify abnormal traffic patterns from mobile stations and access points. Localization techniques such as proximity, triangulation, and scene analysis are common components of wireless IDS to help network administrators determine the location of rogue access points and mobile stations.

Some commercial Wi-Fi location systems such as Ekahau and Airspace RF Fingerprinting have been developed using this approach scene analysis techniques. Ekahau (http://www.ekahau.com/) does not employ any propagation or triangulation methods that suffer from radiowave multipathing, scattering, and attenuation effects. Instead, a scene analysis method called *site calibration* is used for collecting radio network sample points from different site locations. Each sample point contains a statistical summary of the received signal intensity (RSSI) and related map coordinates. By analyzing these data with Bayesian network techniques (explained in Section 7.1.7), Ekahau is said to achieve a positioning accuracy up to 1 m. The limitation of this approach is that the map has to be updated whenever the underlying scene changes with respect to the selected featured being presented. In addition, building a map of features by sampling at many places is not an easy task. Hence, a balance has to be achieved between map granularity and map creation overhead. Researchers have been using robot to conduct this tedious work [4].

7.1.2.3 Proximity

Proximity-based location sensing techniques do not offer direct quantitative position of an unknown object. Instead, they only provide proximity information such as serving base stations of the unknown object. The main idea of these techniques is essentially binary distance measurement to known reference points in proximity. For example, by comparing signal strength at a set of wireless LAN access points from a wireless LAN device, a proximity base location sensing system is able to tell which access points are closer to the devices than others. Because those wireless access points were previously mapped to physical locations, the system can localize and track the device with respect to the signal scope of base stations encountered. Clearly the accuracy of such systems relies on the granularity of the base stations in the proximity determination. If a location-based application employs a symbolic location representation model, then proximity-based localization techniques can be used to localize an unknown object with respect to symbolic locations of reference points in proximity.

7.1.3 Global Positioning System

The global positioning system (GPS) was briefly introduced in Chapter 3. In short, the GPS system consists of a total number of 24 GPS satellites orbiting the Earth, operated by the U.S. Department of Defense. GPS satellite signals can be used by anyone free of charge. At one time, the signals were modified so nongovernmental users could not obtain the greatest accuracy possible. This "selective availability" was discontinued by the federal government in 2000; however, for military purposes, "selective deniability" can still be used to degrade signals received by civilian GPS units in a war zone without affecting military GPS units. The European Union has developed a plan to launch its own GPS system (the Galileo position system) by 2008. A GPS receiver performs location estimation by measuring the time difference of arrival of signals from four GPS satellites. Prior to becoming available to the general pubic, GPS has been widely used in global navigation and in providing synchronization for cellular networks. Because the cost of a GPS receiver chipset continues to drop, GPS is generally considered the best choice for outdoor location sensing. For one thing, aside from working as a standalone mobile device, a GPS receiver chipset can be comfortably embedded into a cell phone or can be packed into a small expansion card. We believe that eventually every smart phone will have an onboard GPS chipset. In addition, many cars now have been equipped with a GPS navigation system.

A GPS satellite circles the Earth twice a day at an altitude of 20,200 km (about 12,600 miles). Each GPS satellite has an atomic clock that keeps extremely precise time. To perform location calculations, a GPS receiver must receive signals from four GPS satellites.

The receiver then calculates a pseudorange, which is the time difference between its local clock and the time when a signal is sent from a satellite. By multiplying the speed of the radio signal and the pseudorange, the receiver can locate itself onto a sphere corresponding to each of the four satellites. The GPS receiver does not need to be equipped with a high-precision clock. As long as the receiver's local clock is stable in the short term, differences between the time points when GPS satellite signals are received can be measured quite accurately and will eventually yield the location of the GPS receiver.

A GPS receiver can calculate its geographic location in terms of latitude, longitude, and altitude, although latitude and longitude are more likely to be used. An even more accurate system is Universal Transverse Mercator (UTM), a topographical rectangular grid consisting of 60 zones of 6 degrees of longitude; coordinates are expressed in meters east of the zone origin and north of the equator. For example, a location of N 40° 05.425 W 075° 07.035 can be converted into UTM to produce 18T E 490004 N 4437799.

The localization accuracy is measured by the distance deviation (*e.g.*, 1 to 5 m). Another way to evaluate the accuracy of a GPS is *precision*, or the number of times a GPS receiver can give a claimed accuracy. Because a GPS receiver must obtain multiple signals from different satellites in order to calculate the first result, there is a significant delay called time to first fix (TTFF), which is commonly in the range of 20 to 40 sec. Network-assisted GPS (A-GPS) uses a large number of networked receivers to assist regular GPS receivers. These networked receivers constantly collect GPS satellite signals, and a regular GPS receiver can request GPS data from these networked receivers instead of GPS satellites.

The global positioning system can generally fulfill the need for accurate positioning in outdoor settings and can provide worldwide coverage. Due to the nature of satellite signal reception, it does not perform well indoors. In addition, because localization (which is highly computational intensive) is done on the GPS receiver, power consumption of a GPS chipset can be a problem that could impede the adoption of GPS-enabled smart phones.

7.1.4 Cellular Network Triangulation

For outdoor location sensing, aside from GPS, a cellular network with a large number of fixed base stations is another excellent, well-established infrastructure. In the case of distance triangulation (lateration), the position of a cell phone can be determined by the distances to three adjacent base stations. When angle triangulation is used (angulation), two base stations are enough. In either case, location information for the fixed base stations in the network is needed in addition to distance measurements. The TOF of radio signals is often used in cellular localization.

Cellular network-based location sensing solutions utilizing triangulation techniques include cell ID, enhanced observed time difference (E-OTD), uplink time of arrival (UL-TOA), and angle of arrival (AOA). Cell ID is the simplest location technique and is essentially a proximity-based localization solution—a cell phone is positioned within the range of a cell. In E-OTD, a network of location measurement units (LMUs) is needed to ensure that the base stations are synchronized to send measurement signals to a cell phone. Time differences between the received synchronized signals are used to calculate the location of the cell phone. UL-TOA takes a slightly different approach in that the time differences of uplink signals (from a cell phone to three adjacent base stations) are measured; consequently, UL-TOA requires more strict time synchronization among base stations. AOA employs angulation instead of lateration. The location of a cell phone can be determined by the intersection of two lines between the cell phone and two base stations. AOA is relatively simple in computation. A disadvantage of AOA is that it requires smart antennae to be installed on existing base stations. More than two base stations are used for greater accuracy when signal reflection and diffusion are factors.

7.1.5 Indoor Location Sensing

Indoor location sensing is the new frontier of next generation wireless systems. In contrast to outdoor location sensing, indoor location sensing systems are more versatile. Infrared, ultrasound, and a number of radiofrequency-based systems have been demonstrated with varying accuracy and cost. Because of the diversity of indoor location-based applications, a system may choose one or more location sensing technologies to fulfill the need for location accuracy, localizing speed, and economic cost.

7.1.5.1 Diffuse Infrared

The first indoor location sensing system was the infrared-based active badge system [5] developed by Olivetti Research, Ltd. (ORL) and AT&T Research Labs. An infrared badge attached to a person has a tiny transmitter that can send a unique identifier to a surrounding networked infrared sensor in the room every 10 sec. The active badge system provides room-grained locations of badges with a deployment of one networked sensor per room in a building. Newer active badges have been equipped with onboard processors and bidirectional infrared capabilities. Figure 7.2 shows the four generations of active badges. Fixed infrared sensors collect data from active badges

Figure 7.2 Active badges. Bottom left is the first generation; bottom right, the second; top left, the third; and top right, the current (white) one. The current active badge has a 48-bit code, bidirectional infrared capabilities, and an onboard 87C751 microprocessor. (Courtesy of AT&T Laboratories Cambridge, © Copyright 2005 AT & T Corporation, http://www.uk.research.att.com/thebadge.html.)

and forward the data to a central database that tracks the location of badges within the room.

The active badge system is a low-cost indoor location solution that utilizes proximity techniques to provide symbolic location information. Its disadvantage is possible interference of the diffuse infrared signals by environmental infrared light, such as sunlight. Moreover, the line-of-sight requirement of infrared transmission makes it impractical to be used in real-time, accurate location sensing systems. Because infrared has a transmission range of up to several meters, a large room may require multiple networked infrared sensors, resulting in dense deployment of such sensors to cover the entire building.

7.1.5.2 Ultrasound

Ultrasound refers to high-frequency sound waves that are beyond the range of human hearing. Ultrasound can be used to perform TOF measurement to accurately determine the distance between an ultrasonic tag embedded in a mobile object and a known point of reference. Ultrasonic location sensing thus can provide an accuracy of several centimeters, and the sound wave can travel as far as 10 m. Active Bat [6] was developed by the same group at AT&T that worked on the active badges. A bat is a small ultrasonic transducer that can respond to a control radio signal and emit ultrasonic pulses to a grid of ceiling-mounted ultrasonic sensors, which will perform TOF calculations and forward data to a central station for further processing. Triangulation techniques are used in determining the position of a bat. The Active Bat system is able to locate bats within 9 cm of their true position 95% of the time. The Cricket location support system [7], developed by researchers at the Massachusetts Institute of Technology (MIT), is a similar approach, but it uses ultrasonic receivers rather than emitters, which are embedded in an object to be positioned. A ceiling-mounted ultrasonic infrastructure of emitters is used, but the triangulation calculation is done at the object. Cricket also uses radiofrequency control signals for time synchronization. Cricket can accurately delineate $4 \times 4\text{-ft}^2$ regions within a room using lateration and proximity techniques.

Ultrasonic sensors can also be used in a self-contained fashion. For example, a sensor is an ultrasonic transceiver, which means that it can act as an emitter and an echo detector at the same time. It measures the duration between a transmission and the reception of an echo. Initially, the system is trained to record time durations when no object is within a specific range of the sensor. This procedure is also called *calibration*. The system then begins to constantly emit and receive short pulses and monitor time durations. When an object such as a person or a part on the pipeline moves close to the sensor and within the range of the trained setting, the sensor is able to report it. Such systems are widely used in gate control, container movement detection, distance and level sensing, automatic light control, etc.

Ultrasonic location sensing generally offers much more accurate range estimation than other types of wireless location sensing approaches, but the cost to establish an ultrasonic location sensing infrastructure is sometimes prohibitively high. The merit of Cricket is that it does not require a grid of ultrasonic sensors, thus it offers greater scalability than Active Bat; however, the overhead of timing and processing ultrasound pulses for TOF computations and radiofrequency signals may become unacceptable when power consumption is of a great concern.

7.1.5.3 Radio Frequency Identification

Radio frequency tags can be used to determine the distance between an RFID reader and a tag by measuring the power level received at the reader. To improve accuracy, reference tags can be placed in known positions to assist position computation of an object. An example of RFID location sensing system is LANDMARC [1]. The LANDMARC system uses a number of active RF tags (RF tags that has a radio transceiver and a button-cell battery to power the transceiver) as reference tags and some RF readers to form a positioning infrastructure. An RF reader has a signal range up to 50 m and can identify 500 tags in 7.5 sec with a collision-avoidance mechanism. Signal strength is denoted by eight incremental power levels, which in turn can be used to deduce physical distance in a free space. Every active tag used in the system, including the reference tags and tags attached to

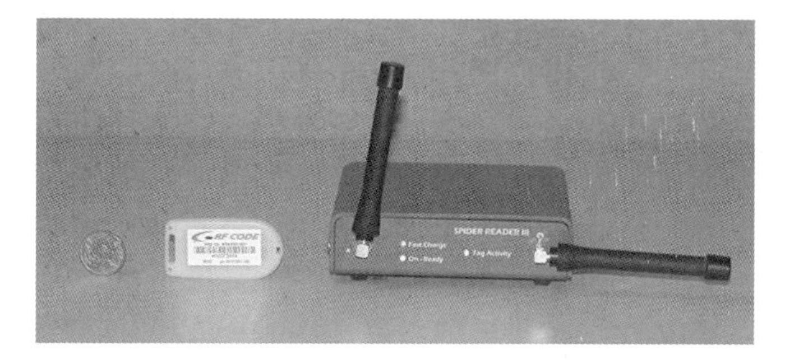

Figure 7.3 Tags and readers used in the LANDMARC system. (Dr. Ni's presentation on indoor location sensing.)

the objects to be positioned, emits a signal with a unique ID every 7.5 sec at a frequency of 303.8 MHz. Figure 7.3 shows an RF reader and a tag used in the LANDMARC system.

Radio frequency identification location sensing systems require a number of reference tags to be placed in known locations in an indoor environment. In the LANDMARC prototype, reference tags are organized into a three-dimensional grid array. LANDMARC does not directly derive physical distance from signal strength using triangulation techniques because in an indoor setting the distribution of received power levels is irregular in space and highly dynamic from time to time due to obstacles such as walls, cubicles, and the movement of human bodies. Instead, LANDMARC uses a novel algorithm that utilizes power levels of all tags to localize an unknown tag using several reference tags in its proximity. Because the reader can calculate a Euclidian distance (E_i) in terms of the received power level between the unknown tag and each reference tag i, the position of an unknown tag can be computed based on the known positions of those selected nearest reference tags, each carrying a distinctive weighting factor in the computation of the position estimate of the unknown tag. Empirically, a reference tag with a small E value will have a large weighting factor. Those dynamically chosen reference tags for an unknown tag effectively help offset environmental dynamics. Environmental factors contributing to the variations of RF signals will not introduce significant error to location determination, as those nearest reference tags of an unknown tag are very likely to be subject to the same effects. Using four RF readers and one reference tag per square meter, LANDMARC can accurately locate an unknown tag such that the largest error is 2 m and the average is about 1 m. As a cost-effective solution to indoor location sensing, LANDMARC does not require a large number of expensive RFID readers, and the system may scale well using more inexpensive RFID tags.

A major limitation of LANDMARC is the fairly unstable and inconsistent behaviors of tags. For example, two tags placed into identical position may report quite different power levels. Latency in position determination is another problem. The inherent power-level interrogation performed by the reader is fairly slow; a reader may spend up to 1 minute to scan all eight power levels to detect tags within the range. Both problems call for improvements in the RFID hardware.

7.1.5.4 802.11 Wireless LAN

802.11 wireless LAN-based indoor location sensing systems employ scene analysis techniques or lateration techniques in a standard IEEE 802.11 network. RADAR [2], developed by Microsoft, is the first effort in this direction. A signal map must be built prior to object localization. This can be done in two steps. First, signal strength at multiple wireless LAN base stations and some selected reference points must be measured, as shown in Figure 7.4. Then, using signal propagation models and empirical analysis, any specific point in the given area can be mapped to a signal strength value. The scene analysis implementation of RADAR can localize an object within a 3-m range with 50% probability, while the signal strength lateration implementation of RADAR only offers an average of 4.3-m accuracy. The same idea can be easily extended to allow full distribution of signal maps in the entire space, using robotic to conduct predefined repetitive measurement [4]. Another approach uses very high frequency (VHF) omnidirectional ranging (VOR) base stations to perform AOA computations for an unknown object in a wireless LAN [8]. VOR is a type of radio-based location system primarily used in aircraft navigation.

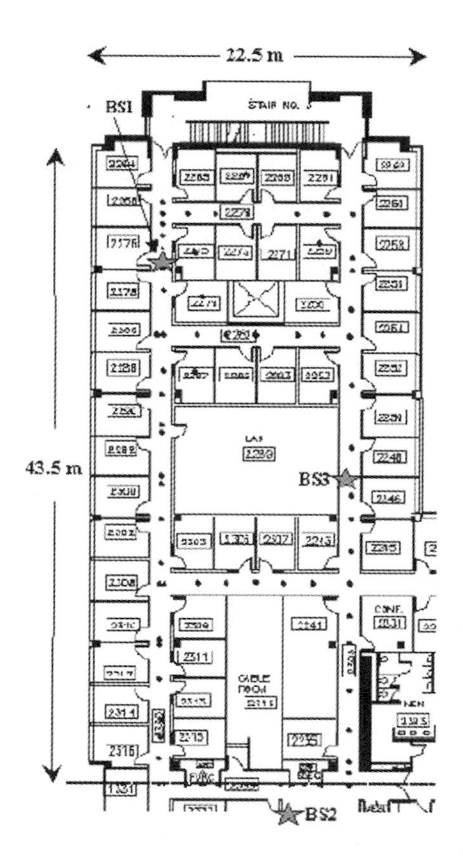

Figure 7.4 RADAR signal strength map [2]. Black dots denote where signal strength is measured. Stars indicate the locations of base stations. (Excerpted from Microsoft RADAR paper. © Copyright 2005 IEEE.)

VOR base stations on the ground constantly broadcast VHF radio signals (30–300 MHz), 360° in azimuth, that carry the identification of the station in Morse code and angle to it, thus helping aircraft to navigate the airspace. Although VOR is being replaced by GPS in aircraft navigation, it may find its way to indoor location sensing. A wireless LAN base station can be customized to enable angle and range measurement such that wireless LAN mobile devices can localize themselves based on signals from multiple VOR base stations.

Because wireless LANs have become ubiquitous network infrastructures in many places (recall the outcome of increasingly popular war driving activities), outdoor location systems can use those wireless LAN access points as reference points to provide localization service. Place Lab (http://www.placelab.org/) is such a project. It is designed to predict the location of a computing device via the known positions of the access points detected by the wireless LAN device on the computing device. The positions of those access points, both indoor and outdoor, can be obtained from a locally cached database on the device. In rural areas where 802.11 coverage is very low, Place Lab uses cellular base stations, fixed Bluetooth devices, and 802.11 access points. Because of the coarse granularity of reference points and elimination of the calibration procedure used in indoor scene analysis localization, the accuracy of Place Lab is about 20 to 25 m, falling between that of a GPS (8–10 m) and the 100-m accuracy of cellular base stations. The clear advantages of this approach are its ubiquity and privacy protection, at the expense of some level of accuracy. GPS location systems only work in an outdoor environment, as GPS signals cannot penetrate most buildings. 802.11 signals do not have such problems. On the other hand, in systems such as Place Lab, location information is completely locally maintained, exposing no location-based privacy to the network. This is in sharp contrast to location systems based on cellular base stations where localization is done on the network side, and users have to count on the network provider not to divulge their location information.

7.1.5.5 Ultra-Wideband

Ultra-wideband (UWB) is the latest sensing technology in this domain. Because UWB signals spread over a very wide bandwidth of more than 500 MHz, this technology can provide highly precise distance measurement. UWB-based indoor location sensing systems can provide great accuracy at a level of a few tens of centimeters and a good range up to 50 m. The power consumption of UWB location sensing is comparatively lower than that of other wireless sensing technologies. Ubisense (http://www.ubisense.net/) is a commercial system developed by the team who created the Active Bat ultrasonic location sensing system at AT&T Cambridge. It consists of UWB receivers and a number of UWB tags, each attached to an object or a person. The cost of such systems remains a major problem.

7.1.5.6 Bluetooth

Bluetooth is yet another low-cost wireless technology for indoor location sensing. Operating in the 2.4-GHz ISM band, Bluetooth is now a *de facto* standard wireless interface on almost all cell phones, smart phones, and PDAs. Like wireless LAN location sensing, received signal strength (RSS) measurement can be used to localize a Bluetooth device. In contrast to wireless LAN, Bluetooth interface can be built in the form of small tags that can be easily attached to an asset or a person. In addition, each Bluetooth device has a unique ID, which makes identification easier than with other wireless sensing technologies. Bluetooth location sensing is able to achieve an accuracy of several meters in an open space. In some environments, such as a hospital or a hotel, where in-room object localization is desired, Bluetooth can be combined with infrared location sensing within a room (*e.g.*, Tadlys' TOPAZ™).

7.1.6 Wireless Sensor Location Sensing

Wireless sensor networks (WSNs) may also make use of location information provided by low-cost, small-size sensors or objects for location-aided operations or location-based applications. The localization in wireless sensor networks is considerably different from that in other wireless networks. Because wireless sensors networks are designed to be configuration free, they have no such localization infrastructure, such as base stations in wireless LANs. Because it is too expensive to equip each sensor with a GPS (at least for now), there might be only a few or even no reference points with known positions (namely, beacons) in a wireless sensor network; hence, a sensor must cooperate with each other to localize itself and some other interested sensors. Note that, due to the inherent constraints of networked sensors, only a small range of distance measurement can be performed at each sensor. Sometimes a collection of these measurements along with several beacons is not sufficient to mathematically determine the physical positions of all the sensors involved.

Essentially, the sensor network localization problem is a variant of the underlying graph theory problem, the *graph realization problem*, which is defined as follows: Given a graph *G* and a set of distances between some vertices in the graph, compute a realization of *G* that represents the locations of those vertices in Euclidean space. Based on geometric theories of this problem, it has been proved that the realizability problem of a weighted graph is strongly NP hard [9]. In practice, the goal of precise localization of wireless sensors is often replaced by position estimation within an acceptable error range. Various distance measurement techniques, such as radio and ultrasound, and localization techniques, such as triangulation and proximity, can be used in wireless sensor networks, but all are subject to the resource constraint nature of sensors and self-organized characteristics of the wireless sensor networks.

Overall, wireless sensor localization systems can be roughly divided into three types:

- *Range-based approaches*, whereby accurate distance measurements are used for geometric computation of the positions of the sensors. Only a small number of beacons are deployed in such systems. Using lateration or more sophisticated multidimensional scaling (MDS), sensors close to beacons can be localized. By recursively performing this computation, the system is able to localize most sensors.
- *Range-free approaches*, whereby proximity rather than accurate geometric distance is used to localize a sensor. Systems of this type can be further classified into two categories: area-based and hop-based. The first detects and denotes the position of a sensor within a known geographic area previously identified. The latter uses the number of hops along a path between two sensors to compute estimates of sensor positions. Range-free approaches do not offer the same level of accuracy as range-based approaches do, but they do not suffer the "localization incapability" problem. Range-free

approaches can be combined with range-based approaches; that is, range-free techniques can be used to ensure that a wireless sensor network is localizable and to compute a rough estimate, and then range-based techniques can be used to precisely localize the sensors.

- *Angulation-based approaches*, whereby AOA measurements are used to compute positions of sensors.

It is worth noting that many techniques used in wireless sensor localization systems are also applicable to general mobile *ad hoc* networks where a fixed location infrastructure is not likely to exist or can be built. While sensors cannot use GPS as the outdoor location sensing solution, a mobile *ad hoc* network may be able to use GPS for every mobile node such as PDAs and cell phones. For both outdoor and indoor location sensing in *ad hoc* networks, triangulation and proximity techniques can be used in association with various wireless radio technologies.

7.1.7 Localization Analysis in Mobile Computing

Recent advancements in location-aware computing have shed some light on utilizing statistical analysis schemes for localization in hopes of addressing two of the most challenging problems in mobile location sensing: the intrinsic inaccuracy of distance measurement, and position estimate uncertainty due to the resource constraints of mobile devices. These schemes aim at leveraging vast amounts of location-related data generated from many types of sensors to facilitate various location-based applications. These schemes are particularly important in a location sensing system that employs *sensor fusion*, a scheme that uses multiple wireless location sensing technologies or systems simultaneously to improve the accuracy and precision of location estimates.

Bayesian filtering is a probabilistic inference technique based on the context of observed data. It is used in many disciplines of computer research, such as spam filtering, information retrieval and logistics, and location sensing. Note that the following discussion of Bayesian filter requires the reader to have some basic knowledge of probability and statistics. You can certainly skip the remaining section as long as you understand that some statistical techniques can be used to improve location sensing accuracy.

When Bayesian filtering applies to location sensing [10], the position of an unknown object is defined as a *state*. A state at time t is represented by the random variable x_t. Location sensors provide observations about the state. A probability distribution over x_t, $Bel(x_t)$, is then used to represent the uncertainty of position estimation based on the observed data. $Bel(x_t)$ is defined by the posterior density over the x_t conditioned on all observed data z_1, z_2, ..., z_t, available at time t:

$$Bel(x_t) = p(x_t|z_1,\ z_2, \ldots, z_t)$$

Because the computation complexity of $Bel(x_t)$ grows exponentially over time, Bayesian filtering assumes the system is Markovian, which implies that an object's location at time t depends only on the previous state, x_{t-1}. An excellent illustration of Bayesian filtering in location estimation is shown Figure 7.5 [10]. Suppose a location sensing system is designed to provide an estimate of a person's position with observed data from a sensor carried by the person as she walks down a hallway. The sensor generates a signal when it is close to a door, but it cannot distinguish different doors.

In Bayesian filtering, $p(z|x)$—namely, the perceptual model—describes the probability of obtaining observation z given that the person is at position x, and the belief immediately after prediction and before the observation is the *predictive belief*, or $Bel^-(x_t)$. Initially, because no sensory data are available, a uniformed distribution over possible location $Bel(x_0)$ is shown in Figure 7.5a. As the person moves close to the first door on the left, the sensor sends a "door found" signal, resulting in high probability at places close to those three doors on the perceptual (observation) model $p(z|x)$. Note that even places not close to the doors have nonzero probability. Without previous observations, the current $Bel(x_t)$ is the same as $p(z|x)$, as shown in Figure 7.5b. When the person walks to the right, the Bayes filter

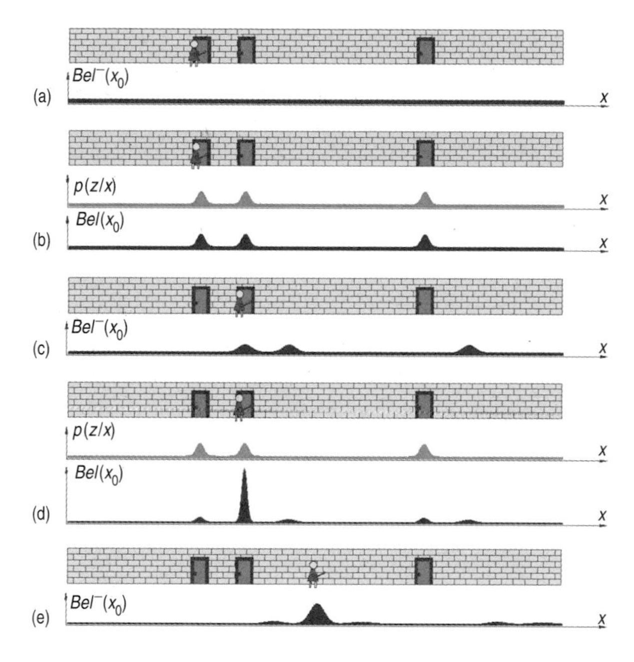

Figure 7.5 A one-dimensional illustration of Bayesian filtering in location sensing [10]. © Copyright 2005 IEEE.

shifts the belief in the same direction due to the Markov assumption and smoothes it out to account for inherent uncertainty in motion estimates. In Figure 7.5c, a predicted belief with three peaks, $Bel^{-}(x_t)$, is shown but each peak is lower than that in Figure 7.5b. This is a predictive belief because the next sensor observation (when the person moves to the second door) is not yet available. In Figure 7.5d, the person is at the second door, and the sensor again sends a "door found" signal. The observation model $p(z|x)$ once again assigns high probabilities to those places next to the doors. The current predicted belief, $Bel^{-}(x_t)$ is now aggregated with the probability distribution from the observation model, resulting in a sharp peak at the second door in the current belief, $Bel(x_t)$. Then, in Figure 7.5e, the person continues to move, and the belief in Figure 7.5d shifts to the right and smoothes at the same time.

The illustration shows conceptually how recursive state estimates of Bayesian filtering can be applied to location estimation. For a

Bayesian filter location sensing system, a *perceptual model*, $p(z|x)$, is needed prior to location estimation. This model can be considered a scene of feature maps used in the scene analysis scheme. The feature in the map of the perceptual model is the likelihood of observing a sensor measurement z (*e.g.*, 2-m ultrasound measurement) given an object's state (position) x in a room. Construction of the map can be performed by autonomous robots. As mentioned earlier, the principle of sensor fusion is to combine multiple location sensing technologies of different error distributions deployed in the same geographic space; therefore, to apply Bayesian filtering to a sensor fusion system, a perceptual model of the underlying space is required for each location sensing technology.

The second step is to select a *filter*. As a probabilistic framework, Bayesian filtering does not define specific filters, such as the statistical representation of belief $Bel(x_t)$. Implementations of Bayesian filtering include Kalman filters, particle filters, multihypothesis tracking (MHT), and dynamic Bayes nets, among others. These filters offer various levels of accuracy, sensor variety, and implementation cost. For example, let us take a look at particle filters [11]. Particle filters use a set of weighted samples called *particles* to compute beliefs. The weights of states are updated and normalized according to sensor observations. Depending on the selected filter, Bayesian filtering can be carried out by computing over observation data and the perceptual model (the map) from multiple types of sensors. The two stages of Bayesian filtering are prediction and filtering. In the prediction stage, one's location is predicted based on the previous belief and knowledge of dynamics such as walking speed. In the filtering stage, a sensor observation is used to correct the prediction, generating a current belief. A sensor fusion scheme using infrared and ultrasound with particle filtering was presented by Fox *et al.* [10]. A similar idea has been proposed for mobile sensor network localization [12].

7.1.8 Localization Techniques Comparison

Table 7.1 provides a comparison of existing localization techniques described in this section. When designing a location-based system,

Table 7.1 Comparison of Localization Techniques.

Localization Solution	Accuracy	Range	Cost	Applications
GPS	8–10 m	Global coverage	GPS receiver or chipset	Military, air/marine navigation, traveling at remote sites, etc.
A-GPS	1–5 m	Urban areas	GPS chipset embedded in cell phones	Outdoor civilian applications such as driving navigation or location tracking
Cell ID	Hundreds of meters depending on the cell size	Cellular network coverage	A function of cellular networks; no cost to a user	Value-added wireless services such as local searches or social networking
E-OTD	100–200 m	Cellular network coverage	Requires LMUs (Location Measurement Unit) to be installed on base stations	Value-added wireless services
UL-TOA	100–200 m	Cellular network coverage	Requires LMUs to be installed on base stations	Value-added wireless services
AOA	100–200 m	Cellular network coverage	Requires smart antennae to be installed on base stations	Value-added wireless services
Infrared	Proximity-based (5–10 m)	Within a room	Low cost	Asset and human tracking inside a building
Ultrasound	5–10 cm	Within 10 m	High cost of ultrasonic devices	Industrial control, building entrance control, etc.
RFID	1–3 m	Less than 50 m	Lower than ultrasound	Human localization and tracking
Wireless LAN	1–10 m	100 m	Inexpensive wireless LAN equipment	Indoor asset and human tracking

(continued)

Table 7.1 (Continued)

Localization Solution	Accuracy	Range	Cost	Applications
UWB	10–50 cm	50 m	High cost of UWB devices	Accurate indoor asset and human tracking
Bluetooth	1–5 m	30 m	Inexpensive Bluetooth devices	Indoor asset and human tracking
Wireless sensor network	1–10 m, depending on the WSN	Depends on deployment of the WSN	Wireless sensors are becoming inexpensive	Outdoor and indoor object localization tracking, applications of WSN networks
Hybrid approaches	1–5 m	Depends on the wireless technologies used	Depends on the wireless technologies used	Accurate and low-cost asset and human tracking

aside from technical factors such as location accuracy and location determination delay, one must also take the application scenario into consideration, as different application scenarios may impose special design constraints including cost, usage pattern, the use of existing wireless infrastructure, and so on.

7.1.9 Building Location-Based Services

Location based services are not limited to mobile computing. Even those using wired computers to access the Internet may sometimes want to find location-related information, such as the nearest movie theater in town or the locations of gas stations along a highway. This implies that the person knows the place in question or his or her current location. Location-based services offer more value to businesses and consumers when they are combined with mobile wireless technologies in the mobile computing arena. The fundamental issues in building location-based services in a mobile wireless environment are listed as follows:

- *Localization*—Find and track locations of someone or some object even if no explicit identification is available
- *Location-related database*—Obtain and incorporate location-related information from a geographic information system (GIS) or any other data sources
- *Service delivery*—Deliver location-specific and user-specific context-aware information to mobile users

It is important to understand that a good localization solution cannot solve all the problems. More important is how to make use of location information for a new dimension of mobile applications and services. The following presents an overview of likely solutions to these three issues.

7.1.9.1 Localization

Much of this section has focused on localization technologies utilizing light, sound, and radio, coupled with a variety of geometric or statistic techniques. It is likely that self-initiated (users want to determine their current location) outdoor location-based services will most likely rely on GPS to deliver localization services. Power consumption of a GPS chipset will continue to decrease. For network-initiated (a network passively discovers and tracks an object's location) outdoor location-based services, base stations of a wireless network such as a cellular network or a campus-wide wireless LAN can certainly help monitor the location of a mobile device in an on-demand fashion. Value-added services such as point of interest (POI) searches, real-time traffic updates along with maps and navigation service, and person/object tracking will become commonplace very soon.

For indoor localization, the solution is likely to be case by case. In other words, the choice of location sensing technologies and location estimate methods will depend on the specific requirements of a particular application. We have introduced several research projects using one or multiple location sensing technologies at different levels of granularity. Proprietary wireless LAN-based indoor localization systems such as Symbol Spectrum24™ and Ekahau's real-time location system (RTLS) have been developed for various application scenarios, including shopping assistance, asset tracking, animal and pet control, patient tracking, proximity-based person localization, etc. Many of these commercial systems are self-contained in the sense that they are vertical systems incorporating all essential elements of a location-based system, which makes interoperability a huge problem when more such systems are being used. To this end, the development of open standards and platforms that allow pervasive interoperable mobile access to location-based services is underway.

Commercial localization systems such as Tadlys' TOPAZ™ (http://www.tadlys.com/) are currently available on the market. The TOPAZ system combines Bluetooth and infrared to track patients and assets in hospitals. The system consists of a number of Bluetooth/infrared access points connected to numerous location servers via a local area network. Patients and assets are identified by tags, which are Bluetooth/infrared transceivers that can be sensed by access points. Each access point can be configured to provide accurate localization using Bluetooth or binary in-room localization using infrared. Localization solutions combining RFID and wireless LAN, such as AeroScout (http://www.aeroscout.com/) and PanGo Locator (http://www.pangonetworks.com./locator.htm), are becoming mainstream solutions to asset tracking. Wireless LAN-enabled active RFID tags can be positioned in real time by a location infrastructure consisting of a number of specialized location receivers and standard wireless LAN access points.

7.1.9.2 Location Information Database

An image of an indoor, two-dimensional floor can be generated using autonomous robots that move along walls and obstructions in rooms and hallways. Substantial information, such as room IDs and cubicle numbers, must be added to form an overlay on top of the geospatial map. In addition, a wide array of POI (Point of Interest) labels can be added to the map as well. Depending on the location-based application, POI may have different meanings. For example, an indoor location-based application deployed in an office building may provide real-time locations of POIs such as printers, copiers, fax machines, rooms of HR people, conference rooms, wandering IT people, and even restless managers, each identified by a unique tag. The map and associated POIs constitute the location information database. To facilitate future changes in the underlying environment, the database must provide a means to allow flexible location configuration and POI updates.

Outdoor location information database are generally two-dimensional or three-dimensional maps or specialized spatially referenced database used in GIS (Geographic Information System). Various types of static datasets, such as road networks, terrain mapping, urban mapping, satellite and aerial mapping, ocean torrents, land cover, and demographic data, can all be tied together geographically to provide a spatial context for an application. Publicly accessible GIS data can be obtained from government agencies and some nonprofit organizations. For example, the U.S. Census Bureau provides an online American FactFinder system that is a source for population, housing, economic, and geographic data. The U.S. Geological Survey (USGS) provides national elevation, land cover, and hydrographic datasets for free download. Commercial GIS providers such as MapPoint and ESRI offer a large variety of visualized raw datasets and corresponding analyses. Digital maps providers such as NAVTEQ and GDT supply maps to well-known online map services such as MapQuest, Yahoo Maps, and Google Maps. Digital maps optimized for mobile devices are often bundled with in-car or handheld GPS navigation systems. These maps have highly distinguishable street markings and can be coupled with the voice-based navigation of GPSs. Figure 7.6 shows an overview map and a route map on a PDA device.

Aside from static databases, a dynamic database that has received much attention in the realm of location-based services is traffic data. Some web-based map services have started to provide current real time traffic information visualized on maps for major U.S. cities.

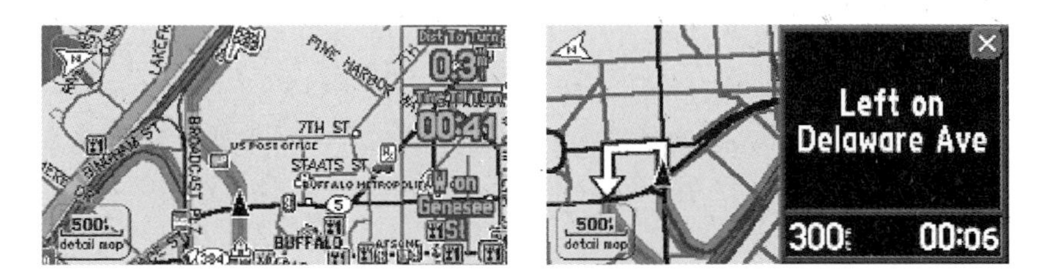

Figure 7.6 Maps on a PDA. (Courtesy of Garmin, Ltd., © Copyright 1996–2005 Garmin, Ltd.)

The same service is also available on any mobile devices that has a wireless connection of some kind. Local departments of transportation often provide traffic data online. For example, the Washington State Department of Transportation (WSDOT) has made up-to-minutes traffic data for the Seattle area available to the public free of charge. WSDOT collects the traffic data via wire loops that link thousands of sensors implanted in roadways to detect traffic conditions. The sensors are located beside the roadway every half mile. When a vehicle passes over a loop, it is counted, and the time the vehicle spends over the loop is measured. Data collected by these loops are transmitted every 20 sec to the WSDOT Traffic Systems Management Center. The raw traffic dataset is published on an FTP server (webflow.wsdot.wa.gov) and updated every minute. Because the service is free of charge, anyone who wants to obtain the latest traffic information can build an application to retrieve the dataset and render it into a form of traffic maps. An example of such applications, CEFlow for Pocket PC, is shown in Figure 7.7. In combination with GPSs, these applications have the potential to offer more intelligent navigation assistance based on current traffic condition.

7.1.9.3 LBS Delivery

As for any other wireless applications for the general public, a standard for location-based services (LBS) will be the key to their success. The standard must address data exchange specifications between different parties in providing location-based services for mobile users. These parties include mobile service providers, location service providers, GIS providers, mobile device manufacturers,

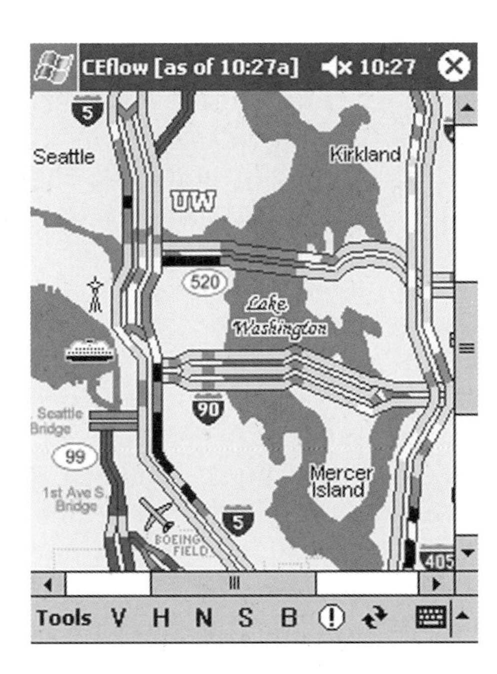

Figure 7.7 A traffic map of the Seattle area produced by CDFlow. (Map and data: © Copyright 2003 Washington Department of Transportation. Software: © Copyright 2005 © Ken Alverson and John Lambert.)

and independent software providers. Of course, one party may assume multiple roles. Furthermore, the LBS standard should allow location systems from different vendors to interoperate, enabling free roaming and flexible service access and accounting.

The Open GIS Consortium (OGC) is such an effort. Founded by a number of universities, governmental research agencies, and some companies in 1994, OGC has gained support from companies such as Oracle, MapInfo, ESRI, and AutoDesk. The vision of OGC, "A world in which everyone benefits from geographic information and services made available across any network, application, or platform," perfectly fits into the more general vision of next-generation mobile computing, according to which everyone with any mobile devices on any platform is able to benefit from mobile services anywhere, anytime. OGC ratified the OpenGIS location services (OpenLS) specification 1.0 in 2004.

Given a continuous feed of location data from a mobile service provider or an independent location service provider, and with a location information database from digital map vendors, a location-based service can be built and delivered to the end user via an Internet connection over a range of wireless technologies. In addition to Open GIS, Microsoft MapPoint is another platform capable of providing maps, navigation assistance, proximity search, and location-related business intelligence in the form of programmable XML web services. A MapPoint system interfaces to a wireless service provider's location service, augmented with enhanced data such as business logic information and POIs, and delivered to mobile users. For specific business needs with respect to LBS in an enterprise, a MapPoint location server acts as the central control to link mobile users, MapPoint web services provided by Microsoft, and location services provided by wireless services providers. A MapPoint location server has three core components: MapPoint location server web service, MapPoint location server database, MapPoint location server providers. MapPoint location server web services offers a set of web services that a mobile application can use. The database maintains location-related data of users and wireless service providers, and the server providers are software interfaces to wireless service providers that specify how location information is retrieved and updated between the two parties.

7.2 Mobile Messaging

Voice communication via cell phones in public places is very often annoying to other people. We have all heard someone talking too loudly on a cell phone on a bus, on an airplane, in sports stadiums, in a movie theater, in a classroom, and so on. On many occasions, people need to communicate with others in a non-intrusive way, other than by direct voice communication. Sometimes people prefer text-based communication to direct voice communication. What is needed is a means of communication that is not annoying to others yet allows real-time effective communication. Mobile messaging is the solution to this problem.

7.2.1 Short Message Service Applications

Text-based short message services (SMSs) offered by mobile service providers have enjoyed phenomenal growth worldwide. As a quick and reliable way to deliver text information, SMS provides cost-effective person-to-person communication and provider-to-consumer value-added services. Picture messages are on the rise due to the unexpected popularity of camera phones. With the advent of 3G cellular services on the horizon, multimedia messages carrying video clips are likely to follow the trend. These traditional messaging services have been leveraged by content providers or service providers to facilitate an assortment of applications. For example, a phone user may subscribe to an SMS-based news and weather service. In some countries, using SMS to spread jokes and ring tones among friends is commonplace and has generated huge revenues for content providers and mobile network operators. In addition to plain messaging exchange, cell phone or smart phone users can use similar services for social networking, such as mobile online introductions, localized dating services, friend searches, community building, and so on. These applications enhance social interactions among users by leveraging mobility and location-based services provided by the underlying mobile wireless infrastructure such as cellular networks, Wi-Fi, and Bluetooth, while messaging is just a way to pass information between a user and a back-end system and between users.

Another new dimension of mobile messaging service is the convergence between wireless network messaging (SMS, EMS enhanced message service, multimedia message service [MMS], and e-mail service provided by BlackBerry) and computer network Instant Messaging (IM) services such as Yahoo instant messaging, MSN messaging, AOL Instant Messenger (AIM), Jabber, Internet Relay Chat (IRC), and ICQ (I Seek You). Considering the massive user base of those messaging services and the fact that IM's short messages are a very effective means of communication over low-bandwidth wireless links on resource constraint mobile devices, mobile messaging services that integrate cellular messages and computer IM have the potential to become extremely popular applications in the foreseeable future. With all these enabling messaging technologies, an entirely new set of mobile applications and services is emerging. Below is a list of existing and potential mobile messaging applications:

- IM chat with a mobile device, enabled by mobile service providers on selected smart phones such as T-Mobile sidekick
- Mobile IM integration with enterprise applications such as customer relationship management (CRM) and enterprise resource planning (ERP) (mobile enterprise applications are discussed later in this chapter)
- Subscription-based information services in the form of mobile messages delivered to mobile devices, such as stock alerts, weather alerts, news update, wireless advertisements, shipping notifications, movie show times and tickets, electronic receipts, traffic reports, or whatever content a mobile user is interested in
- Mobile message-based voting services and television gaming services in which mobile users send text messages to vote or participate in games presented on a television program, for example
- Interactive mobile messaging services provided by content providers, whereby a mobile user sends a request for information and receives a response message from the content provider; Google SMS search service is an example of this type

We introduced cellular messaging services in Chapter 3; this section concentrates on computer IM and issues surrounding the provision of integrated mobile messaging.

7.2.2 Instant Messaging

Unlike traditional mobile messaging services, instant messaging (IM) is a real-time interaction between two parties or among a group of participants via a computer network, which could be the Internet, an enterprise network, a campus network, or even a small LAN. Throughout this section, we use the term *principle* to refer to people, groups, or software in the real world outside of the IM system that use the system as a means of coordination and communication, as defined in RFC 2778. A principle interacts with the IM system via a set of functional components called user agents, each allowing one type of interaction summarized in Table 7.2.

As shown in Table 7.2, an IM system must provide two basic services: presence service and enhanced messaging service. Presence service can be further divided into presence query, presence notification, and presence update, which together allow principles to be aware of the status of others (*e.g.*, online, offline, busy, on the phone) within the IM system. Enhanced instant message service offers a variety of communication methods. An Internet instant messaging system usually consists of a number of directory servers that maintain the states of active principles. IM client applications must first register at the directory server in order to become active principles.

Table 7.2 IM System Services.

Service	Interaction	Description
Presence service	Presence query	A principle queries a directory server for the presence status of others
	Presence notification	A server notifies subscribed principles of presence changes of interested principles
	Presence update	Principles notify a server of their presence change, including initial registration
Enhanced instant messaging service	Text chat	Principles send and receive text messages in a session
	Voice chat	Principles talk to each other in a session
	Multiplayer gaming	Principles engage in a coordinated gaming session
	Application sharing	A user shares the display of a local application with others

The directory servers implement presence service and principle profile management. Principles must register at one directory server when logging in. Generally, a directory server also performs authentication of principles during initial registration.

To provide enhanced messaging services, some IM systems choose to have centralized servers (which could be on the same physical machines of a directory server) to act as gateways forwarding data back and forth between principles. To enable group communication such as message meetings, directory servers have to be involved in conducting message broadcasting among all principles in the group. This type of IM system includes Yahoo Messenger, MSN Messenger, and AIM. Other IM systems, such as IRC and Jabber, let principles exchange data in a peer-to-peer fashion (*i.e.*, directly communicating at the application layer without a central server). The centralized systems are advantageous in providing group communication that otherwise would put a burden on individual principles in sending and receiving messages over multiple connections. This is significant with regard to providing mobile messaging service due to the resource constraints of mobile devices. Additionally, centralized systems provide the logging and auditing of instant messages required by many companies and government agencies. On the other hand, peer-to-peer IM systems are more scalable in serving a large number of principles at the same time because the bottleneck at a directory server, also known as the single point of failure, has been eliminated.

These IM systems are designed for casual use by the general public, but many people use these systems for business purposes to avoid phone calls or face-to-face verbal communication in work environments. Some IM providers and independent software vendors see this as an opportunity to offer enterprise IM systems for three reasons. First, enterprise IM can reduce the cost of telephone usage, especially for companies with branches in many locations. Second, well-confined enterprise IM provides enhanced security to prevent sensitive business information from being divulged. Third, enterprise IM allows close computer-based collaboration that voice or e-mail cannot achieve, and it can be further integrated into other enterprise applications such as ERP or CRM to improve productivity and reduce operational costs. Indeed, enterprise IM allows mobile messaging for mobile workers who must interact with back-end enterprise systems or other mobile workers. We will return to this issue later.

We will use MSN Messenger as an example to discuss IM architecture. As shown in Figure 7.8, the MSN Messenger architecture consists of a notification server (NS), switchboard (SB), and client application. An MSN network can have many notification servers and switchboards. An NS is a directory server implementing presence service, whereas an SB is the communication proxy for each messaging session. A client application first connects to an NS, which in turns directs the client to an SB if an IM session is being requested. Note that the figure does not show the communication between the NS and SB. Presence services are provided by NS1 and NS2, which collectively manage inter-NS communication to obtain global principle states. Each of the principles has to connect to an NS, and communication between principles goes through the SB.

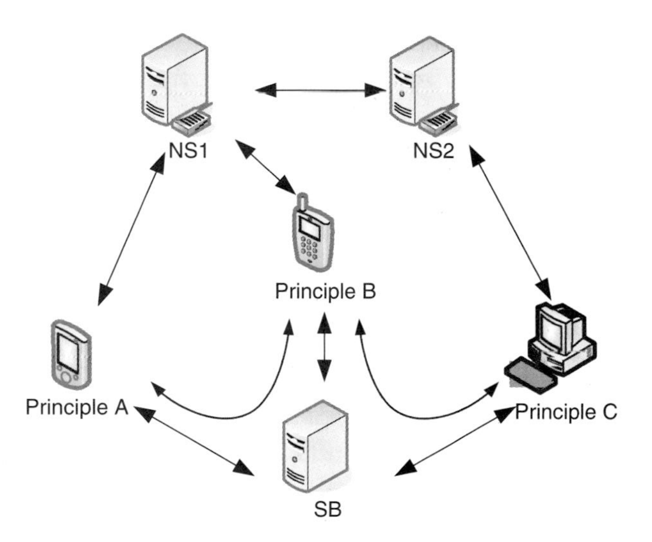

Figure 7.8 MSN Messenger architecture.

7.2.3 SIP/SIMPLE *Versus* XMPP

The major problem keeping IM from becoming a pervasive mobile service is a lack of interoperability among popular IM systems, most of which employ proprietary protocols and have their own user bases. An exception is Jabber, which is designed based on an open protocol called the extensible messaging and presence protocol (XMPP). AOL has opened up AIM presence and community services to partners who are interested in integrating AIM into their Internet applications, but the AIM protocol is still a proprietary protocol. Because cross-system IM is generally not possible (Note: In October 2005, Microsoft and Yahoo announced that MSN Messenger and Yahoo Messenger will interoperate in 2006), it is common to see multiple IM client applications on a single computer or mobile device. As is the case in promoting other technologies, IM standardization is crucial to enabling interoperability among different systems. There used to be some industry consortiums, such as IMUnified and FreeIM, which were founded by a number of companies in this business hoping to develop technical specifications for interoperation among member systems; however, these do not exist now due to a lack of interest. On the other hand, IETF has formed two working groups to define open IM protocols: the SIMPLE working group and the XMPP working group. It is very likely that IM systems will move toward one of these two protocols.

7.2.3.1 Session Initialization Protocol

Session initialization protocol (SIP) is a signaling protocol for multimedia session management developed by IETF. The principal application scenarios of SIP include voice over IP, IM, and audio/video conferencing. SIP defines a series of message exchanges that allows users to establish, modify, and terminate multimedia sessions, including telephone calls, audio/video transmission, instant messaging, and presence update. In addition to voice call management, SIP provides personal mobility, which refers to the ability to reach a party under a single, location-independent address even when the user changes terminals (*e.g.*, using different cell phones or computers with different IP addresses). Simple SIP session can be established between two parties by request/response message exchanges on top of a chosen transport (default is UDP (User Datagram Protocol)). A proxy server can be used to assist SIP session setup between two parties, as shown in Figure 7.9. A device supporting SIP must implement an SIP user agent (UA), which maintains the state of ongoing SIP sessions. Three SIP servers are defined: proxy server, redirect server, and registration server. The proxy server is used when two SIP user agents cannot or will not talk directly to each other; it is responsible for routing messages among clients. The redirection server allows a server to push routing information for a request back to a client; thus, subsequent message exchanges during the ongoing session will no longer involve the server. Redirection servers effectively reduce the load on the proxy servers. The registration server maintains presence information regarding subscribed SIP clients.

SIP often works with the session description protocol (SDP), a protocol that defines how multimedia content is described. One thing worth noting is that SIP, as a signaling protocol, does not perform actual data communication. For voice-over-IP applications utilizing SIP, the underlying transport layer protocol is the real-time transport protocol (RTP). The Third Generation Partnership Project (3GPP) has elected to use SIP as the signaling protocol in universal mobile telecommunications system (UMTS)/wideband CDMA (WCDMA) networks.

When SIP is introduced to IM, two models are capable of providing IM with SIP. One is the paging model, in which IM messages

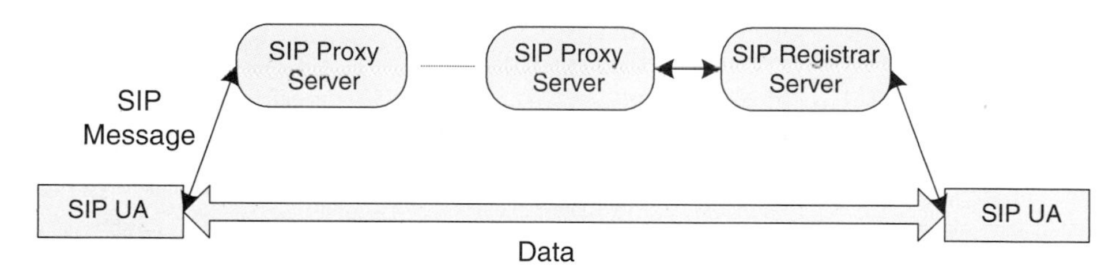

Figure 7.9 Session initialization protocol (SIP).

are piggybacked as the payload of signaling messages. The second is session model, which uses SIP to negotiate IM session parameters but relies on another protocol for data exchange. The first model generates significant overhead for the messages being sent, and the system may not scale well as SIP itself does not provide congestion-control mechanisms. Thus, the session model was chosen when IETF began work on the SIMPLE protocol.

7.2.3.2 SIP for Instant Messaging and Presence Leveraging Extensions

SIP for Instant Messaging and Presence Leveraging Extensions (SIMPLE) is an SIP extension for instant messaging applications developed by the IETF SIMPLE working group. It uses HTTP-like plain-text messages carried by SIP to perform presence service and negotiate IM session between two entities. The addressing scheme in SIMPLE is quite flexible in that it is independent of the user's location and the device being used. As an extension of SIP, SIMPLE has three logical entities: presence user agent (PUA), presence agent (PA), and watcher. It also offers the following new methods for instant messaging and presence services, SUBSCRIBE, MESSAGE, and NOTIFY. A user (the "presentity") may connect to several PUAs to propagate presence information. A PA is in charge of managing presence information uploaded from PUAs. A watcher is a user who is interested in someone else' presence information. The SUBSCRIBE method enables a watcher to send a subscribe request message. The subscribe message contains the URI of the desired presentity. The subscribe request is routed in the SIP network and eventually reaches the PA that serves the desired presentity. After authentication and authorization of the watcher, the NOTIFY method is used by a PA to inform a watcher of the presentity being requested.

The logical entities of SIMPLE's presence service are depicted in Figure 7.10. The MESSAGE method permits a user to send IMs to a presentity in paging mode described as follows. When a user decides to send an instant message to another, or an "instant inbox," the user agent of the sender formulates and issues an SIP MESSAGE request destined to the intended receiver, which is identified by an "address

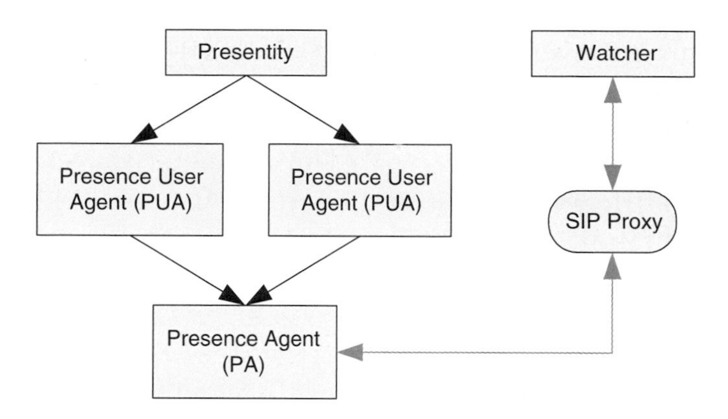

Figure 7.10 SIMPLE presence service.

of record" (such as joh@example.com) or a device address of the receiver if the sender is aware of that. The MESSAGE request carries the message content as its payload. The MESSAGE request may traverse a set of SIP proxies before reaching the destination, during which time hop-by-hop request routing is performed by SIP. A response will be generated and routed back to the sender in the same way. Another mode for exchanging instant messages is the session model. SIMPLE makes use of the message session relay protocol (MSRP) to establish an MSRP session dedicated to message exchange. MSRP introduces two messages: VISIT and SEND. The VISIT message and SIP INVITE message are used to establish an MSRP session, whereas an MSRP SEND message carries the content of an IM message.

SIMPLE supports both peer-to-peer and proxy-assisted data communication. In particular, for real-time audio and video communication, one client first attempts to make a peer-to-peer connection with the other. If that fails due to firewalls or network address translation (NAT) on either side, proxy-assisted communication method will be used.

7.2.3.3 Extensible Messaging and Presence Protocol

A competing protocol also developed by IETF is XMPP (Extensible Messaging and Presence Protocol), which is based on an open source instant messaging specification called Jabber. Unlike SIMPLE, which operates on top of the general-purpose SIP, XMPP is an XML-based application layer protocol designed solely for instant messaging and presence management. XMPP specifies the use of the transmission control protocol (TCP) rather than UDP. An XMPP address, also called Jabber Identifier (JID) contains a set of ordered elements including a domain identifier, node identifier, and resource identifier, in the form of <user@domain/resource>. XMPP strictly follows the client–server model to provide presence service and IM service. This implies that a client must register at a server before joining the network, and messages exchanged between two clients are always forwarded between one or more servers along the way.

Two specifications of XMPP, XMPP-Core, and XMPP-IM, were approved by IETF in early 2004. XMPP-Core defines XML schemas for the representation of real-time communications and presence information, whereas XMPP-IM defines XMPP stanzas (a piece of XML) used for communicating IMs and presence information in the form of XML streams. Three types of stanzas have been defined: <message/> for message exchange, <presence/> for presence information delivery, and <iq/> (infor/query) for generic query message. A client may use a single connection to the server for serialization of XML stanzas of multiple IM sessions.

A major difference between XMPP and SIMPLE is that XMPP is a more complete protocol that has been implemented in Jabber, while much of SIMPLE remains as design guidelines and is open to debate. Even SIMPLE-based commercial enterprise IM systems such as Microsoft's Live Communication Server (LCS) and IBM's Lotus Sametime cannot operate with each other. SIMPLE's advantage over XMPP is SIP's capability to support signaling for voice and video sessions. Because future IM will surely become more closely integrated with telephony and other multimedia communication methods, SIP/SIMPLE seems to hold a better position. For specific text-based IM, XMPP offers higher scalability. Table 7.3 highlights key differences between these two approaches.

Because SIM/SIMPLE and XMPP are not compatible with each other, interoperability problems must be addressed. One way to

Table 7.3 SIP/SIMPLE *Versus* XMPP.

Category	SIP/SIMPLE	XMPP
Data communication method	Peer-to-peer and proxy assisted	Client–server based
Presence service	Presence agent	XMPP server
Message queuing	Proxy server	Not available
Protocol	Repurposed SIP signaling protocol	XML-based data communication protocol
Message overhead	Large because of SIP header	Small
Implementations	Implemented in several commercial systems, which arc not fully compatible	Jabber and a range of open source clients
Mobile communication support	SIP/SIMPLE has been adopted by 3GPP	XMPP can operate on top of underlying wireless technologies

achieve interoperability is to have both protocols mapped to the same IM abstract semantics—namely, the common profile for instant messaging (CPIM) defined in RFC 3860. Both SIMPLE and XMPP have provided such mapping as part of the specifications. Another approach is to directly map IM and presence service from one to the other on some gateways that resides at the border of the two networks.

7.2.4 Mobile Instant Messaging

We consider the legacy messaging services of mobile wireless networks such as SMS and EMS to be basic forms of IM, because they permit text messaging delivered in near real time, but the message content is very limited in both type and size. As IM moves into the mobile world, these services will likely be replaced by convenient, feature-rich mobile IM services enabled by specially designed IM systems within a mobile wireless network.

IM in mobile computing possesses some unique characteristics that separate mobile IM from traditional IM. First, presence in mobile

IM implies broader meanings in addition to availability, such as location, available communication means, and current activity. It would be ideal if a presence server were able to obtain extensive information about a user from mobile wireless core networks. Second, 2.5G and 3G data services are touted as "always-on" as long as the mobile device is powered on. Thus, a user is always ready to receive SMS. This contradicts the notion of presence in traditional IM, where users have more flexible control over their status information. Third, mobile wireless connections generally cannot supply the same magnitude of link bandwidth and reliability as wired Internet does, thereby requiring optimized design of message delivery in a mobile IP system.

A near-term approach to mobile IM is to provide traditional IM as part of 2.5G and 3G data services. For example, several wireless service providers offer AOL messaging service or MSN messaging service via general packet radio service (GPRS) or cdma2000 1x data connections. Although this approach generally works, it does not take into account the three fundamental issues mentioned above. For one thing, these quickly migrated mobile IM systems may not scale well when message traffic continues to grow at a fast pace because the system does not distinguish mobile wireless clients from wired network clients. Mobility presents another problem that must be solved by a mobile IM system. Additionally, mobile spamming (spimming) will cost subscribers money if the problem is not addressed properly. In a word, the inherent advantages and limitations of wireless links call for a long-term solution, a complete redesign of IM systems to meet unique requirements of mobile IM services. In the following, we introduce Wireless Village as architectural designs in this category.

7.2.4.1 Wireless Village

The Wireless Village (WV) [13] is an early effort to introduce IM to mobile wireless networks. It was developed by the Instant Messaging and Presence Services (IMPS) group at the Open Mobile Alliance (OMA). OMA is an industry association formed by a number of mobile wireless service providers, device manufacturers, information technology companies, and content and service providers. The goal of OMA is to foster the development of mobile service enabler specifications that support the creation of interoperable end-to-end mobile services.

Wireless Village aims to offer the following core features in the context of IM in an IMPS system: presence, instant messaging, group, and shared content. Additional features such as access features and server interoperability features can also be implemented in association with connected mobile wireless networks. Wireless Village employs an architecture that can be closely coupled with 2G or 2.5G mobile wireless networks. Two types of elements are defined in Wireless Village: WV client and WV server. Any mobile station implementing the protocol is considered a WV client. A WV server consists of five elements: presence service element, instant messaging service element, group service element, content service element, and service access point. A WV server interfaces to WV clients and other WV servers through its service access point. Wireless Village defines a set of protocols for communication between servers, WV embedded client, and mobile core networks, as shown in Figure 7.11. The client–server protocol (CSP) provides for WV-embedded clients to access serving WV servers. The server–server protocol (SSP) defines interserver communication within the same WV network or between

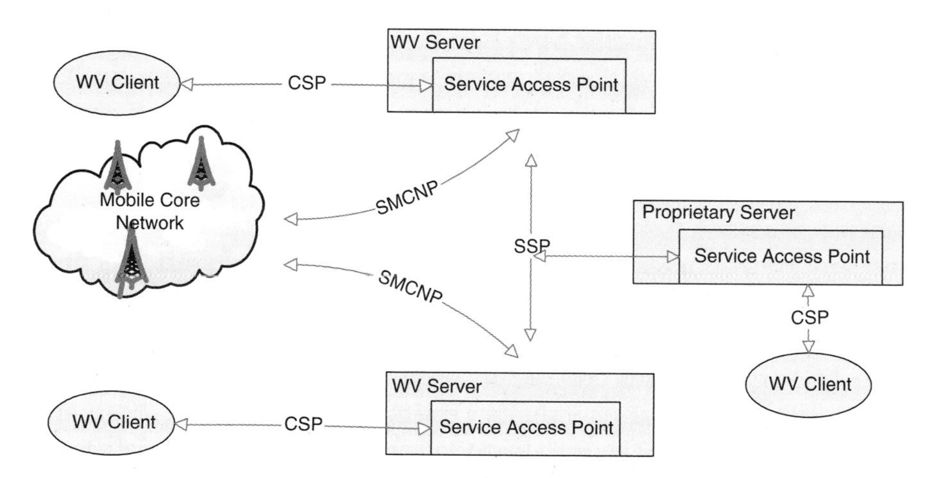

Figure 7.11 Wireless Village architecture [13].

different service providers. SSP allows a proprietary IMPS service to be connected to a WV network via a service access point. The server mobile core network (SMCNP) protocol essentially allows WV servers to access mobile wireless core networks, thereby integrating WV mobile messaging service into the existing service offerings of a mobile wireless network.

Wireless Village is a loosely coupled approach in the sense that it requires few internal modifications to legacy mobile core network. Wireless Village can be implemented in conjunction with 2G or 2.5G wireless networks as "add-on" elements to the legacy network. As for 3G and 4G wireless networks, SIP/SIMPLE has been chosen by 3GPP as the IM signaling protocol.

7.2.4.2 Wireless IM Server Issues

In a mobile IM system, IM directory servers and proxy servers serving a mixture of wireless clients and wired clients must be optimized to facilitate efficient and reliable message delivery over wireless links. For example, message transport for mobile IM may not perform well if specific characteristics of underlying mobile wireless networks are not taken into account. Here is a scenario. Suppose during an IM session a mobile device experiences a handoff or network connectivity problem that causes message delivery failure. The IM server will retransmit the message when the connection is restored. There is a chance that while waiting for a pending ACK (Acknowledgement), the IM server may simply continue to send text messages to the mobile device. This behavior is often determined by the Nagle algorithm in many TCP implementations as defined in RFC 896. The Nagle algorithm effectively reduces the number of small TCP segments by aggregating them into a large segment, thus improving data transmission efficiency for highly interactive applications. According to the Nagle algorithm, only one outstanding segment is allowed. While waiting for the ACK of this outstanding segment, subsequent small segments will be aggregated into each one, but any full-size segment gathered by the TCP layer will be transmitted immediately.

The Nagle algorithm is enabled by default in many TCP implementations. For example, Jabber server enables Nagle algorithm by default; however, in an interactive mobile IM session, the Nagle algorithm may not perform well [14] if mobile devices are engaged in high mobility. While the IM server is waiting for a pending ACK, for immediate message delivery, the Nagle algorithm should be disabled such that the IM server can continue to send text messages (thus the number of outstanding segments is larger than one) and possibly resend the original message, which will be reordered at the recipient. However, disabling the Nagle algorithm (using the TCP_NODELAY option) increases the number of small TCP segments, thereby reducing data delivery efficiency over precious wireless link bandwidth. The solution is to have the IM application rather than TCP control when to send messages based on an application layer time-out mechanism. This is again an example of cross-layer design issues in the mobile computing paradigm. For details regarding IM server implementations, please refer to Leggio and Kojo [14].

7.3 Mobile Multimedia Streaming

Multimedia streaming is a method of delivering synchronized multimedia content such that the recipient can begin to play the content during the course of the transmission. Multimedia streaming is further divided into two types: on-demand streaming and live streaming. In on-demand streaming, the content (data file) as a whole is generated at a server and is delivered it to a client upon request. Many websites offer on-demand video news, movie trailers, etc. In live streaming, however, the content is a datastream that is being generated and delivered to a client simultaneously. Live streaming is often used in live broadcasting of an event on the web (webcast). Multimedia streaming is by definition unidirectional in that the stream is sent from a content server to the client only. Some applications such as video telephony and video conferencing require two live streams for both directions between two parties.

Traditional mobile wireless networks are primarily voice based because of limited wireless link bandwidth, which makes it impossible to deliver high-quality video content. This situation is changing as 2.5G and 3G cellular systems and other wireless data technologies such as wireless LANs and wireless metropolitan area networks (MANs) begin to proliferate, rendering effective delivery of multimedia steaming content, especially video content. In light of the rapidly growing mobile user community and the need for rich content over high-resolution, color cell phones, mobile service providers are exploring the opportunity to offer a new set of value-added applications that combine multimedia streaming, location-based systems, mobile commerce, and instant messaging. Following are some examples of such applications:

- Mobile video message and video call
- Mobile video IM and conferencing
- Mobile news, sports, and music video clips
- Mobile television broadcasting, such as Verizon's VCast service
- Mobile online music
- Mobile live broadcasting of events
- Mobile video monitoring in surveillance
- Mobile remote control

For mobile multimedia applications, the data delivery methods can vary significantly. In addition to unicast (one to one), multicast (one to a group of nodes forming a multicast tree for message delivery), broadcast (one to all nodes in a network at the same time), and anycast (one to a set of nodes but only one of them is chosen at any given time) can also be used. These methods can be implemented at the network layer where packets are routed in the network infrastructure or at the application layer where the application itself handles group management and data transfer. For example, if routers and hosts in a network support IP multicast, than multimedia streaming can be realized using multicast in favor of performance; otherwise, application layer multicast can be utilized to leverage the benefit of multicast at the expense of potentially high communication overhead resulting from a mismatch between overlay topology and underlying physical network topology.

7.3.1 Multimedia Primer

A challenge of mobile streaming is to deal with the heterogeneity of wireless technologies, multimedia protocols and formats, and the terminals—mobile devices. Many types of mobile wireless technologies are in use in the wireless world, and they differ tremendously in terms of data rate, bit error rate, link reliability, and power consumption, as well as the business models based on the various technologies and operating costs. And more emerging wireless technologies are on the way. Interoperability is a huge problem, even between devices using the same technology, because of those proprietary systems and protocols. On the other hand, it is very common to see competing standards, file formats, or specifications supported by different interest groups, effectively leading to a significant segmentation of the market. This is particularly true in network-based multimedia streaming systems.

As shown in Figure 7.12, the three basic components of a multimedia delivery system are encoder, player, and server. Encoders are multimedia authoring tools that generate files or streams according to specific formats (codecs). For mobile video services on 3G networks, MPEG-4 is the obvious choice (explained below). A multimedia server or content server delivers the content, in the form of a file created offline or a stream generated online, to a client player, which is capable of communicating with the media server conforming to some protocols and playing the received content correctly. A player usually supports a collection of codecs it can render locally. It is commonplace to see commercial solutions that are based on proprietary or open codecs, integrating encoders, players, and a media server and supporting a range of data transmission protocols.

When providing mobile multimedia services, codecs and players for desktop computers on the wired Internet must be optimized for low bandwidth, small screen size, and low computing power in a

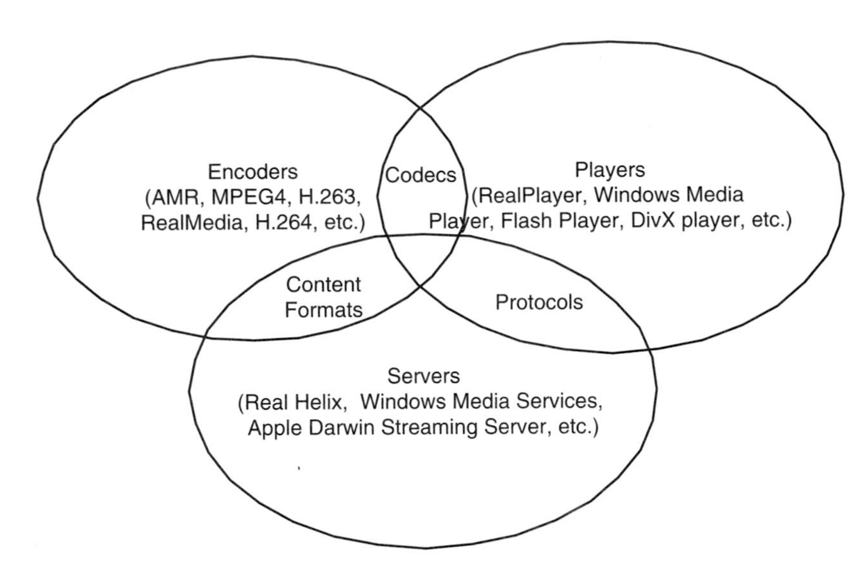

Figure 7.12 Three components of a multimedia delivery system. (© Copyright 2005 RealNetworks: Inc. Courtesy of rtsp.org, http://www.rtsp.org/2001/faq.html.)

mobile environment. Aside from proprietary codecs such as RealMedia codecs and Windows Media codecs, other widely used codecs for mobile multimedia services are MPEG-4, Mp3, AAC, and AMR:

- MPEG-4, designed by the Motion Picture Expert Group (MPEG), is the solution to high-quality video at low bit rate, primarily for low-bandwidth delivery. MPEG-1 (CD quality) and MPEG-2 (DVD quality) are not suitable for mobile multimedia streaming systems because their imposed data rates are too high to fit into wireless links. MPEG-4 encodes video, audio, and three-dimensional objects described using virtual reality modeling language (VRML).
- Mp3 (MPEG-1/2 Audio Layer 3) is an open standard of highly compressed audio codec with very little loss of sound quality. The bit rates of mp3 range from 128 to 384 kbps, and the compression ratio is 10 to 1 or 12 to 1. Typical mp3 songs are several megabytes in size.
- Advanced audio coding (AAC), also developed by MPEG, provides high-quality audio at low data rates. AAC was designed to

replace mp3. Four profiles are defined in AAC: low-complexity profile, main profile, sample term prediction, and long-term prediction (LTP). MPEG-4 employs the LTP profile as an audio codec.

- AMR (Adaptive Multi-Rate) narrowband and AMR wideband are low-bit-rate codecs standardized by the European Telecommunications Standards Institute (ETSI). They operate at a bit rate ranging from 4.75 to 12.2 kbps. AMR-WR (wideband) operates at 6 to 36 kbps for mono audio, and 8 to 48 kbps for stereo audio. AMRs have been adopted by GSM and 3GPP as the voice codec to be used in packet-switched streaming (PSS) and multimedia broadcast/multicast service (MBMS) (explained later in Section 7.3.3).

Table 7.4 outlines frequently used network protocols for multimedia data transmission. Table 7.5 shows popular multimedia solutions from different vendors. Because of their collective dominance in the

Table 7.4 Network Protocols for Multimedia Transmission.

Protocol	Description
Real-time protocol (RTP), RFC1889	An end-to-end network protocol for transmitting real-time data on-demand or in interactive services. It consists of two parts: data transmission and a control subprotocol called RTCP. RTP is transport independent, meaning that it can be used on top of UDP or TCP.
Real-time streaming protocol (RTSP), RFC2326	An end-to-end network protocol framework for transmitting streaming content from a live feed or in a file. It is designed to work on top of RTP while providing controls over multiple delivery sessions and means of delivery such as multicast as well as unicast.
Progressive Networks Media/Audio (PNM/PNA)	A proprietary protocol developed by RealNetworks. PNA uses TCP for player control messages, and UDP or TCP for audio data. It has been replaced by RTSP.
Session definition protocol (SDP), RFC2327	A standard format for session description, including session name and purpose, media description, codecs, transport protocols, bandwidth requirement, and user information. SDP can be used in different multimedia related network protocols such as RTSP, SIP, and HTTP.

Table 7.5 Summary of Multimedia Systems.

File Format	Video Codecs	Audio Codecs	Player	Encoder	Protocols in Players
RealMedia (.rm, .rmvb, .ra)	RealVideo	RealAudio	Real Player	RealProducer	RTSP/RTP, PNA, and HTTP
Windows Media (.wmv, .wma)	Windows Media Video	Windows Media Audio	Windows Media Player	Windows Media Encoder	MMS, RTSP/RTP, and HTTP
QuickTime (.mov, .qt)	MPEG-4 and Sorenson Video	AAC and AMR	Apple QuickTime Player	QuickTime Encoder	RTSP/RTP
MPEG-4 (.mp4)	MPEG-4	AAC	Generally supported by all players	Generally supported	RTSP/RTP, PNA, MMS, and HTTP
MP3 (.mp3)	N/A	Mp3	Generally supported by all players	Many commercial encoders available	RTSP/RTP, PNA, MMS, and HTTP

computer multimedia arena, it is very likely that a mobile version of these systems will encroach on the mobile wireless world.

One way to address the issue of diversity in mobile streaming is to employ a network-independent signaling scheme capable of dealing with various media formats and transmission control between a media server and a mobile device. An example of this approach is 3GPP's packet-switched streaming (PSS) specification and end-to-end solutions to multimedia streaming over UMTS/WCDMA networks. The 3GPP2's effort in this regard is multimedia streaming service (MSS).

7.3.2 Packet-Switched Streaming in 3GPP

The 3GPP PSS specification defines a framework of interoperable multimedia content delivery over UMTS networks. It specifies a logical architecture as well as a set of multimedia streaming services that integrate video, audio, images, and formatted text in various formats. Figure 7.13 depicts those network elements involved in a 3G PSS specification. Streaming clients are players on mobile terminals. According to UMTS specifications, an external IP network is connected to a UMTS core network via the Gi interface (Details of UMTS networks are described in Chapter 3). Content servers in an IP network store static multimedia content and generate live streams, both of which can be delivered to streaming clients. Content cache can be implemented on proxy servers placed in IP networks to improve streaming performance. Portals are websites of media content. A mobile user can browse those web links to media content on portals and then send a request for particular content to the content server. User and terminal profiles indicate the user's preferences and terminal capabilities with respect to a list of media formats and supported media quality specifications. Notice that, in the simplest form, only streaming clients and a content server are required to provide

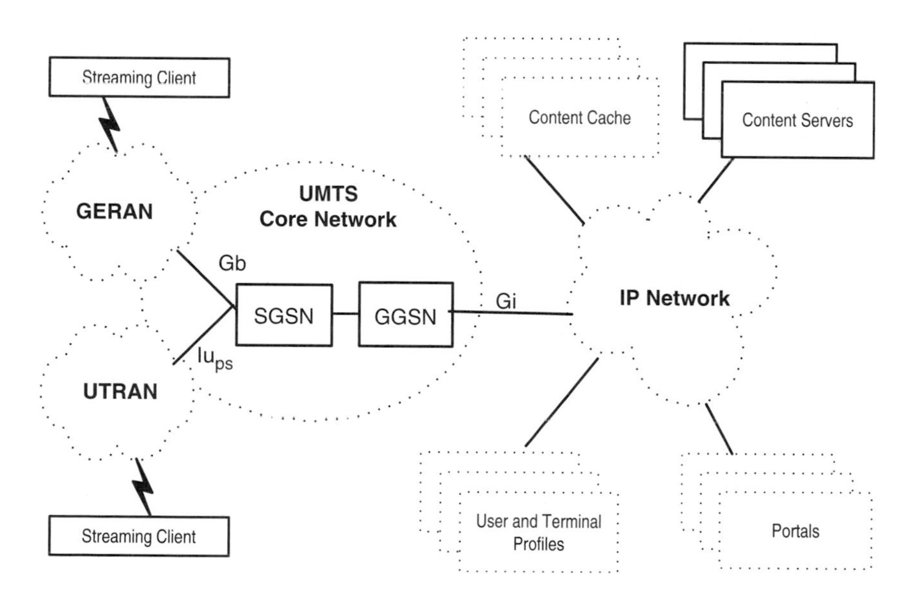

Figure 7.13 Network elements involved in a 3G packet-switched streaming service [15]. (© Copyright 2005 3GPP.)

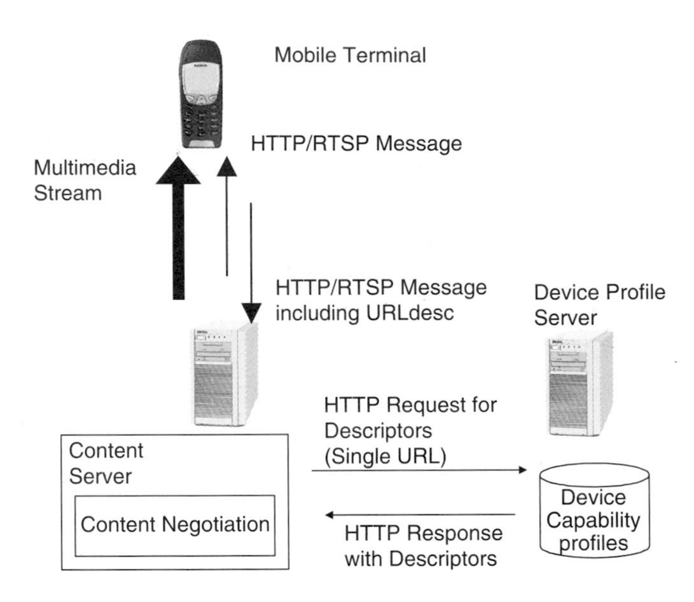

Figure 7.14 Content negotiation in PSS [15]. (© Copyright 2005 3GPP.)

PPS services. An example of the negotiation procedure of streaming control is illustrated in Figure 7.14.

A mobile streaming client initiates a streaming session by sending an HTTP/RTSP message containing an RTSP URL to a content server. If a mobile device profile server is in use, the content server will contact it to obtain a descriptor of the requested device in question, then the media server and the steaming client will exchange several RTSP messages to determine transmission parameters that are suitable to the client's capability, a procedure referred to as *capability exchange*. For example, based on the link bandwidth of the client, the media server can determine the appropriate bit rate of video streams to be delivered to the client.

3GPP PSS makes use of existing network protocols to support mobile streaming in wireless environment. The protocols and their supporting applications are summarized as follows [16]:

- Session establishment: RTSP
- Capability exchange: RTSP and HTTP

- Session control: RTSP and SDP
- Real-time media transport: RTP
- Static media and scene description transport: HTTP and TCP

Scene description consists of spatial layout of multiple types of content on a display, as well as synchronization information. This piece of information is often used in creating and delivering interactive multimedia presentations that incorporate video, audio, images, and text, described using the synchronized multimedia integration language (SMIL, http://www.w3.org/TR/SMIL2/), which is an XML-based presentation language developed by the World Wide Web Consortium (W3C). Static media includes still images, bitmap graphics, vector graphics, text, timed text, and synthetic audio, whereas real-time media could be one or more of the following: video, audio, and speech. The codecs of these media types specified in 3GPP PSS are shown in Table 7.6.

7.3.2.1 DRM in PSS

The 3GPP PSS specification has included digital rights management (DRM) schemes to control digital content dispensation in a mobile streaming application. DRM is generally defined as capabilities of a multimedia system or a network to protect copyrighted digital content against illegal copying and distribution. DRM is considered an essential component for multimedia content delivery in wired and wireless applications. For example, online music stores often provide free trials of one or two songs from an artist's album to attract perspective purchasers. When the music content is streamed to a client for trial, the system must ensure that it is not possible for the client to save the content locally, which would facilitate further distribution of the content via some file share systems such as peer-to-peer networks. To this end, DRM solutions such as Microsoft Windows Media DRM and RealNetworks Helix DRM are able to provide secure media packaging, license management, and a range of business models including purchase, rental, exchange, subscription, etc.

Packet-switched streaming specifies the use of OMA DRM extensions, which consists of two portions. The first is RTP payload

Table 7.6 Codecs in 3GPP PSS.

Type	Codec
Speech	AMR narrowband
	AMR wideband
Audio	Enhanced aacPlus
	AMR wideband
	MPEG-4 AAC-LC
	MPEG-4 AAC-LTP
Synthetic audio	Scalable polyphony MIDI
	Mobile DLS
	Mobile XMF
Still images	JPEG and JFIF
Bitmap graphics	GIF and PNG
Vector graphics	SVG Tiny 1.2
	ECMAScript
Text	XHTML Mobile Profile
	SMIL
Timed text	Defined by PSS
3GPP file format	.3gp file defined by 3GPP
Video	H.263 Profile 3 Level 45
	MPEG-4 Visual Simple Profile Level 0b
	H.264 (AVC) Baseline Profile Level 1b

encryption using a form of 128-bit AES with a content encryption key (CEK) for each type of payload in a session. The second is an integrity protection mechanism-based on the secure real-time protocol (SRTP) defined in RFC 3711. Content providers, streaming servers, and clients are all involved in providing integrity. Using a preshared CEK, a streaming server can derive an SRTP master key and an integrity key and use them to protect SDP attributes of the content and the encrypted content. The message authentication code (MAC) tag appended per packet is based on HMAC-SHA1 and truncated to 80 bits to reduce computation overhead during verification.

7.3.3 Mobile Broadcasting and Multicasting

3GPP PSS essentially defines a unicast mobile streaming service between a streaming server and a client. This model works fine if the number of streaming clients at any given time is not large such that the streaming server is able to handle all sessions being delivered; however, like some other server-centric network systems, when a large number of clients is accessing the streaming server during an event, the server is likely to be overwhelmed as it cannot handle a huge number of individual connections. One way to accommodate many clients at the same time is to use load-balancing techniques on servers. The downside of load balancing is increased complexity and cost of the system. Another way is to employ broadcast and multicast schemes instead of unicast. Mobile broadcasting is particularly used in pushing recorded or live television programs to mobile devices from a cellular network or satellites. Note that point-to-multipoint mobile streaming can also be done by broadcasting or multicasting. Two broadcasting-based systems that primarily target video broadcasting are DVB-H and MBMS.

7.3.3.1 Digital Video Broadcasting for Handheld

Digital video broadcasting for handheld (DVB-H) is an extension of digital video broadcasting for terrestrial digital video broadcasting (DVB-T) developed by the ETSI and finalized in 2004. DVB-H enables simultaneous transmission of multiple television, radio, and video channels to mobile devices. Note that DVB-H does not use any mobile wireless network infrastructure for content delivery. Instead, it relies on a separated, single-frequency wireless infrastructure—namely, a DVB network and an integrated wireless receiver on each participating mobile device to deliver data rates of up to 2 Mbit/sec in an 8-MHz channel. DVB-H can be integrated into a mobile wireless network so mobile service providers, along with broadcasters, can supply value-added, broadcasting-based services to subscribers in addition to unicast-based mobile streaming services. Codecs used by DVH-H are MPEG-4 for video and AAC for audio. Video streams are modulated using QPSK or 16QAM and are multiplexed using COFDM. To reduce the power consumption of the receiver, DVB-H provides a time-slicing mechanism that allows a receiver to receive bursts of data in a noncontinuous fashion. DVB-H defines three types of receivers: type A (128 kbit/sec) for a video resolution of QCIF (Quarter Common Intermediate Format), 180 × 144), type B (384 kbit/sec) for CIF (Common Intermediate Format, 360 × 288), and type C (2 Mbit/sec) for CIF.

7.3.3.2 Multimedia Broadcast/Multicast Service

Multimedia broadcast/multicast service (MBMS) is a 3GPP standard for mobile broadcasting service. Before MBMS, 3GPP had developed the cell broadcast service (CBS) for text broadcasting. Unlike DVB-H, MBMS uses dedicated channels in existing and upcoming mobile wireless network infrastructure. The data rate specified by MBMS is 64 to 384 kbps. MBMS uses IP multicast to delivery content to a number of large user groups simultaneously. MBMS introduces a new component, the broadcast multicast servicecenter (BM-SC), to the 3GPP packet domain infrastructure. Residing between a content provider and a GGSN, BM-SC acts as the data source in a domain to serve locally connected clients. The gateway GPRS support node (GGSN) is responsible for IP multicast group management and routing packets to corresponding serving GPRS service nodes (SGSNs), which will perform authentication, authorization, and mobility management of multicast participants. An SGSN involved in MBMS will also maintain MBMS user environment context along with its packet data protocol (PDP) context.

3GPP2 has developed a similar service, the broadcast-multicast service (BCMCS), for cdma2000 networks. Five components have been introduced into the BCMCS architecture: broadcasting service node (BSN), BCMCS controller, BCMCS content server, BCMCS content provider, and an optional multicast router. A content server receives content streams from the content provider and sends them to BSNs. A BMCS controller manages session information needed by the radio access network and content servers. BSNs, content servers, and PDSNs are entities responsible for performing IP multicast of content streams. Aside from mobile broadcasting services by 3GPP and 3GPP2, some countries have developed their own mobile broadcasting standards. For example, Japan has been using integrated services digital broadcasting (ISDB) for mobile television services. Korea has a similar system in operation (T-DMB).

Providing mobile broadcasting services is not simply a technical challenge. Apart from tackling system performance and integration issues, service providers and content providers must find a good business model to determine their investment on network infrastructure, technology licenses, and operational cost. The pricing of premium content offered by mobile streaming services must be flexible to accommodate the needs of both business professionals and consumers.

7.4 M-Commerce

M-commerce (or mobile e-commerce or wireless commerce) is a set of services and applications operating in mobile wireless networks that allow consumer-oriented transactions and business transactions to be performed via mobile devices. In its simplest form, m-commerce can be considered an integration of e-commerce and wireless web. M-commerce services and applications rely on the underlying converging wireless infrastructure to deliver information and to enable interaction with high reliability and security. In addition, enterprise information systems must be optimized to serve mobile transactions. Parties involved in this supply chain include mobile service providers, content providers, financial institutes, and mobile users.

M-commerce is set to take off as a result of the widespread use of cell phones, wireless-enabled PDAs, and smart phones, and there is an obvious need for real-time, ubiquitous mobile access on these mobile devices. Consider the success of e-commerce on the Internet. Despite the Internet bubble, e-commerce has become a hugely profitable business sector. Business to client (B2C) services such as online shopping and online investment have seen amazing growth. Business to business (B2B) and business to employee (B2E) applications continue to gain substantial ground in many companies as a way to improve productivity and reduce operational cost. With the proliferation of mobile devices and ubiquitous mobile access, it would be quite reasonable to expect a dramatic growth of m-commerce services and applications.

7.4.1 M-Commerce Applications

Online shopping is an excellent example of E-Commerce on the Internet. In a mobile wireless world, however, due to the inherent limitations of mobile devices, simply porting desktop e-commerce applications to mobile is not likely to work. First of all, the form factor of a mobile device makes it impossible for users to view very sophisticated pages or utilize user interfaces commonly used by desktop applications. Second, mobile security presents some distinctive challenges when sensitive financial data are being exchanged within or between m-commerce applications. Not only the wireless infrastructure but also the mobile device housing the mobile application client should be highly secure in terms of data confidentiality and integrity during transmission or in storage. Third, because wireless links connecting to the front end (mobile application client on a mobile device) are inherently not as stable as wired network connections, techniques such as offline processing in disconnected mode along with timely data synchronization are often used. This inevitably complicates the design of m-commerce systems.

E-commerce is mostly regarded as transaction-related operations and functionality enabled by computers and communication networks, whereas m-commerce may provide both enhanced business information access (type I) and mobile transaction support (type II), as shown in Table 7.7. Examples of type I m-commerce are mobile stock quotes and sports news subscription service, in which subscribers pay some fee to receive requested real-time information for business or consumer needs. Type II m-commerce tends to be more complicated as it entails mobile transaction such as mobile payment, mobile banking, and bidding or online shopping using mobile devices. Which type of m-commerce is being used largely depends on two interrelated factors: business/consumer needs and technological support. Nonetheless, not all forms of e-commerce can currently be

Table 7.7 Examples of M-Commerce Applications, where Type I Refers to Enhanced Business Information Access and Type II Refers to Mobile Transaction Support.

m-commerce Application	Type	Notes
Mobile catalog and advertising	I	Information about products, services, and promotions; coupons; and advertisements are pushed to mobile devices on-demand or via location-based services. NTT DoCoMo's iMode is an example of this service.
Mobile information service	I	Users can subscribe to receive weather, news, sports updates, flight and bus schedules and status, maps, traffic updates, and location-based information delivered in the form of messages (including SMS, EMS, and MMS), web pages, and multimedia streams and clips to mobile devices or in-car telemetrics systems.
Mobile entertainment	I and II	Users can receive premium content (audio and video clips) in a mobile streaming application. Online paid music download services are likely to be directly accessible from mobile devices sooner or later, and consumer electronics such as Apple's iPod will surely have wireless capability in the near future, if they do not already do so.
Mobile shopping	II	Users can use mobile devices to view and order goods. Traditional payment methods such as credit cards or phone bill combined payment can be used.
Mobile payment and ticketing	II	Mobile devices can make payment at automatic vending machines and ticket machines. They can also be used as digital wallet or digital ID in business transactions.
Mobile auction	II	Users are able to receive real-time information about online auctions and place bids from mobile devices.
Mobile online banking	II	Users use mobile devices to manage bank accounts, pay bills and credit card balances, transfer funds, etc.
Mobile online investment and brokerage	I and II	Real-time financial information, news, and stock quotes can be streamed to a mobile device. Users can also place orders directly from a mobile device.
Mobile gaming	I and II	Users can play single-player or multiple-player mobile games on a mobile device wirelessly. Games can also be ordered for download.

(continued)

Table 7.7 Continued

m-commerce Application	Type	Notes
Mobile e-learning	I and II	A mobile device can be used in an educational or business training environment to enhance interactions and information access. Students with a mobile device can take quizzes, send questions to instructors, participate in online discussions, and obtain customized assistance based on their progress. Instructors can use mobile devices to collect information about topics and questions and guide individual students.
Supply chain and inventory management	II	Enterprises may use mobile devices to track goods and products wirelessly and to enable close interaction between employees for better decision making.
Mobile enterprise applications	I and II	Enterprise applications such as CRM, ERP, IT helpdesk, and corporate portals can support mobile terminals, providing real-time data access for such areas as pest control, shipping and delivery services, insurance, police forces, and the military.
Mobile conferencing	I and II	Users can make use of built-in video cameras and microphones for video conferencing.

conducted on mobile devices due to their intrinsic limitations. Even if future mobile devices eventually solve those technical problems, there might be no substantial need for some types of e-commerce services using a mobile device. As mobile computing evolves to become more pervasive, we will surely see many forms of type I and type II m-commerce services and applications. The underlying operational platform tends to more be versatile; aside from wireless web, other existing mobile applications such as IM and location-based services can also become an integral component of m-commerce architecture.

A list of m-commerce applications is provided in Table 7.7. We consider any mobile application that relates to some sort of business operations an m-commerce application, even if no transaction is conducted. In reality, this broad categorization of m-commerce means that almost all network-based operations with a content provider or service provider can be considered an m-commerce application, except mobile web surfing from a mobile web browser. Note that the listed items are generally not self-contained, meaning that some may require others to realize a specific functionality.

7.4.2 M-Commerce Architecture

An m-commerce service architecture consists of three basic functional elements: user application software on mobile devices as clients, mobile wireless network infrastructure, and m-commerce servers as data providers or service enablers. Each element may be highly heterogeneous. M-commerce clients may operate on a variety of mobile devices with distinct capabilities and form factors, and the mobile wireless network infrastructure may be a single wireless network such as a 3G cellular network or a hybrid of multiple wireless networks at different scales. Depending on the underlying m-commerce service, an m-commerce server may be a web server associated with an application server optimized for mobile services, a multimedia streaming server, an AAA server responsible for imposing security and billing, an enterprise server for business transactions, etc. To improve the performance of m-commerce services, proxy servers for different purposes are often used as well.

Figure 7.15 provides a conceptual overview of m-commerce services. An m-commerce client device must have at least one type of wireless access to a wireless network infrastructure such as cellular networks, wireless LAN, WiMax, or Bluetooth. M-commerce generally requires high-end cell phones and PDAs with a fast mobile processor, large storage, and a large display. Converged smart phones are seemingly good candidates for m-commerce client devices. The network can be either base station based or *ad hoc* based. On the server side, all servers connect to a wired network infrastructure such as an enterprise network and the Internet via an Internet service provider (ISP). An m-commerce service will be delivered to a client device in the form of a mobile application client, requiring mobilemiddleware that encapsulates low-level support by different platforms. Similarly, m-commerce servers must operate on the same

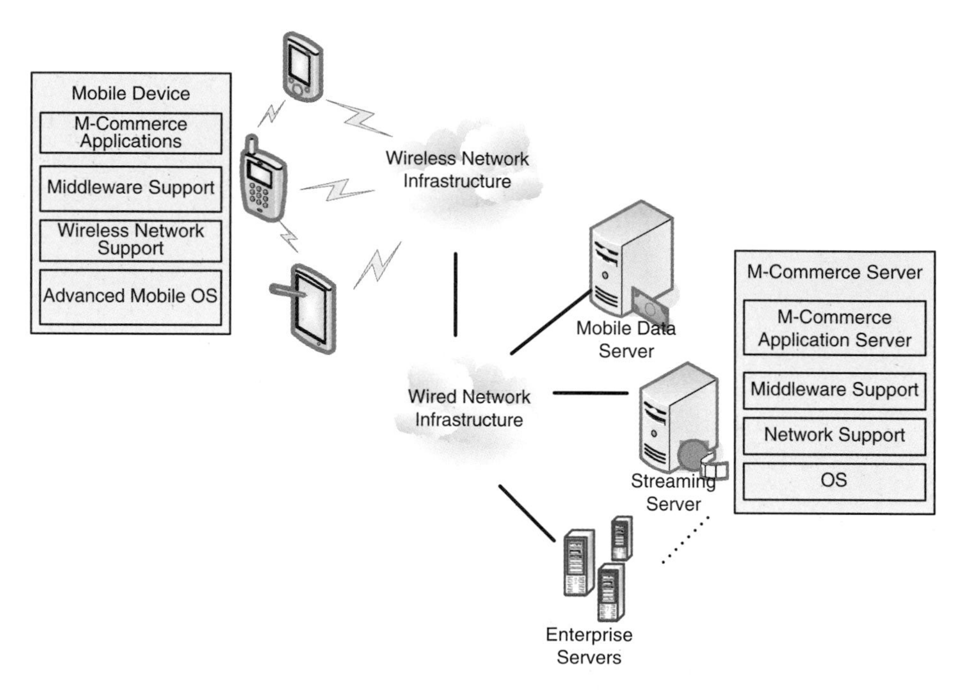

Figure 7.15 M-commerce applications overview.

middleware. For example, the wireless application protocol (WAP) can be considered a middleware for a range of m-commerce services that utilize wireless WAP as the underlying software platform and user interface to deliver content and services.

The issue of providing high-performance, user-friendly m-commerce services has presented numerous challenges. Examples of m-commerce mobile client issues include user interface design for usability and extensibility, applications and middleware for low power, high performance and reliability, and device security. Network infrastructures issues include localization support, mobility support, heterogeneous mobile device support, and mobile network security. The network infrastructure issues are transparent to m-commerce providers and have been extensively discussed in previous chapters. Thus, we will only address those issues here in this section. M-commerce server issues are concerned with integrating m-commerce service into existing back-end network and server infrastructures and choosing the most appropriate software platform to deliver a service (which in turn determines the platform on the client side). Indeed, many of these issues are general mobile application design issues and do not exist solely in the m-commerce domain. For example, user interface design for mobile devices is a key issue for nearly all mobile applications, including SMS and IM. Some issues such as mobile security have been discussed in general in Chapter 6; therefore, in the following text we will focus on those issues closely related to business transactions and issues that give rise to particular m-commerce problems. Examples of real systems will be introduced along with the discussion.

7.4.3 M-Commerce Client Issues

Suppose a user is nearly convinced of the economic and technological advantages of a specific m-commerce service and has decided to give it a try using a mobile device. What could possibly cause the user to give up? A complicated, nonintuitive, inconsistent user interface and slow, error-prone, often disconnected, unreliable applications might be some of the most obvious reasons. In addition, device security could be another important problem in m-commerce. Ten of thousands of PDAs get lost every year in the United States, and most are not well protected from nonauthenticated use. In a number of cases cell phones and smart phones have been hacked wirelessly. We will look at these issues in this section.

7.4.3.1 Mobile User Interface Design

User interface (UI) is a crucial issue for mobile applications. In fact, it may be the #1 problem hampering the acceptance of mobile applications by prospective users. Because the display of a mobile device is comparatively smaller than that of a desktop computer and the input methods are largely restricted, the design of a good mobile user interface is much more difficult than that of a desktop application user interface. Direct migration of a desktop user interface to mobile will not work well in most cases because of the complexity of desktop application user interfaces. In addition, simply trying to shrink and partition a desktop application user interface to multiple pages on a mobile device is not likely to give users a comfortable experience. Mobile UI design requires rethinking from the very beginning of an application design, with the goal of providing a highly convenient, time-saving, and device-independent experience. Researchers have chosen seven design elements of customer interfaces (7Cs) that together provide guidelines to mobile user interface design [17]. Table 7.8 presents a brief introduction to these design elements.

The UI design prescription varies from device to device. For example, mobile applications on PDAs and cell phones cannot use the same UI design guidelines, as PDAs generally have much larger displays and employ pen-based input instead of the multiple-key T9 method commonly used on cell phones. Thus, UIs on PDAs bear fewer constraints than on cell phones. Smart phones tend to have a more powerful input method, using a tiny keyboard instead of a phone keypad, effectively making input somewhat easier than on traditional cell phones. As for form factor, earlier smart phone models were more like traditional cell phones; however, as the need for value-added mobile applications has become more evident, newer smart phones have begun to have larger screens and full keyboards or pen-based text input to accommodate sophisticated applications. These models are often referred to as PDA phones, as they apparently lean toward the PDA side with respect to UI design. Nevertheless, the trend of convergence among communication devices, computers, and consumer electronic devices will continue to affect mobile UI design in many aspects.

7.4.3.2 Middleware

One way to deal with the heterogeneity of mobile devices used in an m-commerce application is to utilize middleware that hides the difference between various models of mobile devices and provides a uniformed platform for mobile applications. In the context of network services and applications, middleware refers to software acting

Table 7.8 Seven Design Elements of Customer Interfaces for M-Commerce.

Design Element	Description	Guide to Mobile Interface Design
Context	The functional and artistic design of a website or a standalone UI	Easy-to-read font (big enough, legible)
		Simple layout of controls
		Easy navigation of linked pages
		Broad but shallow menus
		Multiple pages instead of scrolling
Content	The resource presented by a UI	Short sentences
		Cross-link of related information
		Audio if possible
Community	User information sharing (mostly for retail)	Allow text and instant messaging
		Display customer rating next to the item
Customization	UI tailoring to devices or personal profiles	Allow customized item selection and layout
		Easy log-in
Communication	Human computer interaction	Lists instead of keyboard input
		Wizard-like dialogs
		Short video clips if possible
Connection	Linkage between sites and applications	Quick “Home” link
		Reduce cross-site links
Commerce	Presentation of commerce information	Big buttons
		Reminders before transactions
		Status updates at will

as a middle layer between applications and the underlying operating system, which is capable of providing unified software development interface over various low-level wireless network transport protocols and hardware platforms. Another use of middleware is to act as a general software agent between a client and server applications, thereby allowing developers to focus on business logic rather than communication details while implementing distributed transactions. Mobile middleware specifically abstracts a range of wired and wireless technologies and mobile operating systems, thereby allowing for the same user interface to be deployed on heterogeneous mobile devices. WAP is the most well-known mobile middleware, and it allows web-based m-commerce applications to be independent of the underlying hardware platform. Other well-known mobile web platforms include i-mode and BREW.

A broader definition of mobile middleware may as well include any programming languages, data coding standards, or common runtime that collectively enable interoperability between applications on different platforms. In this sense, Java J2EE, XML, and Microsoft's .Net Compact Framework can all be considered mobile middleware. All of these technologies were covered in Chapter 4.

7.4.3.3 Mobile Client Security

M-commerce security issues can be divided into two categories: mobile client security and system security. Mobile client security is mainly concerned with user authentication and data protection in case the device is lost. System security refers to systemwide security mechanisms, protocols, and supporting technologies. General mobile security issues were discussed in Chapter 6. The following is an introduction to mobile client security issues in the context of m-commerce. We leave system security to the discussion on m-commerce systems and services.

The first problem in mobile client security is to authenticate a user to a mobile device. This is often done by prompting a user to enter a custom personal identification number (PIN) or password; however, in reality, many people choose to disable this feature for convenience. Thus, if the mobile device is lost or stolen, anybody can use it and access files on the device. In other cases, even if the authentication feature is enabled on a mobile device, people tend to choose simple numbers or phrases such as birthdays, office phone numbers, or family member names as their PINs or passwords; these can be easy to guess based on the legitimate user's personal information collected by other means (social engineering, for example). In the computer password security domain, it is often suggested that English-word passwords and simple combinations of words and numbers should be considered insecure because they are highly susceptible to password hash-based brute force dictionary attacks using tools such as John the Ripper (http://www.openwall.com/john/) and L0phtCrack(http://www.atstake.com/products/lc/). Given a hash of some passwords, an attacker can recover some simple plaintext passwords by trying all possible combinations of words and numbers. This is also a problem in mobile security. Passwords used on many PDAs and smart phones are easily guessed because they are too short and simple. In addition, challenge-and-response authentication protocols in wireless networks such as Cisco LEAP in 802.11 and WPA-PSK in 802.11i are also vulnerable to dictionary attack if the password or pass code is too short or poorly chosen. A rule of thumb is to enforce a strong password policy on a mobile device or in a wireless network that require users to choose sufficiently long passwords consisting of characters from different sets including English uppercase and lowercase characters, non-alphanumeric characters, and digits.

Many types of biometric systems have been developed to eliminate the use of passwords for authentication. By comparing a stored feature of legitimate users and the one a user submits, a biometric system on a mobile device can determine if access should be granted in case of verification. More challenging is biometric identification, in which a mobile biometric system can be used to collect features of a user and map it to a unique identity. An example of biometric identification system is the U.S. BioVisas program, in which two fingerprints of every visa applicant are collected to be used for later identification, if necessary. Fingerprints and voiceprints (voice patterns) are by far the most widely used features in computerized biometric systems. In a fingerprint biometric system, a person's fingerprint image is digitized and compared with samples stored in a database. Some laptop computers have a built-in fingerprint scanner and some related software tools to allow fingerprint-based authentication. The problem with fingerprint biometric systems is that a fingerprint does not always generate the same sample data. For example, a cut in the finger, surface oils, and finger position may affect the scanned image, resulting in inaccurate matching result.

The principle challenge of mobile biometrics is the computational overhead incurred by the feature matching procedure, as it is a CPU-intensive task even for desktop computers; therefore, feature matching is often accomplished using low-power dedicated chips. To this end, the leading biometric technology is again fingerprint recognition. The size of a fingerprint reader, mostly in the form of solid-state sensors, continues to shrink, making it possible to embed them in mobile devices. Companies such as AuthenTec and Cogent Systems have developed low-cost, small-size embedded fingerprint systems; for example, the AuthenTec EntrePad™ AES3400 fingerprint scanner is a 6.5-mm square sensor array. Cogent System's BioStrip technology has been used on some HP iPAQ PDAs. These two examples represent the two basic types of fingerprint system that use either a firm-pressure (touch) scanner or a swipe (slide) scanner. The first requires a person to press a finger onto a capture pad, whereas the latter requires a rapid swipe of a finger over thesensor face. Figure 7.16 shows these two types of fingerprint scanners, as developed by Fujitsu.

Fingerprint biometric systems present a cost-effective way to realize identity authentication not only for replacing PINs or passwords on a mobile device but also for integration of a user's identity into m-commerce applications. For example, in a mobile banking application, when prompted to enter account information, a user may simply swipe a finger over a fingerprint scanner, and the application is able to map the fingerprint to an account. In this case,

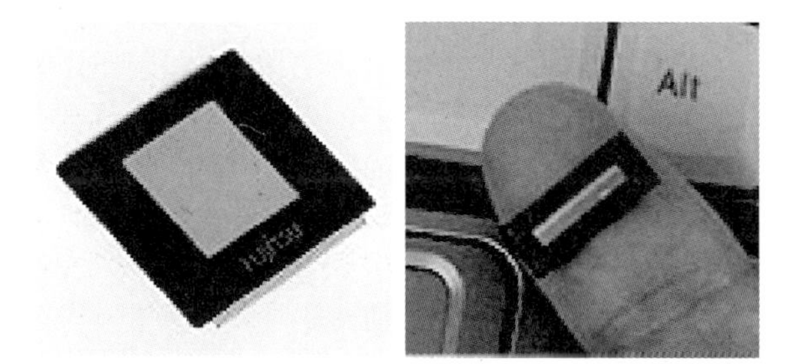

Figure 7.16 Two types of fingerprint scanners for mobile devices. (© Copyright 2005 Fujitsu. Courtesy of Fujitsu.)

the fingerprint acts as a secure combination of account number and password that can be sent to a remote m-commerce application server for authentication.

Aside from fingerprint biometric systems, other types of biometric systems employing various features of a person, such as voice recognition, facial recognition, iris recognition, retina recognition, and hand geometry may also be used on mobile devices. Each of these systems involves some level of intensive training to build a dataset of stored features for verification. Voice (speech) recognition is another means of verification in addition to text-based input. Nuance, IBM's ViaVoice, and Microsoft's speech recognition utility embedded in Windows systems are the major products in this category. In the mobile client security domain, voice recognition leverages the features of a person's voice for authentication. An example of such systems designed for mobile devices is NeuVoice Audiometrix™.

Facial recognition involves three-dimensional image recognition of facial features and is more sophisticated than fingerprint recognition. As more mobile devices have built-in cameras, it is very likely that facial recognition technology for mobile devices will become an integral security function to authenticate persons in the vicinity of the mobile device.

Iris recognition and retina recognition use features of a person's eyes as identity: both iris (the colored tissue in front of the lens and behind the cornea of the eye) and retina (a light-sensitive tissue at the back of the eyeball) are unique to each person and never change. Retinal scanning requires users to position their heads closely and accurately enough so the image of retina can be taken properly. Iris scanning is much more unobtrusive, only requiring the user to stand within a short distance from the scanner. Because of their superior accuracy in terms of very low false rejection and false acceptance and their relatively high cost, these two techniques have been primarily used in highly secured environment such as banks and military stations. It is not clear whether these technologies will be used on mobile devices in the foreseeable future.

A slightly different biometric technique from the above-mentioned methods is hand geometry recognition, which takes the shape of a person's hand for verification rather than identification because hand geometry is not unique among human beings.

It is also possible to combine multiple low-cost biometric methods to improve the accuracy of the recognition system. These so-called multimodal biometric systems are particularly advantageous compared to single modal systems when a feature of a person cannot be collected as identification. BioID, developed by HumanScan, is an example of such systems. BioID uses face (static features), voice, and lip movement (dynamic features) to provide identity authentication. Multimodal biometric systems are generally targeting desktop security applications and are not suitable for mobile security for the time being.

7.4.4 Network Infrastructure Issues

M-commerce systems require the underlying network infrastructure to supply a wide range of network services essential to the realization of the system, as well as reliable and efficient delivery of m-commerce services being offered. For example, in order to provide different service levels for live multimedia streaming subscribers, the system requires the mobile wireless network to be able to delivery multiple levels of video quality to corresponding subscribers. Mobile localization in association with GIS is a general requirement for many value-added location-based m-commerce applications such as mobile advertising and logistics. Below is a list of requirements for m-commerce:

- *Seamless integration of heterogeneous wireless network technologies*—A mobile user will be able roam across different types of wireless networks with the same identity. Ongoing m-commerce service sessions should not be affected by handoff. IP is seemingly the foundation of the framework of integration. This is a general network requirement for m-commerce applications.
- *Mobile localization support for LBS*—Either the network or the mobile user, after proper authentication and authorization,

is capable of tracking the location of an object using some localization techniques such as GPS, cellular base stations triangulation, wireless LAN, and so on. A user's location information must be well protected in the system to ensure privacy.

- *High quality and efficient mobile multimedia delivery*—Given the limited bandwidth of wireless networks (as compared to wired networks) and the mobility of mobile devices, a mobile multimedia delivery system such as a sports news streaming system or a video messaging system must use high-performance, adaptive network protocols and codecs that are specially designed for mobile environments.
- *Quality of service across multiple wireless networks*—M-commerce services can be differentiated to impose multiple service level guarantees. Bandwidth, maximum latency, and packet loss are three major items in describing the quality of service. This is again a general network requirement for m-commerce applications.
- *Secure in-network data transmission and routing*—The network must provide data confidentiality and data integrity at the radio layer, link layer, and network layer. All m-commerce applications rely on these low-level security services to function in a highly susceptible mobile environment.
- *Group communication and ad hoc communication support*—Apart from infrastructure-based unicast wireless communication, the network should allow broadcast, multicast, and anycast when needed. *Ad hoc* communication should also be supported as a way to quickly establish a self-organized communication infrastructure. Applications requiring these support include mobile IM, mobile gaming, live media broadcasting, and *ad hoc* collaboration

Very often these issues are not strictly originating from m-commerce applications but from a more general network design perspective; however, if we view m-commerce as a platform upon which different parties perform business operations in addition to communication, we will realize the significance of these issues because the underlying platform indeed has a huge impact on the business world. For details of these issues, please refer to Chapter 3 and previous sections of this chapter.

7.4.5 M-Commerce Systems and Services

We now turn to some m-commerce systems and services to show how m-commerce may help traditional business operations. From an m-commerce system designer's point of view, the key question is how to make the system appeal to perspective consumers and enterprises. For consumers, an m-commerce application must be easy to use and capable of allowing business transactions to be performed more conveniently. For enterprises, an m-commerce application must be able to improve business productivity and reduce operational cost. Of course, technological issues surrounding building an m-commerce system to provide a value-added service are equally important. Figure 7.17 provides a diagram of the m-commerce application landscape. Note that we use mobile payment (m-payment), mobile banking, and mobile enterprise as three types of applications in the m-commerce arena. Mobile payment and mobile banking are

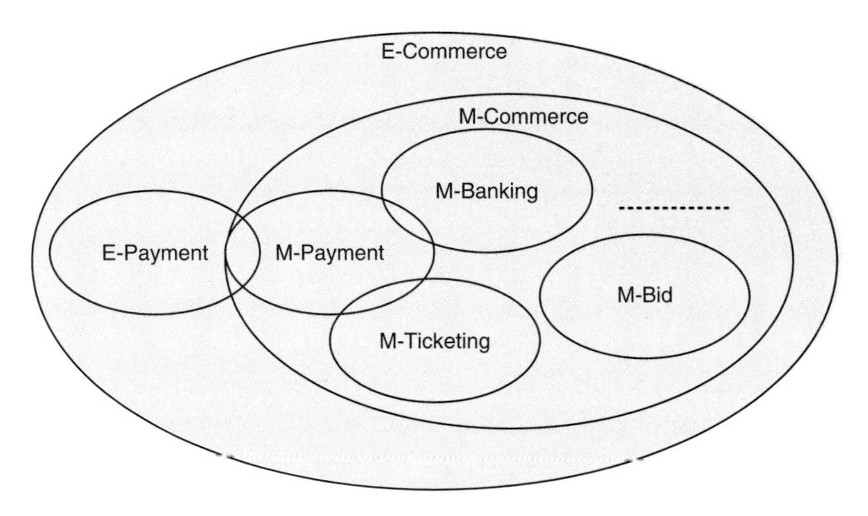

Figure 7.17 M-commerce services landscape.

related to each other with some overlapped services. M-commerce applications and services fall into the general electronic commerce domain, which encompasses a much broader range of electronic business operations such as electronic payment services and B2B, business to consumer (B2C), and consumer to consumer (C2C) transactions, such as bidding on eBay.

7.4.5.1 M-Payment

M-payment refers to the process of using a mobile device to perform an electronic payment between business parties that both have access to some mobile wireless networks. M-payment is not merely a mobile extension of electronic payment over the Internet; it possesses some unique features that require special attention. For example, electronic payment on the Internet normally performs authentication based on the user's submitted identity, whereas in m-payment the payment system can be closely integrated with operation of the mobile wireless network such that authentication can be conducted with a user's preestablished identity. Because of this convenience and the advantage of enabling electronic payment everywhere at any time, m-payment is believed to become one of the most promising m-commerce applications. There has been some standardization effort to define m-payment interfaces to allow interoperability among various parties and systems.

Depending on the overall approach to m-payment, the two types of m-payment system architectures are macro m-payment and micro m-payment. Macro m-payment systems typically involve financial institutions such as credit card companies and banks, as well as consumers and mobile service providers, in a payment operation. The amount of payment is thus at least several dollars to cover the operational costs of such a system. Examples of macro m-payments include mobile online shopping and mobile brokerage. Micro m-payment, on the other hand, only deals with payments of very small amounts of money. A micro m-payment service can be provided by a mobile service provider, a special micro m-payment provider, or a wireless-enabled payment device within proximity of the mobile device. Because the payment is too small to cover operational costs, online retailers usually do not provide such products or services. Examples of micro m-payments include online music download services in which each song is usually less than $1 and blogging services in which bloggers are paid by the number of visitors to the page. Web-based micro m-payment providers such as Peppercoin and BitPass claim to support payments as low as 1¢. Proximity-based micro m-payment systems, such as using cell phones to pay for beverages at vending machines or to purchase mass transit tickets, have been in operation for several years.

The goal of a mobile payment transaction is to transfer values from a customer (the payer) to the merchant (the payee). Other parties involved in this procedure are one or more mobile service providers that operate the wireless network infrastructure; financial institutions such as banks, credit card companies; and stock brokers. There are two logical roles a financial institution must assume for each payment: acquirer and issuer. An acquirer is the financial institution representing the merchant, and the issuer represents the consumer. In a typical credit-card-based m-payment, a consumer communicates with a merchant's system from a mobile device via a wireless network. Upon receiving confirmed transaction details including credit card information from the consumer, the merchant forwards this information to the acquirer, which in turn transfers transaction details to the issuer within a closed, highly secure, financial network. The issuer verifies the transaction details and begins to transfer funds to the acquirer. When the fund transfer is complete, both the merchant and the consumer will be informed.

An m-payment can be performed in the absence of a financial institution. In this case, the consumer's account maintained by the mobile service operator will be charged, and an associated fund transfer will be reflected on the phone bills of the two parties. There might be another party—namely, an independent mobile commerce provider such as PayPal—that serves as a broker for merchants interfacing mobile service providers and financial institutions.

For proximity-based m-payment, the consumer's mobile device is able to communicate with a point-of-sale device using a short-range wireless technology such as infrared, wireless LAN, RFID, Bluetooth, or built-in contactless smart card. The consumer simply waves the mobile device (*e.g.*, cell phone) to the point-of-sale device, which connects to a back-end m-payment system that uses the consumer's identity (*e.g.*, cell phone number) to retrieve funds from the consumer's account at the mobile service provider. Table 7.9 explains the roles of these parties.

Figure 7.18 illustrates a general m-payment architecture, including wireless network-based m-payment and proximity-based m-payment. Because several parties are involved in this big picture, the key to the success of a mobile payment service, and perhaps any other m-commerce applications, is the cooperation between these involved parties in promoting the application and generating value from it. Mobile service providers have the basic network facilities in place and are looking for applications that take advantage of them; merchants may consider m-payment or m-commerce as a way to generate more sales and to enlarge their customer basis; financial institutions view m-payment as an additional method to increase their volume of transactions or reduce operating costs; and independent m-commerce providers want to enlarge their merchant base while promoting their payment services among consumers. Along the value chain, how can the revenue be shared such that each party is able to justify their investment in the m-payment system? It would appear as though the mobile service provider must turn over most of the funds collected from consumers to merchants in order to get a

Table 7.9 Parties Involved in M-Payment Applications.

Parties	Description
Consumer	Initiates a payment using a mobile device
Mobile wireless service provider	Operates the underlying network in which payment is performed across different parties
Financial institutions	Performs fund transfer as informed by consumer and merchant
Merchants	Provides goods or services and allows electronic payment via mobile wireless networks
Independent m-commerce provider	Represents a collection of merchants and receives payment from consumers

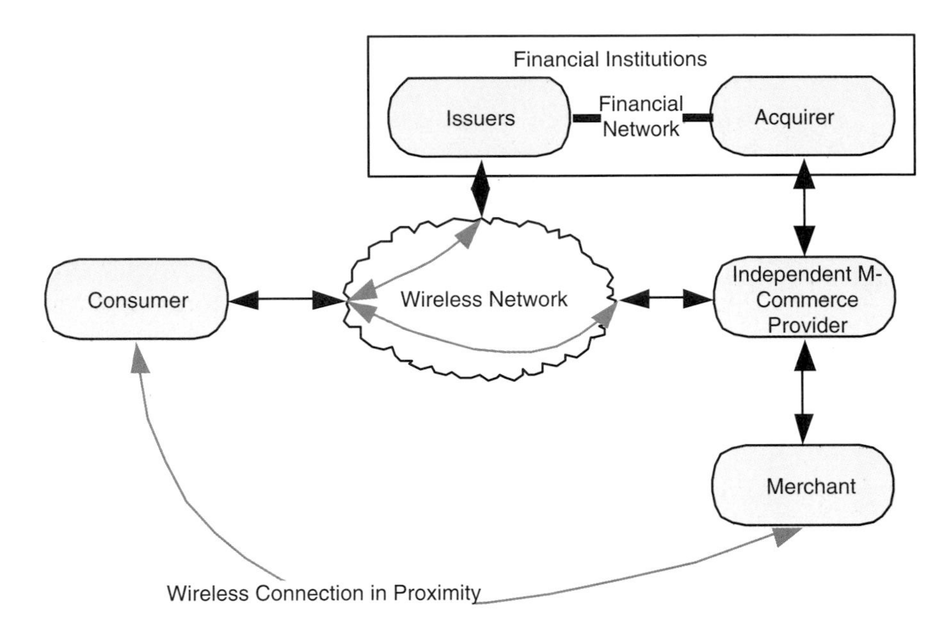

Figure 7.18 M-payment architecture.

large quantity of merchants to sign up. For example, NTT DoCoMo's iMode service splits funds paid by consumers at a ratio of 9:1, whereas merchants (in this case, mobile content providers) take 90%.

The government also plays a significant role in m-payment advancements. Any m-payment services must comply with specific regulations concerned with financial transaction security. In some countries, mobile service providers and independent m-commerce service providers must obtain a license from the government in order to provide m-payment services.

When designing an m-payment system, the five main technical considerations are [18, 19]:

- *Security*—Security is the most important issue in m-payment. Security issues are the primary concern of consumers and merchants. Every step of an m-payment transaction must be tightly secured. In addition to the mobile client security described above, the system must ensure mutual authentication, data confidentiality, data integrity, and nonrepudiation of a transaction. The system must be trusted by consumers and merchants and must be able to protect their identities. In some cases, the system should support user-controlled anonymous payment to protect a user's privacy. Security mechanisms such as digital signature, digital certificate, application layer encryption, and authentication can be incorporated into an m-payment system.

- *Usability*—Not only should a mobile interface be user friendly but also the network service provided by the system must be highly dependable. In addition, the system should allow a consumer to easily customize the service to facilitate convenient, on-the-go payment using a range of mobile devices. We introduced user interface design guidelines earlier. On the server side, to improve system reliability, server farms or geographically located content distribution systems can be used.
- *Performance and reliability*—From a consumer's point of view, an m-payment must not take too much time to complete. The system should deliver guaranteed performance under a heavy load. Aside from automatically utilizing higher-data-rate wireless links on a mobile device in a heterogeneous wireless environment, data compression should be enabled whenever possible throughout the system. Furthermore, an m-payment system must tolerate intermittent signal loss so as to guarantee correct payment transactions.
- *Cost*—An m-payment system must be a cost-effective solution compared with traditional payment methods. It should become an integral part of the legacy billing system. As confidence in the system grows, the increase in revenue should exceed the costs incurred by the additional users.
- *Interoperability*—An m-payment system should be independent of mobile devices and not rely on a specific wireless technology to deliver to a mobile device. Different m-payment systems must be compatible to extend the scope of service and user base of

a single system. Several industry consortia have been formed to promote international standardization of m-payment services, the payment process, and software platforms. The most influential consortia are led by mobile service providers or financial institutions, such as SimPay (http://www.simpay.com), Mobile Payment Forum (http://www.mobilepaymentforum.com), the M-Commerce and Charging (MCC) charter of Open Mobile Alliance (http://www.openmobilealliance.org/tech/wg_committees/mcc.html), and Mobey Forum (http://www.mobeyforum.org/). Because m-payment is still in the initial stages of development, none of these standardization efforts has gained substantial support globally.

Many m-payment systems have been developed and put into operation. Some have been adopted by a fairly large number of users. The value or fund being used in these systems is either tokens, such as the Meest M-Token system, or real money, such as Nokia's M-Wallet WAP application; both can be prepaid or postpaid. Tokens are symbolic credit associated with an account maintained at a mobile service provider or an independent m-commerce provider. Because a token is usually mapped to a small fraction of real money, it is well suited for micro payment whereby the charge of a product or service access could be very small. For example, a consumer may use prepaid tokens to download sport news clips or music clips from a content provider that serves as an independent m-commerce provider as the same time.

Using real money for m-payment effectively transforms a mobile device into a wireless wallet. The account may be managed at a bank, a credit card company, a mobile service provider in the form of phone bills, or an independent service provider. Payment functionality can be implemented at the client side using a chip card such as the SIM card or a separated payment card embedded into a mobile device, or at a central payment server and a client-side application as a virtual card. Both micro and macro m-payment can use real money to transfer funds, but in the case of micro m-payment, it is impractical to make every low-value payment with a bank account or credit card account; therefore, the mobile service provider or an independent m-commerce service provider must be able to act as an intermediary broker to perform payment aggregation and dissemination for users, and make money from doing so.

Unlike wireless network m-payment systems, proximity-based m-payment systems are more diverse in terms of the payment methods and underlying transaction procedures. For example, cell phones with attached RFID tags or smart cards can be used to initiate purchases at automatic vending machines, at a POS, or at an automatic ticketing machine. ZigBee, IM, ultra-wideband, and nearfield communication (NFC) can also be used in the context of m-payment as a means of wireless proximity communication. Another way to use built-in wireless capability is online shopping without credit cards. The payment system simply sends a digital receipt to the user, which can be verified at a store when the user pays for the product and picks it up. NTT DoCoMo, KDDI, and J-Phone have provided such services. The digital receipt in these services is a barcode that can be read at most stores.

Wireless web technologies such as WAP and iMode are widely used in m-payment systems as a cross-platform application protocol to reach the consumer. For example, the well-known PayPal service provided by eBay uses WAP to allow peer-to-peer payment as well as bank fund transfer. In addition, many existing m-payment systems use SMS as a vehicle for communication between a consumer and a merchant or as a means of payment receipt or an electronic ticket, such as the MoPay (http://www.mopay.co.za/) system, the Contopronto system (http://corporate.contopronto.com/), and the StreetCash system (http://www.streetcash.de).

7.4.5.2 Mobile Banking

Mobile banking and mobile payment are emerging mobile commerce applications utilizing a mobile wireless infrastructure as well as mobile devices for business transactions. Unlike m-payment services where financial institutions may be completely bypassed by mobile service providers, consumers, and merchants, the mobile banking service is well centered at the integration of mobile services and traditional financial services at a bank or a credit card company. Mobile banking service does not necessarily result in payment transaction; rather, it allows account holders to manage their accounts conveniently from a mobile device independent of place and time. Mobile banking, as well as Internet banking, is a cost-effective way to provide banking services in addition to traditional telephone banking services and over-the-counter banking services, as it does not require any tellers and transactions are secure and automatic. Like m-payment, a mobile banking service is often delivered in the form of SMS, Java/BREW applications, or a wireless web application over secured wireless links. Because only an account holder and the bank (or credit card company) are involved in a mobile banking service, the overall architecture of mobile banking is generally much simpler than that of a mobile payment system involving more parties along the value chain.

Typical service of mobile banking include:

- View account statement and transaction details.
- Transfer fund between accounts in real-time.
- Pay bills and confirm checks.
- Trade stocks with enhanced notification and brokerage.
- Make proximity-based payments with a bank account.

As mentioned before, mobile banking and mobile payment are closely related when a bank account is being used for a payment transaction; therefore, the technical requirements for m-payment are also applicable to mobile banking: security, usability, performance, cost, and interoperability. The banking industry has made a substantial standardization effort for mobile banking. The Mobey Forum (http://www.mobeyforum.org) and the European Committee for Banking Standards (ECBS) and two examples of such. The Mobey Forum, a consortium of more than 30 financial institutions and financial technology providers, aims at promoting open architecture and non-proprietary secure and user-friendly mobile banking and payment services. It has adopted a "preferred payment architecture" for local payment in which consumers use a personal trusted device (PTD), such as a cell phone, to withdraw cash from an ATM or pay at a POS. ECBS defines a European electronic banking standards framework for building a variety of banking services over the Internet and mobile wireless networks. ECBS also published a standard for electronic payment initiator (ePI) and implementation guidelines to allow flexible representation and interoperation of online payment.

7.5 Mobile Enterprise

Mobile enterprise applications refer to a set of corporate applications that assist employee in performing business practice in many aspects, including fast data integration and decision making, real-time communication, and close control over goods and services. Mobile enterprise applications present a tremendous opportunity for enterprise application designers to take advantage of mobile computing in business operations. The goal of general enterprise applications is to enable business process automation in various scenarios. Table 7.10 outlines widely used enterprise applications in the corporate world. Note that all enterprise applications must support both back-office (within the corporate site) and front-office operations (at field sites or customer sites).

7.5.1 Mobile Enterprise Overview

Among these enterprise systems, ERR is generally the most sophisticated and requires intensive on-site customization and optimization when it is deployed in a company, whereas CRM and KM may simply work off the shelf. ERP essentially provides a data layer platform upon which a broad set of business processes can be interconnected and automated, such as bills of materials (BOMs), purchasing, accounts payable, manufacturing resource planning (MRP), sales and marketing, inventory management, etc. SCM is closely related to ERP with regard to sharing inventory and accounting information.

Table 7.10 Summary of Enterprise Applications.

Enterprise Application	Description	Mobile
Enterprise resource planning (ERP)	A highly customizable suite of applications for electronic processing of business operations; also referred to as an e-business package	ERP mobile client can be used to access a mobile web portal.
Customer relationship management (CRM)	A software solution that manages a customer database in association with sales, marketing, and product development information	CRM mobile client is used to access the CRM portal.
Supply-chain management (SCM)	A software solution that automates manufacturing, shipping, delivery, and purchasing of goods and services within and between companies	SCM can utilize RFID-based pallet and product tracking.
Collaboration	A set of communication tools that enables a variety of person-to-person communications and group collaboration methods, such as e-mail, group calendar and scheduler, enterprise instant messaging, voice over IP, and video conferencing	Collaboration includes mobile IM, mobile multimedia applications, wireless Internet applications, etc.
Knowledge management (KM)	A software solution to the management, analysis, and systematic presentation of information, work items, knowledge, and documents contributed by employees in a company	Acquisition and retrieval of knowledge can be done via a mobile device
Enterprise application integration (EAI)	The integration of various legacy and new enterprise applications within a company and between different companies	EAI should be transparent to mobile clients.

SCM is primarily concerned with improving the flow and efficiency of the supply chain, which can be broken down into five stages: supply chain planning, sourcing, manufacturing, delivery, and return. CRM and partner relationship management (PRM) are stand-alone applications focusing on maximizing the benefits of customer assets or partner assets. Three types of business processes are involved in CRM: sales, which uses customer data for sales planning, sales analytics, and account management; marketing, which uses customer data to plan and manage campaigns and promotions; and service, which primarily deals with customer services. Customers can directly interface to CRM to perform self-guided operations. Obviously CRM and ERP share some functionalities as well as customer data. KM is not as common an enterprise application as ERP, SCM, and CRM. All these systems can be exposed in some way to a collaboration platform that enables effective communication in a virtual working environment; yet, all these systems, including the collaboration platform and legacy enterprise systems, could possibly be integrated into a more strategic, enterprise-wide solution using some middleware. Between companies, web services are often used to enable interoperability.

The back end of enterprise applications always sits in a wired network, whereas the front end could be running on wired desktop computers or mobile devices within the enterprise network or on remote computers and mobile devices on the Internet via virtual private networks (VPNs). Enhancing these enterprise applications with mobile features could extend the reach of the application, thereby improving work productivity and reducing operational cost. The simplest form of mobile enterprise applications is supplying a mobile client that allows a mobile worker or a customer to access back-end systems anywhere, anytime, as indicated in Table 7.10. Furthermore, for each type of enterprise applications, a new set of mobile modules that enable augmented collaboration and intelligent business processing can be integrated into an enterprise system. Figure 7.19 provides an overview of the enterprise application environment augmented with mobile technologies. Two types of applications for mobile wireless technologies could be used in an enterprise business environment. One is to build wireless networks and corresponding applications that directly engage in business processes such as manufacturing, floor planning, and inventory management, as part of the entire enterprise information and business processing system. Examples of this type of mobile application are wireless sensor networks in a manufacturing plant, location-based asset tracking in a hospital using wireless LAN or Bluetooth, and RFID-based inventory management in a warehouse. The other type of application is

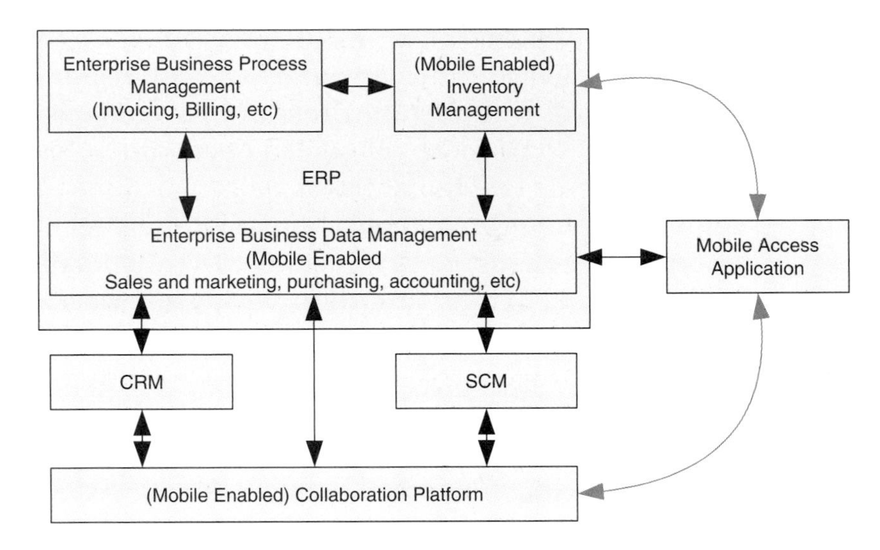

Figure 7.19 Mobile enterprise applications.

to provide optimized mobile client portals on a mobile device for employees working in the field or at customer sites; for example, onsite field engineers can use mobile devices to directly access a back-end database in real time to obtain job details and schedules and to communicate with other field workers on other sites. These two types of mobile applications essentially "mobilize" an enterprise system from the internal to the external.

7.5.2 Mobile Enterprise Applications

Below is a summary of potential mobile applications for enterprise systems. These applications are by no means exhaustive, but they do illustrate the diverse ways mobile enterprise applications can help a company to conduct business:

- *Inventory management*—Traditionally, inventory management has relied on manual scanning of a UPN (Universal Personal Number) code on pallets or products, which is time consuming and error prone. As RFID technology matures, low-cost RFID tags attached to pallets and products can be polled wirelessly by a nearby RFID reader. These RFID readers are furthered connected to server via a wired network, where raw tag data are mapped into detailed product information, aggregated, and sent to the central inventory system. A reader can also be attached to a mobile device to permit convenient point-and-check on the move. This scheme is actually a binary form of a proximity-based localization system. This may not suffice when accurate location of a product is needed, such as in a huge warehouse. Using some RFID tags placed at fixed locations as reference tags, it is possible to track the locations of product tags in real time. As mentioned earlier in the location-aware computing section, RFID can be combined with wireless LAN to provide real-time locating and monitoring (RTLM) for indoor asset management. Both ERP applications and SCM applications of a company or a retail store can be enhanced with RTLM to improve productivity (*e.g.*, discovering shortages of materials, placing orders, confirming orders). Technical requirements of such systems include sufficient localization accuracy, long battery life of tags, seamless integration with other enterprise systems, and easy deployment and upgrade.

- *Onsite processing*—Field engineers, salespersons, and mobile workers at a remote site may take advantage of mobile client applications to access back-end enterprise systems in real-time. For example, by using a mobile device that wirelessly connects to a job scheduling system, a field engineer can be dynamically assigned to a work item and can retrieve detailed information about the work item without going back to the office. While a phone call can be an option for quick and direct communication, it is generally difficult, if not impossible, to exchange detailed information such as numeric data and pictures in a phone call, not to mention the manpower of operators necessary to answer phone calls. When the work item is done, the engineer may update the status of the work item in the back-end database, and check out the next item. By using a multifunction device such as a smart phone, the engineer can also send rich-media content of the work item to the back-end server, including pictures, audio description, instrumental measurements, and video clips. The procedure does not require any manual intervention on the server side. For salespersons and mobile workers who travel to a customer's site, a mobile client application running on a mobile device will allow them to conduct CRM-related business practices onsite, such as collecting and updating customer data, providing quick responses to customers' inquiries, checking product availability, generating real-time quotes for products or services, and onsite invoicing. Because of the enhanced quality of service offered by the salesforce, customer loyalty is likely to be improved. Another benefit of onsite processing is the increasing granularity of enterprise workflow, leading to agile and well-justified decision making. Onsite processing can be done with a single mobile portal, which is a unified mobile interface to back-end enterprise applications on a mobile device. The principle challenge to facilitating mobile-enhanced onsite processing is providing a range of sophisticated interoperable enterprise business processes on a mobile device of small form factor, unreliable wireless capability, and limited computing power. Obviously, a well-designed enterprise application integration (EAI) will make this a lot easier at the back end, thus the front end—the much simplified mobile access application on a mobile device—only has to interface with EAI rather than individual enterprise applications.

- *Onsite collaboration*—Field engineers, salespersons, and mobile workers may also use the wireless capability of a mobile device to communicate with colleagues back at the office or on other sites. Aside from e-mail, voice mail, and phone calls, a wide range of communication methods can be used on a mobile device, including IM, voice over IP, video streaming, and video conferencing. A major difference between these tools and consumer-oriented collaboration tools is that they are well suited for enterprise environment and can be integrated into other enterprise applications. For example, location-based presence service in mobile IM is particularly useful for a user who wants to find out the current work status of other colleagues. Imagine an enterprise IM application running on a mobile device that shows the geographic location of colleagues and their work status. This would improve job scheduling as well as work item tracking. Field engineers, mobile workers, and salespersons in a conversation can quickly refer to the same piece of business data retrieved from a backend enterprise system, thereby improving communication efficiency and work productivity. Besides, enterprise security is a major difference that separates mobile enterprise collaboration tools with the counterparts for the mass public. In order to deal with the peculiarities of mobile enterprise applications, a company usually imposes strict policies for the establishment of secured computing infrastructure as well as user usage. An example of mobile security in the enterprise is the discovery of rogue wireless access points. A rogue wireless access point is an access point installed without authorization and thus does not conform to enterprise network security policies. A rogue wireless access point could be a serious security problem because it exposes a supposedly well-protected enterprise network. Enterprises often use wireless intrusion detection systems along with positioning techniques to approximate the location of rogue access points and then use a mobile device to pinpoint them. Some wireless intrusion detection systems even combine location with authentication to deny access to stations that lie outside a predefined physical perimeter, such as AirTight SpectraGuard (http://www.airtightnetworks.net/) and Newbury Watchdog (http://www.newburynetworks.com).

7.6 Wireless Application Gateway

Mobile applications running on a mobile device typically connect to an enterprise system or an e-commerce system through a wireless application gateway. A wireless application gateway, as the term implies, is essentially the access server for mobile devices. When the underlying application is delivered to a mobile device via the web, a wireless application gateway will serve as a wireless web server dedicated to mobile browsers; otherwise, a wireless application gateway acts as a general-purpose back-end server for applications running mobile devices. Network connection between a wireless application gateway and a mobile client could be any kind of wireless networks such as cellular, wireless LAN, Bluetooth, and WiMax, whereas the wireless application gateway and back-end systems are often connected via a wired network.

7.6.1 Wireless Application Gateway Architecture

The functional components of a wireless application gateway are depicted in Figure 7.20. On top of IP is the wireless transport support including datagram transport layer functions, session management, and transaction support. Depending on the purpose of a wireless application gateway, certain business logic pertaining to CRM or ERP, location-based mobility support such as customer search in a geographic area, or end-user profile support can be realized based on wireless transport support. Security mechanisms must be provided along with these supporting functional components as well as with

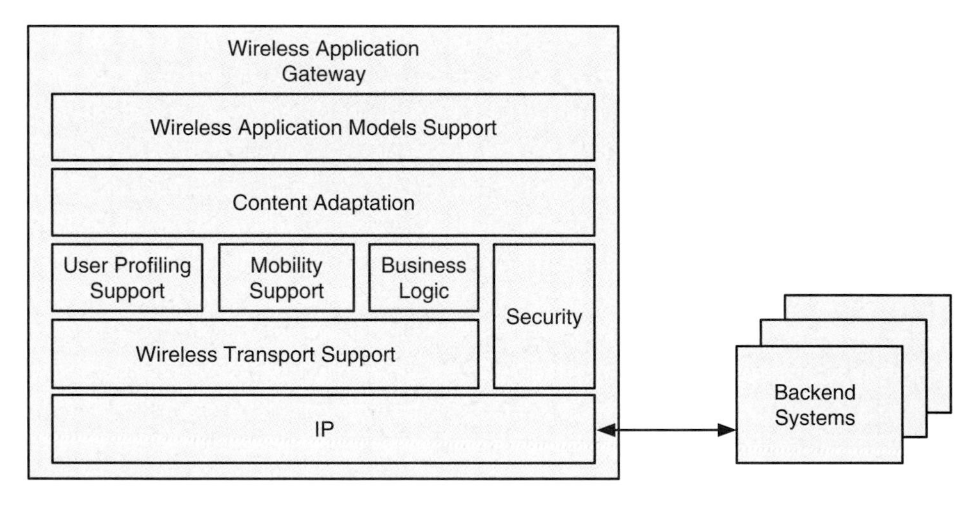

Figure 7.20 Wireless application gateway architecture.

wireless transport layer implementation. Server process output, usually in the form of a markup language such as HTML, WML (in earlier WAP), xHTML (in WAP), or cHTML (in iMode), must be transformed into an appropriate structure before being delivered to a mobile client application. (Refer to Chapter 5 for a summary of mobile markup languages.)

The process of transforming data described in markup languages into a form suitable for display on mobile devices is called content adaptation (explained in Section 7.6.3). Apart from server-side content adaptation, another way to adapt content to mobile devices is to do so on the client side. The topmost layer of the wireless application gateway architecture is a handful of wireless application models representing different messaging services. Note that some of the functional components shown in Figure 7.20 can be grouped into a general protocol suite dedicated to wireless web access from a mobile device. An example of such is the WAP specification, which consists of the following components: datagram transport layer protocol (WDP), security layer (WTLS), transaction layer (WTP), session layer (WSP), and application layer (WAE). Details of these layers of WAP were discussed in Chapter 3.

A wireless application gateway can be regarded as a bridge between a variety of backend systems and a mobile device. In this sense, it is also a bridge between the wireless part and the wired part of an enterprise application being delivered to mobile devices. The wireless transport inherently has high latency, low bandwidth (although it is improving), and a higher bit error rate and is subject to device mobility. This implies that the wireless transport component and the mobility support component must ensure that data generated from either side can be reliably and efficiently transferred to the destination. These components also provide session management support for related applications, as well as data compression to save the bandwidth of wireless links. In addition, power consumption on a mobile device is another big issue, requiring a good partitioning of the application (wireless part *versus* wired part) such that power consumption of a mobile application client can be substantially reduced.

As shown in Figure 7.20, the functions of a wireless application gateway are not limited to dealing with wireless data delivery; the gateway can also provide data processing with regard to some business logic that is mainly concerned with business processes and mobile device users such as salespersons, field engineers, or mobile workers. The business logic implemented by a wireless gateway is not merely a subset of business logic for enterprise systems but is an entirely new design dedicated to mobile applications. For example, when a number of power plant field engineers are assigned to several sites to repair a power outage, they could use mobile devices to pull information about the site and surrounding sites and to be notified when the status of another site changes. Such functions are enabled by the underlying wireless application gateway, which creates a job object for each site and maintains communication between these objects. To achieve this, the wireless application gateway must obtain necessary information from back-end enterprise systems and provide updates when the jobs are done.

7.6.2 Wireless Application Models

The simplest form of a wireless application gateway is a web server dedicated to mobile browsers on mobile devices. Two types of services can be provided by the web server: push or pull. In a push service, a user first subscribes to some content, and the web server is able to push the content onto the subscriber's mobile browser on a regular basis. In a general-purpose wireless application gateway, a user may subscribe to event alert, status update, news feed, and any other information of interest. Depending on the business logic, some of the information may be directly maintained at the wireless application gateway. If some information to be pushed is not locally available at the wireless application gateway, the wireless application gateway will retrieve it from back-end systems. Then it pushes packages of requested information to a mobile application client. In some cases, the information being pushed may just be a brief abstract of a report or a URL. If interested in an item, the user will proceed to pull the details of that item (*i.e.*, send a properly formatted request for the item to the wireless application gateway), which will generate a detailed reply.

The pull service is a request-and-response procedure like HTTP. A user may request a pull service at will. To improve performance, the wireless application gateway may employ prefetching/caching and aggregation techniques [20] for back-end information that is not likely to change very soon over time (for example, data items with fairly large time-out values). The mobile application client, such as a mobile browser, can also maintain a local cache of received data items to further improve performance; therefore, a single data item may have copies at the wireless application gateway and at a mobile application client. To ensure consistency, a data item is associated with some metadata indicating its freshness. For example, the wireless application gateway may perform freshness validation of the data item by querying the serving back-end system for the last modification time or a sequence number of the data item. Comparing this value with the one previously received, the wireless application gateway is able to decide whether or not a reload of the data item is needed. Alternatively, the metadata associated with each data item can be an expiration time. So, instead of asking the back-end system for validation, the wireless application gateway may simply compare the expiration time with its local clock time and perform a reload if needed. The same techniques can be applied to local content cache of a mobile application client. It is worth noting that data updates are not always unidirectional. For example, a mobile application client may update some data item locally when it is used offline, or a wireless application gateway may not always write-through every update due to performance considerations. In these cases, the wireless application gateway must support data synchronization, which may involve multiple mobile devices and multiple back-end systems.

7.6.3 Device-Independent Content Adaptation

Because a web page, no matter whether it is statically or dynamically generated, is usually designed to fit onto a fairly large screen size (usually at least 15 inches), its layout is optimized for high-resolution settings such as 800 × 600 or 1024 × 768. Such optimization is not feasible on mobile devices, which mostly have a QVGA or lower display resolution. Furthermore, form factors differ from device to device. For example, some devices do not support portrait or landscape orientation, and others cannot display certain types of images.

Content adaptation a core functional component of a wireless application gateway is the solution to these problems. Content adaptation refers to a set of techniques that transform web pages, formatted documents and text streams, and structured output from another system into a format optimized for the display of various mobile devices. Content adaptation also implies personalization of content display according to the user's preference. For static web pages, a complete redesign of the entire web portal for mobile applications could be an option, but doing so requires a great amount of time and effort. For dynamically generated content, it would be ideal for the wireless application gateway to directly adapt the output to the display of a specific device. In either case, a wireless application gateway must support dynamic content transformation to address the heterogeneity of web page delivery to mobile devices.

There are three issues in the context of device independent content adaptation. First of all, content adaptation must consider the targeted client's characteristics, which could be described in plaintext embedded in the header of a request from the client or in a preconfigured profile. The W3C Composite Capabilities/Preferences Profile (CC/PP) standard defines a data format for device characteristics and user preferences in the form of XML. Examples of characteristics used in CC/PP include screen size, bits per pixel, color capable (yes or no), keyboard (phone keypad or pen based), image capable (yes or no), supported image formats, and so on. The WAP specification provides a similar mechanism, the User Agent Profile (UAProf).

Second, the content adaptation component of a wireless application gateway must define a set of rules for each device. These rules dictate how a generally defined markup tag should be displayed on that device. Extensible stylesheet language transformation (XSLT) is a language of template rules for the display of XML documents. An XSLT file consists of a list of templates that map an element of a common XML document to an XSL style sheet, which in turns determines the output structure. This is a server-side adaptation approach because the wireless application server stores one XSLT file for each type of device and uses it to generate markups for corresponding devices. It is also possible to put this burden on the client side by using cascade style sheets (CSS) to render an original XML or other types of markup documents on a mobile device.

Finally, the layout of a web page or a structured document must be carefully adjusted to fit into small mobile displays. For web pages, a technique known as *transcoding* is often used to dynamically rearrange blocks of a normal web page into a single column or a series of small subpages, thereby eliminating horizontal scroll on a mobile browser. Opera Small Screen Rendering (http://www.opera.com/products/mobile/smallscreen/) is an example of the single-column approach. The problem of single-column rendering is that the resultant page is too long, which is against the design guideline of web pages. The subpage-based transcoding approach essentially generates an outline of content blocks for a normal web page. The outline can be a text-based table of contents or a visualized thumbnail image in which each block is marked by a different color. Within each block, multiple web content components are rendered vertically to eliminate horizontal scrolling. An example of the subpage-based transcoding approach utilizing thumbnails as index pages is the so-called "page-splitting" technique described by Chen *et al.* [21]. Figure 7.21 shows how the page-splitting technique is applied to a web page. The thumbnail index page is shown on the left, and the subpages are shown on the right. Note that the transcoding procedure can be done on the mobile client side or on the wireless application gateway side, depending on the link bandwidth between them and the computing power of the underlying mobile device.

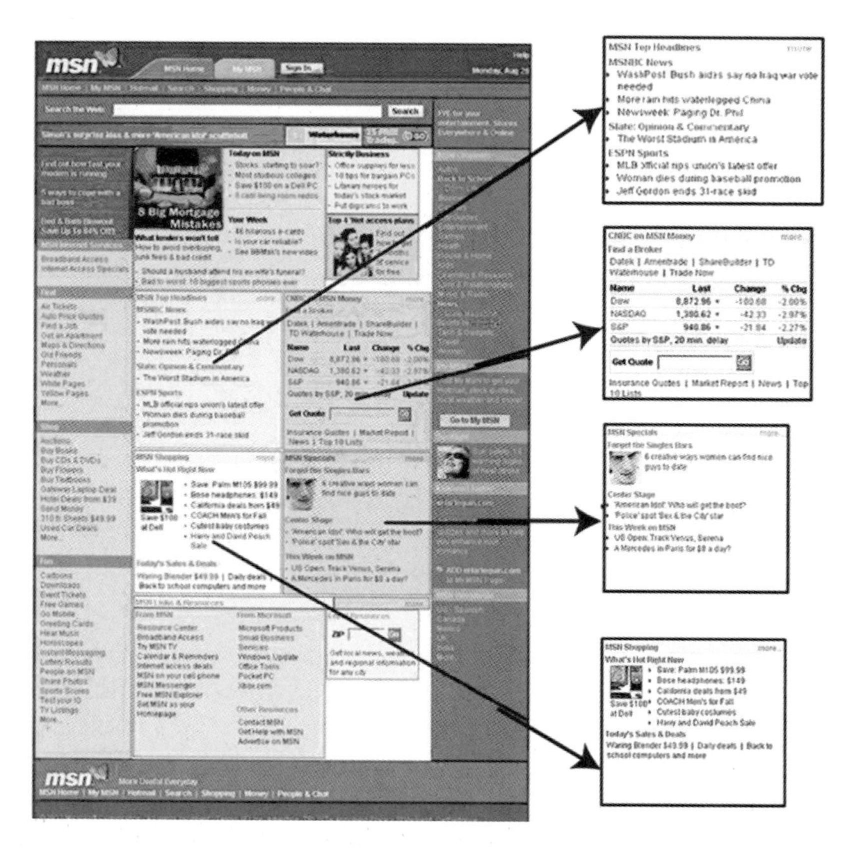

Figure 7.21 Transcoding of web pages using outlined subpages [21]. (© Copyright 2005 IEEE.)

7.7 Telematics

Telematics refers to the combination of wireless communication, onboard electronics, and information technologies in the context of integrated automobile and freight information system. Conventional telematics systems on high-end vehicles use GPS and other in-vehicle sensing devices to remotely track a vehicle, monitor its operation, conduct mechanical or electronic diagnostics, and provide navigation and emergency assistance for the driver. Newer telematics systems make better use of wireless communication capabilities to deliver a wide range of personalized content onto a liquid-crystal display (LCD) or a built-in radio speaker in a vehicle.

7.7.1 Telematics Systems

Below is a list of telematics system either already in use or being developed:

- In-vehicle road assistance, such as emergency service when a car breaks down; lock and unlock doors; provide maps and turn-by-turn navigation.
- Vehicular diagnostics and enhanced safety such as real-time and offline remote vehicle diagnostic services; vehicle maintenance alerts; mobile messaging when a car's alarm is on. Insurance companies can use in-car devices to monitor a driver's behavior and wirelessly transmit to a back-end system when evaluating the driver's insurance policy. The same device can be used to train drivers; for example, when the driver brakes too hard, a chime could ring.
- In-vehicle, value-added, location-based services, such as POI (gas stations, lounging, etc.) searches, restaurant reservations, purchasing tickets, real-time traffic reports, and location-based objects (*e.g.*, person, portable device, car) discovery, monitoring, and tracking. Parents will be able to trace their children in real time when the children are driving and could even be alerted when a child is driving too fast.
- Internet access and vehicular networking, including checking e-mails, web surfing, instant messaging, in-car wireless LAN and Bluetooth networks, wireless tolling, and traffic reporting using built-in sensors. A vehicular network should allow convenient data exchange between devices and should be able to interconnect with wireless sensors and networks passing by.
- Multimedia and entertainment content delivery, such as satellite radio services, satellite television services, online music streaming, on-demand news streaming, audio/video sports broadcasting, and wireless gaming on *ad hoc* networks or via the Internet. A typical in-car stereo system can be converted into a media center.

- In-vehicle mobile business applications, including almost every mobile application in the m-commerce arena that can be deployed onto a common mobile device. Some examples are CRM mobile access, job scheduling and reporting, and mobile collaboration.

Many telematics applications on the list must be made available when the car is parked as well as moving on the road. Depending on the application scenario, a telematics system can operate in the following three modes: driving mode, in which a motorist is driving the car; parked mode, in which the car is parked with the engine off but the power is on so devices such as GPS and radio are still in operation; and shutdown mode, in which the power is off and only battery-powered devices such as alarms are in operation. Each mode essentially imposes various restrictions on the underlying telematics system in terms of power consumption, wireless communication quality and cost, and safety considerations. For example, when a car is parked close to a Wi-Fi hotspot such as a coffee shop or a restaurant, the telematics system should be able to detect the Wi-Fi hotspot and switch to it if it offers better signal quality and link bandwidth than the built-in Internet access method such as cellular data service or satellite data service.

7.7.2 Telematics Issues

A key design objective of telematics systems is that the system should keep the driver focused on the road to ensure driving safety. This means interactions between a telematics system and a driver should distract the driver as little as possible. For this purpose, voice-activated user interfaces are widely used to keep the driver's hands on the wheel and eyes on the road. Speech recognition allows the driver operate the system, go through menus, and input data. Also, text-to-speech conversion technology can be used to receive messages, traffic updates, route information, e-mails, news, and business communications.

Telematics systems usually use a means of wide area wireless communication as the default network connectivity. GPS, satellite data service, or 2.5G/3G cellular services are among the most commonly used telematics wireless networks. A telematics system may also make use of some wireless local area networks in the vicinity for Internet access. In addition, direct intervehicle communication (car-to-car walkie talkie) might occasionally be required by a group of people on the road as a way to reduce communication overhead and economic cost. Intervehicle communication is essentially a mobile *ad hoc* network with vehicles as the moving nodes. These nodes may share several Internet connections managed by some of the member nodes. They can also provide rich collaborations among people in the vehicles, such as multiplayer gaming, conferencing, and application sharing. The major challenge in intervehicle communication is to maintain stable connections between nodes in the mobile *ad hoc* network, as wireless signals are subject to fading, diffusions, multipath effect, and interference from other wireless signals nearby especially when the nodes are moving at high speed.

To build a telematics system, security and privacy issues must be taken into account. Security mechanisms of a telematics system are extremely crucial as they could affect the normal operation of the vehicle and perhaps result in accidents. Although a telematics system can monitor and analyze data obtained from in-vehicle sensors, a virus or a piece of malicious code downloaded to a telematics system must not be able to control a vehicle's electronic or mechanical components. Privacy issues are equally important in the telematics domain. When a telematics system is reasonably priced, many people will want to take advantage of its services but will not want to be monitored or tracked by anyone else on the road or exposed to the release of any personal information.

7.8 Summary

What would be the "killer app" for next-generation mobile computing? Considering the heterogeneity of mobile wireless technologies, it is understandable that people in different interest groups would have a different answer. In fact, any party who has made a significant investment in a wireless network infrastructure should find a way to take advantage of it. To this end, we identify three key factors that will shape future applications of mobile computing.

- Dramatic increase of data rate in various wireless networks
- Continuous deployment of wide area wireless networks
- Evident trend of convergence between heterogeneous wireless networks

Taking these factors into account, we selected seven categories of mobile applications that are likely to proliferate in the next several years and have presented an extensive discussion of these applications. In the broad domain of location-aware mobile computing, a variety of wireless location sensing techniques and systems were introduced, followed by a general analysis of the design of a location-based system coupled with introductions to some commercial localization systems for asset and people positioning and tracking, inventory management, wireless site survey, and even wireless intrusion detection systems. Indeed, wireless localization has become a key building block for existing and future mobile applications.

Mobile messaging as an alternative to voice communication has become a phenomenal success in Asia and Europe. In parallel with the increasing popularity of SMS, EMS, and MMS in 2.5G and 3G cellular systems, mobile instant messaging that provides presence service and enhanced means of communication has begun to gain some ground. A principle challenge to the success of mobile IM is interoperability. For historical and economic reasons, well-known proprietary desktop IM networks are not designed to be compatible with each other. Open IM protocols such as SIMPLE and XMPP are expected to be adopted by major IM operators. It is not clear whether mobile IM service will become a one-to-one clone of such popular IM networks in the wireless world such as mobile AIM or mobile MSN messaging, or if there will be an open, extensible IM architecture across various providers' networks.

Mobile multimedia streaming services are well positioned to leverage the increased data rate of wireless links, and mobile broadcasting services also have the potential to become indispensable offerings. Of all potential mobile steaming applications, mobile television and on-demand video streaming have already become available in some countries. Both 3GPP and 3GPP2 have developed frameworks for multimedia streaming. We believe that, as 3G networks continues to roll out, we will see more high-quality multimedia streaming applications being used by a larger user base, primarily engendered by some novel business models.

For m-commerce and m-enterprise services, further categorization is necessary because it encompasses a very broad range of applications, either consumer oriented or enterprise oriented. All these applications essentially share some general requirements such as mobile client design and network issues, but they also possess unique characteristics that require special consideration when being integrated into a legacy system. The emergence of telematics systems also presents phenomenal opportunities to automobile makers, mobile service providers, independent service providers, and embedded system manufacturers.

Telematics systems are excellent examples of the rise of pervasive computing, which encompasses a new dimension of challenges beyond the reach of current computer systems and networks. Smart phones, as a generally used mobile device in this paradigm, must be able to facilitate those mobile applications in an intuitive and unobtrusive way.

7.9 What's Ahead

Needless to say, there will always be surprises when it comes to the next general mobile applications. Recall that the web was initially designed for European researchers to exchange technical documents. Now it is the platform for Internet computing. Also recall that instant messaging tools stemmed from toy applications that allowed people to chat casually online. Now it is moving quickly into an enterprise computing environment. Also recall that GPS was once used exclusively by the military. Now we are seeing a trend of each smart phone having a GPS chipset.

Indeed, we cannot predict the future, especially from a technologist's perspective; however, in this book of next-generation mobile computing, we have tried to present a complete picture of the mobile computing realm, which essentially is comprised of three types of elements:

- The trend of convergence between communication, computing, and consumer electronics
- Enabling cutting-edge technologies and related researches, system building blocks, established architectures, challenges and approaches, interesting R&D projects, and commercial systems
- Existing and future mobile services, systems, and applications

This is indeed a rapidly changing field. It is impossible to cover all the amazing advancements even for a single wireless technology or emerging mobile application, and some of the systems and technologies may quickly become outdated and be replaced by new ones. However, it was not our intent to enumerate every proposed solution or system within a specific area; instead, we wanted to identify the core idea—either technological or business-related—that could be possibly adopted to solve problems in the near future. For example, the lessons learned from the well-known 802.11 wireless LAN security problems will certainly remind wireless system researchers to pay attention to wireless security in the very first stages of system development. On the other hand, the success of SMS service in Asia and Europe may give some insight into blending a somewhat simple technology with a marketable and effective business model, which is particularly crucial for mobile IM service providers and mobile online gaming service providers.

One of the major objectives of this book is to allow people from different sectors of the mobile wireless world, such as industry research labs, academia, standardization bodies, industry forums, and commercial wireless system designers, to be informed of the latest development in other sectors. For example, wireless Internet service providers (WISPs) that are interested in the integration of Wi-Fi hotspots and cellular networks may want to learn of proposed solutions by academic and research-and-development laboratories. Researchers working on mobile multimedia streaming will need to explore the enabling technologies of in-operation mobile television services and mobile video phone services in some countries. Here, we want to emphasize that mobile technologies developed in laboratories, with no exceptions, must have real-world applications in order to become really useful. This philosophy has dictated the writing of this book—technologies are always introduced in parallel with applications, which in turn creates challenging problems that call for research and development of new ideas and systems.

In summary, next-generation mobile computing will create a fantastic world in which we are able to enjoy unprecedented levels of freedom in communication, computing, and entertainment. The power of convergence of data access and pervasive mobile intelligence enabled by smart mobile devices such as smart phones is the driving force behind this wave of computing. We will surely see many exciting technological breakthroughs and innovations of mobile computing in the next several years. We hope the book can serve as a glimpse into the new frontier.

Further Reading

3GPP packet-switched streaming (PSS) specifications, http://www.3gpp.org/ftp/Specs/html-info/26246.htm.

A good reference on mobile enterprise applications is P. Brans, *Mobilize Your Enterprise: Achieving Competitive Advantage Through Wireless Technology*, Prentice Hall, Englewood Cliffs, NJ, 2002.

An excellent summary of enabling technologies and potential opportunities for telematics systems is presented in Y. Zhao, Telematics: safe and fun driving, *IEEE Intelligent Syst.*, 17(1): 2002.

For a description of vehicular telematics business applications, see Vehicular mobile commerce, *IEEE Computer*, 2004 (http://www.computer.org/computer/homepage/1204/communications/).

For a summary of streaming technology in 3G mobile wireless networks, refer to I. Elsen, F. Hartung, U. Horn, M. Kampmann, and L. Peters, Streaming technology in 3G mobile communication systems, *IEEE Computer*, 34(9):46–52 2001.

For protocol details of XMPP and Jabber, see http://www.xmpp.org and http://www.jabber.org/protocol/.

IETF SIMPLE working group, http://www.ietf.org/html.charters/simple-charter.html; RFC3428 Session Initiation Protocol (SIP) Extension for Instant Messaging, http://www.ietf.org/rfc/rfc3428.txt.

Mobile Payment Forum, http://www.mobilepaymentforum.org.

MPEG home page, http://www.chiariglione.org/mpeg/; MPEG-4 standard, http://www.chiariglione.org/mpeg/standards/mpeg-4/mpeg-4.htm.

OMA Mobile Commerce and Charging Working Group, http://www.openmobilealliance.org/tech/wg_committees/mcc.html.

Wireless Village, http://www.openmobilealliance.org/tech/affiliates/wv/wvindex.html.

References

[1] L. M. Ni, Y. Liu, Y. C. Lau, and A. P. Patil, "LANDMARC: Indoor Location Sensing Using Active RFID," in *Proc. IEEE*

International Conference on Pervasive Computing and Communication (PERCOM'03), Dallas-Fort Worth, TX, 2003.

[2] P. Bahl and V. N. Padmanabhan, "RADAR: An In-Building RF-based User Location and Tracking System," in *Proc. the 19th Annual Joint Conference of the IEEE Computer and Communications Societies (INFOCOM'00)*, Tel-Aviv, Israel, 2000.

[3] J. Hightower and G. Borriello, "Location Sensing Techniques," University of Washington, Computer Science and Engineering, Technical Report UW-CSE-01-07-01, 2001.

[4] A. M. Ladd, K. E. Bekris, A. Rudys, G. Marceau, L. E. Kavraki, and D. S. Wallach, "Robotics-based Location Sensing using Wireless Ethernet," in *Proc. theh 8th Annual International Conference on Mobile Computing and Networking (MOBICOM'02)*, Atlanta, GA, 2002.

[5] R. Want, A. Hopper, V. Falcao, and J. Gibbons, "The Active Badge Location System," *ACM Transaction on Information Systems*, 10(1):91–102, 1992.

[6] A. Harter, A. Hopper, P. Steggles, A. Ward, and P. Webster, "The Anatomy of a Context-Aware Application," in *Proc. the 5th Annual International Conference on Mobile Computing and Networking (MOBICOM'99)*, Seattle, WA, 1999.

[7] N. B. Priyantha, A. Chakraborty, and H. Balakrishnan, "The Cricket Location-Support System," in *Proc. theh 6th Annual International Conference on Mobile Computing and Networking (MOBICOM'00)*, Boston, MA, 2000.

[8] D. Niculescu and B. Nath, "VOR Base Stations for In-Door 802.11 Positioning," in *Proc. the 10th Annual International Conference on Mobile Computing and Networking (MOBICOM'04)*, Philadelphia, PA, 2004.

[9] J. B. Saxe, "Embeddability of Weighted Graphs in K-Space is Strongly NP-Hard," in *Proc. Allerton Conference Communication Control Computer*, 1979.

[10] D. Fox, J. Hightower, L. Liao, D. Schulz, and G. Borriello, "Bayesian Filtering for Location Estimation," in *IEEE Pervasive Comput.*, vol. 2, 2003, pp. 24–33.

[11] A. Doucet, N. d. Freitas, and N. Gordon, *Sequential Monte Carlo in Practice*: Springer-Verlag, 2001.

[12] L. Hu and D. Evans, "Localization for Mobile Sensor Networks," in *Proc. the 10th Annual International Conference on Mobile Computing and Networking (MOBICOM'04)*, Philadelphia, PA, 2004.

[13] OMA (Open Mobile Alliance): The Wireless Village Initiative: System Architecture Model v1.1, http://www.openmobilealliance.org/tech/affiliates/ LicenseAgreement.asp?DocName=/wv/wv_architecture_v1.1.pdf, 2002.

[14] S. Leggio and M. Kojo, "Instant Messaging in Wireless Networks," in *Proc. Fourth Berkeley-Helsinki Ph.D. Student Workshop on Telecommunication Software Architectures*, 2004.

[15] 3GPP TS 26.233 V6.0.0 Transparent End-to-End Packet Switched Streaming Service (PSS) General Description, http://www.3gpp.org/ftp/specs/latest/Rel-6/26_series/26233-600.zip, 2004.

[16] 3GPP TS 26.234 V6.2.0 Transparent End-to-End Packet Switched Streaming Service (PSS) Protocol and Codecs, http://www.3gpp.org/ftp/specs/latest/Rel-6/26_series/26233-600.zip, 2004.

[17] Y. E. Lee and I. Benbasat, "Interface Design for Mobile Commerce," *Commun. ACM*, 46(12):49–52, 2003.

[18] S. Karnouskos and F. Fokus, "Mobile Payment: A Journey Through Existing Procedures and Standardization Initiatives," *IEEE Communications Surveys and Tutorials*, 6(4):44–66, 2004.

[19] Mobile Payment Forum While Paper: Enabling Secure, Interoperable, and User-friendly Mobile Payments, http://www.mobilepaymentforum.org/pdfs/mpf_whitepaper.pdf, 2002.

[20] P. Brans, *Mobilize Your Enterprise*: Prentice Hall, 2003.

[21] Y. Chen, X. Xie, W.-Y. Ma, and H.-J. Zhang, "Adapting Web Pages for Small-Screen Devices," *IEEE Internet Computing*, 9(1):5056, 2005.

Index

C

D

E

H

I

J

N

O

P

S

U

V

W